Test Automation Using Selenium WebDriver with Java

Navneesh Garg

- Selenium WebDriver 4.0
- Learn Automation on a Web Based Application
- Real Life Experiences
- Step By Step Instructions
- Interview Questions Based on Selenium

Selenium WebDriver Step By Step Guide

Test Automation Using Selenium WebDriver with Java
By Navneesh Garg

ISBN - 978-0-9922935-8-1

Publisher: Adactin Group Pty Ltd.

Contents

About the Author

Navneesh Garg

Navneesh Garg is a recognized test automation architect and corporate trainer, specializing in test automation, performance testing, security testing and test management. As a tool specialist, he has worked on a variety of functional automation tools including Selenium, OpenText UFT, Tosca, TestComplete, TestPartner, SilkTest, Watir, RFT, and on varied technologies including Web, Java, Dot-net, SAP, Peoplesoft and Seibel.

His previous book "Test Automation using Unified Functional Testing" is among the bestselling books on HP UFT. This book has consistently ranked among the top 100 testing books on Amazon. It was the first book to be released globally on the latest version of HP UFT.

He is an entrepreneur and founder of several successful IT companies which encompass the Adactin Group, CresTech Software, and Planios Technologies.

As an experienced corporate trainer, he has trained professionals in Selenium and other test tools across a wide range of global clients such as Macquarie Bank, Corporate Express, Max New York Life, Accenture, NSW Road and Maritime Services, Australian Dept of Education, HCL Technologies, Sapient, Fidelity Group, Adobe Systems, and many more. He has training experience in diverse geographies such as Australia, India, Hong Kong and USA.

As a technical test delivery head for his company, he has led and managed functional automation testing and performance testing teams across a wide range of domains, using commercial tools and open source tools. Certified in OpenText UFT, OpenText Quality Center, OpenText LoadRunner, IBM Rational Functional Tester and as a Certified Ethical Hacker, he has designed several high-end automation frameworks including using Selenium and its integrations with tools like TestNG, JUnit, Selenium Grid, Jenkins and ANT.

Preface

My motivation for writing this book stems from my hands-on experience in the IT and testing domain and the experience I have gained as an automation consultant working in numerous complex automation projects.

Selenium, being an open source tool, is gaining huge popularity but still is not conceived as an easy to use tool especially by testers due to a variety of reasons, including tool setup, programming background and support issues. A key objective of this book is showcase in a simple guided way how to use Selenium WebDriver so that we can attain maximum return on investment from using the tool. Not only will we learn how to use the tool but also how to effectively create maintainable frameworks using Selenium.

In my previous book "Test Automation using HP Unified Functional Testing" we had taken a similar step by step guided approach using commercial tool HP UFT which has been excellently received by the testing fraternity.

Scope of Topics

As part of the scope of this book we will cover **Selenium WebDriver (Selenium 4.0) with Java** as a programming language. We will also cover how to use **Selenium IDE** which is a Firefox based Selenium Plug-in for easy record and replay.

We will be using **Eclipse** as the main IDE for creating Selenium WebDriver tests.

No prior knowledge of Java language is required for this book but having understanding of object oriented programming language concepts will definitely help. As part of this book we will be covering **Basics of Java** which would be required to use Selenium WebDriver for beginner users.

In the later section we also show how to configure and use **Selenium Grid** to run parallel tests on multiple browsers and OS configurations.

As part of reporting frameworks, the book will show how to configure and use both custom **JUnit and TestNG reports**.

We will also see how Selenium WebDriver integrates with **continuous Integration** tools like **Jenkins.**

My intent in this book is to discuss the key features of Selenium WebDriver, WebDriver methods and cover all crucial aspects of the tool which help to **create effective automation frameworks**.

The book **does not** have samples or examples on how to use Selenium WebDriver with Python, C# and Ruby languages. The book **focuses** on using **Selenium WebDriver with Java language**.

Key Audience

The target audience for this book are manual functional testers who want to **learn Selenium WebDriver quickly** and who want to create effective automation frameworks that generate positive ROIs to stakeholders.

Salient Features of this Book

This book has been designed with the objective of **simplicity and ease of understanding**.

A major fear amongst functional testers who want to learn Selenium is the fear of programming language and coding. As a part of this we will cover just enough **basics on Java programming language** that will give the readers confidence to use Selenium WebDriver.

This book follows a **unique training based approach** instead of a regular text book approach. Using a step by step approach, it guides the students through the exercises using pictorial snapshots.

Selenium being an open source tool needs quite a few independent components to be installed like Eclipse, TestNG, ANT, etc. This would usually scare testers. In this book we will cover step by step installation and configuration of each of these components.

Another major highlight of this book is a **custom developed Web based application used throughout the book** instead of learning automation on custom html pages with few form fields and links.

Another differentiator is that I have tried to include **many practical examples and issues** which most of the automation testers encounter in day-to-day automation. These experiences will give you an insight into what challenges you could face with automation in the real world. Practical examples cover how to use most of the features within Selenium WebDriver.

It also covers aspects of **Continuous Integration tool; Jenkins** so that Selenium WebDriver scripts can be integrated with the development environment and run on nightly builds.

The book also covers the most **common interview questions** on Selenium WebDriver and automation.

Sample Application and Source Used in Book

The sample application used in the book can be accessed at the following URL: *https://adactinhotelapp.com/*

The source code used in the book can be found at the following link *www.adactin.com/stores/*

Feedback and Queries

For any feedback or queries you can contact the author at www.adactin.com/contact.html or email navneesh.garg@adactin.com

Order this book

For bulk orders, contact us at info@adactin.com

You can also place your order online at adactin.com/stores/

Acknowledgements

I would like to thank my family (my parents, my wife Sapna, my wonderful kids Shaurya and Adaa) for their continued support. Without them this book would not have been possible.

Special thanks to Emily Jones and William Bugan for their reviews and feedback, which immensely helped as I worked on this book.

I would also like to thank my colleagues and clients for the inspiration, knowledge and learning opportunities provided.

1

Introduction to Automation

Introduction

In this chapter we will talk about automation fundamentals and understand what automation is and the need for automation. An important objective of this chapter is to understand the economics of automation and determine when we should carry out automation in our projects. We will also discuss some popular commercial and open source automation tools available in the market.

Key objectives:

- What is automation?
- Why automate? What are the benefits of automation?
- Economics of automation
- Commercial and Open-Source automation tools

1.1. What is Functional Automation?

Automation testing is to automate the execution of manually designed test cases without any human intervention.

The purpose of automated testing is to execute manual functional tests quickly and in a cost-effective manner. Frequently, we re-run tests that have been previously executed (also called regression testing) to validate functional correctness of the application. Think of a scenario where you need to validate the username and password for an application which has more than 10,000 users. It can be a tedious and monotonous task for a manual tester and this is where the real benefits of automation can be harnessed. We want to free up manual functional tester's time so that they can perform other key tasks while automation provides extensive coverage to the overall test effort.

When we use the term "automation", there is usually confusion about whether automation scope includes functional and performance testing. Automation covers both.

- Functional Automation – Used for automation of functional test cases in the regression test bed.
- Performance Automation – Used for automation of non-functional performance test cases. An example of this is measuring the response time of the application under considerable (for example 100 users) load.

Functional automation and performance automation are two distinct terms and their automation internals work using different driving concepts. Hence, there are separate tools for functional automation and performance automation.

For the scope of this book, we will be only referring to **Functional Automation**.

1.2. Why do we Automate?

Find below key benefits of Functional Automation:

1. Effective Smoke (or Build Verification) Testing

Whenever a new software build or release is received, a test (generally referred to as "smoke test" or "shakedown test") is run to verify if the build is testable for a bigger testing effort and major application functionalities are working correctly. Many times we spend hours doing this only to discover that a faulty software build resulted in all the testing efforts going to waste. Testing has to now start all over again after release of a new build.

If the smoke test is automated, the smoke test scripts can be run by developers to verify the build quality before being released to the testing team.

2. Standalone - Lights Out Testing

Automated testing tools can be programmed to kick off a script at a specific time.

If needed, automated tests can be automatically kicked off overnight, and the testers can analyse the results of the automated test the next morning. This will save valuable test execution time for the testers.

3. Increased Repeatability

At times it becomes impossible to reproduce a defect which was found during manual testing. Key reason for this could be that the tester forgot which combinations of test steps led to the error message; hence, he is unable to reproduce the defect. Automated testing scripts take the guess work out of test repeatability.

4. Testers can Focus on Advanced Issues

As tests are automated, automated scripts can be base-lined and re-run for regression testing. Regression tests generally yield fewer new defects as opposed to testing newly developed features. So, functional testers can focus on analysing and testing newer or more complex areas that have the potential for most of the defects while automated test scripts can be used for regression test execution.

5. Higher Functional Test Coverage

With automated testing a large number of data combinations can be tested which might not be practically feasible with manual testing. We use the term 'Data driven testing' which means validating numerous test data combinations using one automated script.

6. Other Benefits

- **Reliable:** Tests perform precisely the same operations each time they are run, thereby eliminating human error.
- **Repeatable:** You can test how the software reacts under repeated execution of the same operations.
- **Programmable:** You can program sophisticated tests that bring out hidden information from the application.
- **Comprehensive:** You can build a suite of tests that cover every feature in your application.
- **Reusable:** You can re-use tests on different versions of an application, even if the user-interface changes.
- **Better Quality Software:** Because you can run more tests in less time with fewer resources.
- **Fast:** Automated tools run tests significantly faster than human users.

1.3. When should we Automate? Economics of Automation

Let us take a scenario. If your Test Manager comes up to you and asks whether it is advisable for your company to automate an application, how would you respond?

In this scenario, the manager is interested in knowing if functional automation will deliver the organization a better return on investment (ROI) besides improving application quality and test coverage.

We can determine whether we should automate a given test if we can determine that the cost of automation would be less than the total cost of manually executing the test cases.

For example, if a test script is to run every week for the next two years, automate the test if the cost of automation is less than the cost of manually executing the test 104 times (2 years will have 104 weeks).

Calculating the **Cost of Test Automation**

$$\textit{Cost of Automation} = \textit{Cost of tool} + \textit{labor cost of script creation} + \textit{labor cost of script maintenance}$$

Automate if:

Cost of automation is lower than the manual execution of those scripts.

The key idea here is to plan for the cost of script maintenance. I have seen a lot of automation projects fail because project managers did not plan for the labor costs involved in script maintenance.

Example Let me give you an example from my personal experience.

I performed some automation work for one of our investment banking clients. We had a five-member team, which automated almost 3000 test cases in about six months time, which included around total 30-man months of effort. At the end of project, we gave the client's testing team a hand-over of the entire automation suite created by our team. Our recommendation to them was that they would need at least a one or two member team to continuously maintain the scripts. This was because there were still functional changes happening to the application and scripts would need maintenance. But since the client project manager had no budget allocated for this activity; they skipped this advice and continued to execute automation scripts. After the first six months of the 3000 test cases, only 2000 test cases were passing, while the rest started failing. These scripts failures were because script fixes were needed due to application changes. The client team was okay with that and continued to execute those 2000 working test cases, and got rid of the remaining 1000 test cases, which were now executed manually. After another six months, only scripts corresponding to 1000 test cases were passing. So they got rid of another 1000 test cases and started executing them manually. After another six months (1.5 years in total), all the scripts were failing, and testing had to move back to manual functional testing.

In the above real-life scenario, the cost of automation and its benefits could have been reaped, if the client had allocated 1-2 automation testers (could have been part-time) to maintain the scripts and had properly planned and budgeted for it.

1.4. Commercial and Open Source Automation Tools

This section lists some of the popular Commercial and Open Source Automation Tools.

Vendor	Tool	Details
OpenSource (free)	Selenium	Open Source tools and market leader in Open Source segment. Primary for WWeb-based automation. Support C#, Java, Python, and Ruby as programming language.
OpenSource (free)	Playwright	Playwright is an open source automation tool used for Angular based web applications and API Automation. It supports Java, C#, Typescript, Javascript and Python language.
OpenSource (free)	Cypress	Cypress is a front-end testing tool that directly operates inside the browser, which enables it to listen and modify browser behaviour at runtime. The language used to develop cypress tests is Javascript.
OpenText	Unified Functional Testing	OpenText UFT (previous version was called QTP) is the market leader in Test Automation in the commercial tools segment. It uses VBScript as the programming language and its ease of use makes it a tool of choice against other competing tools.
SmartBear	TestComplete	Low-cost alternative to other commercial tools with good features for automation. You have the option to program using VBScript, JScript, C++Script, C#Script or DelphiScript language.
Tricentis	Tosca	Tosca is a tool that covers API, exploratory, mobile testing, system integration and regression testing. It even supports performance testing through its integration with NeoLoad. It comes with a wizard for automating test cases and supports programming languages such as Java and C#.

§

2

Training Application Walkthrough

In this chapter we will introduce our customized Web based training application, which we will use as a part of our book.

Key Objectives

1. Training application walkthrough
2. Understand a sample scenario.

2.1. Training Application Walkthrough

As part of this book, we will be working through a custom Web based training application. The reason why we planned to use our custom built Web based application was that this book is focussed around Selenium which only supports Web based applications. Also 80-90% of applications tested and automated in real projects are Web based applications. We will have a much closer and a better understanding of how we need to automate Web-based applications.

Our sample application is a simple hotel booking web application, which has the following key features

- Search for a Hotel
- Book a Hotel
- View Booked Itinerary
- Cancel Booking

Let us browse through the application

1. Launch Chrome and enter URL www.adactinhotelapp.com to see Login page.

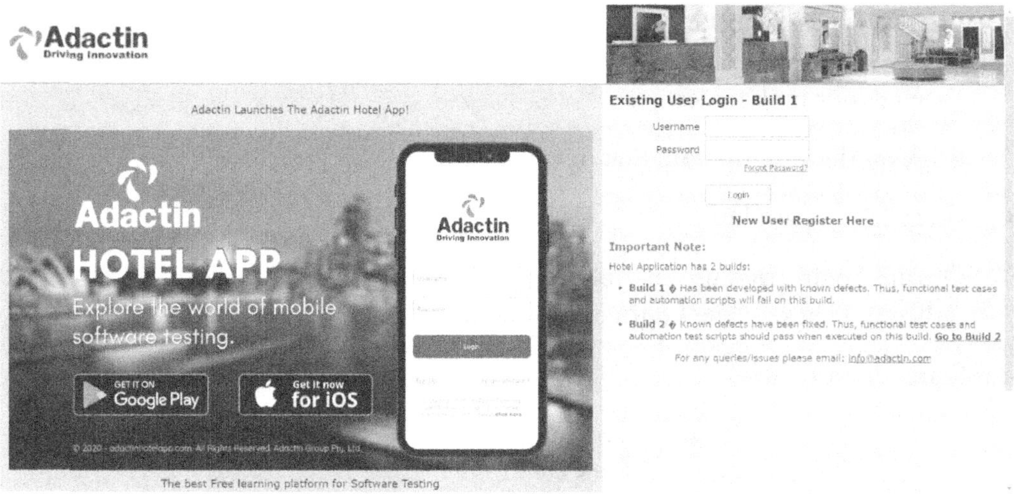

Figure 2-1 – Application Login Page

2. Click on **"New User Register Here"** to go to the Registration page.

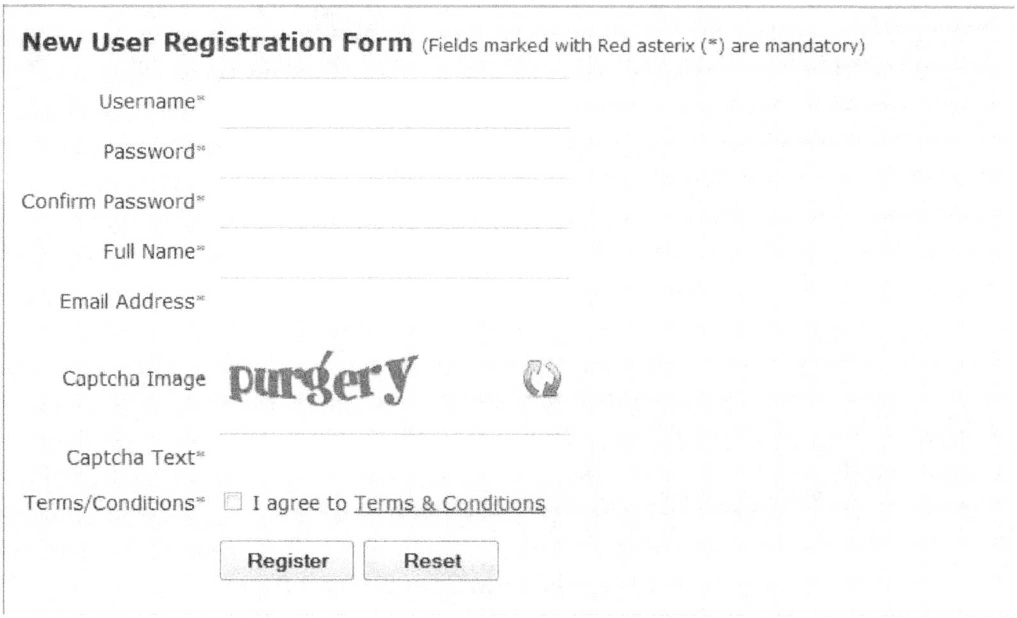

Figure 2-2 – Application Registration Page

3. Register yourself by entering all the fields. Remember the username and password as you will be using this username/password to login to application and remaining part of automation.
4. After you register, an automatic email will be sent to your email-id for confirmation. In case you do not receive the email, re-verify it in junk folder as email might have gone to your junk folder.
5. Click on the confirmation link in email
6. Go to Login page link.
7. On the Login page use the username/password with which you have registered earlier and click on the **Login** button. You will come to Search Hotel Page.
8. Search for a Hotel-
    - i. Select a location, e.g., Sydney
    - ii. Select Number of Rooms e.g., 2
    - iii. Select Adults per Room, e.g.,2
    - iv. Click on **Search** button

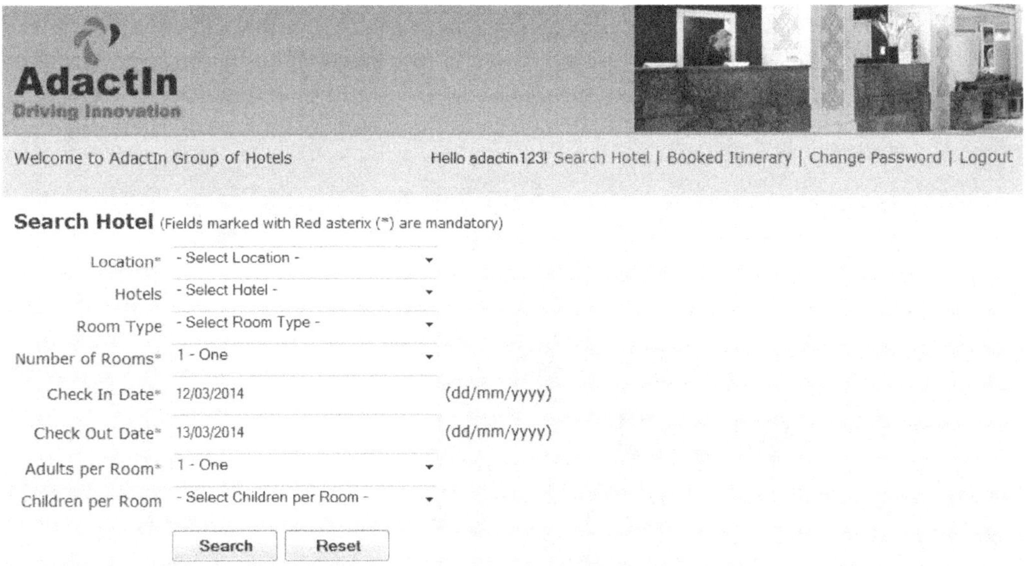

Figure 2-3 – Application Search Hotel Page

9. Select a Hotel-

    i. **Select** one of the Hotel Radio Buttons and click on continue button, e.g., select radio button next to Hotel Cornice

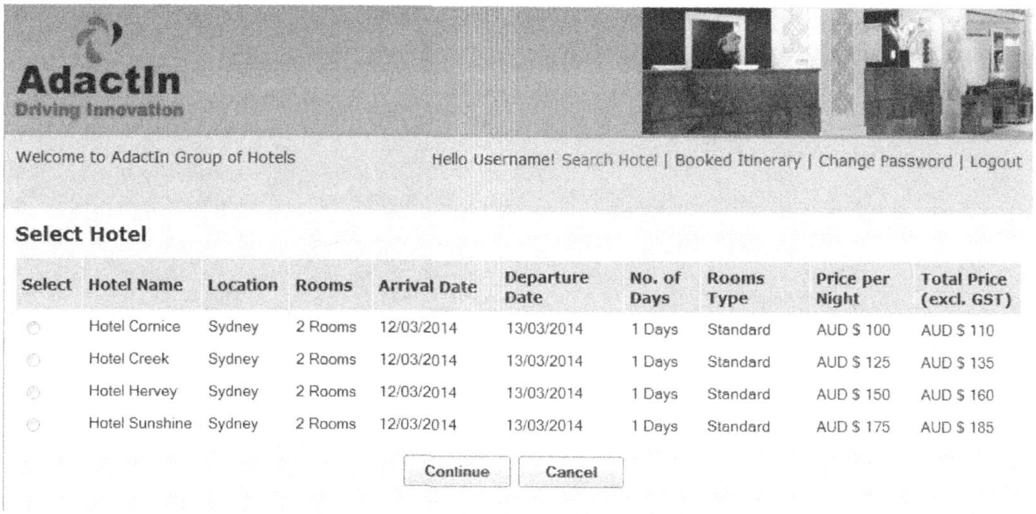

Figure 2-4 – Application Select Hotel Page

10. Book a Hotel-

    i. Enter First Name
    ii. Enter Last Name
    iii. Enter Address
    iv. Enter 16-digit credit Card Number
    v. Enter Credit Card Type
    vi. Enter Expiry Month
    vii. Enter Expiry Year
    viii. Enter CVV Number
    ix. Click on **Book Now**

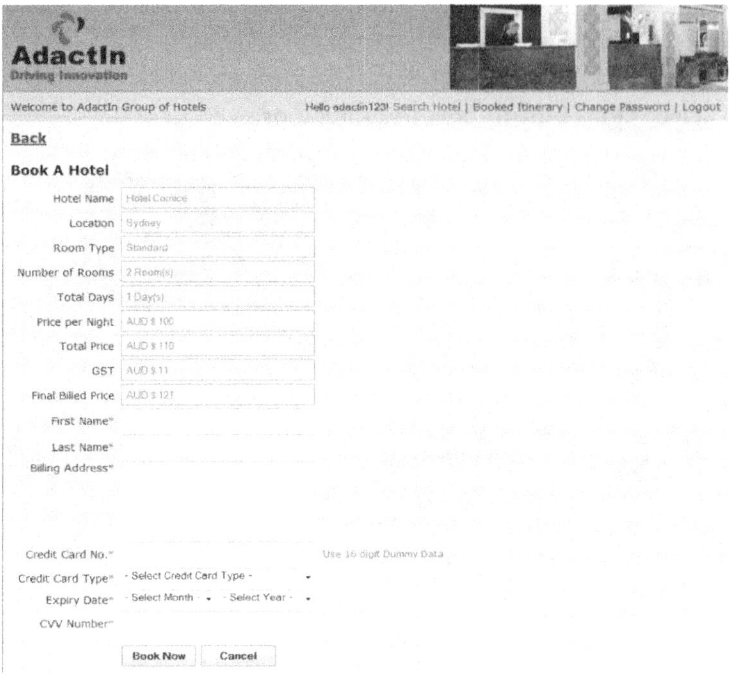

Figure 2-5 – Application Book a Hotel Page

11. After you see booking confirmation, you will notice that you get an Order No. generated.

Figure 2-6 – Application Booking Confirmation Page

12. Copy the Order No. to clipboard. In our case it is 8K1l446G95.

13. Click on **My Itinerary** Button or click on **Booking Itinerary** link at the top right corner of the application. User will go to **Booked Itinerary** Page.

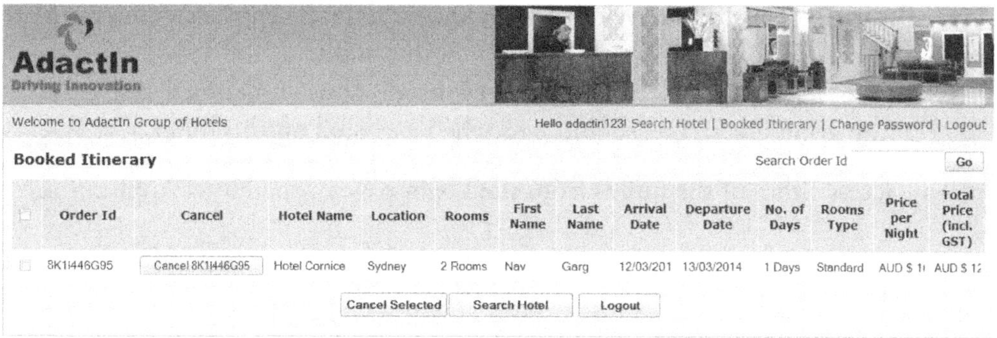

Figure 2-7 – Application Booked Itinerary Page

14. Enter the Order No. copied in the previous step in search order Id field and click on Go button. You will see the order you recently created.

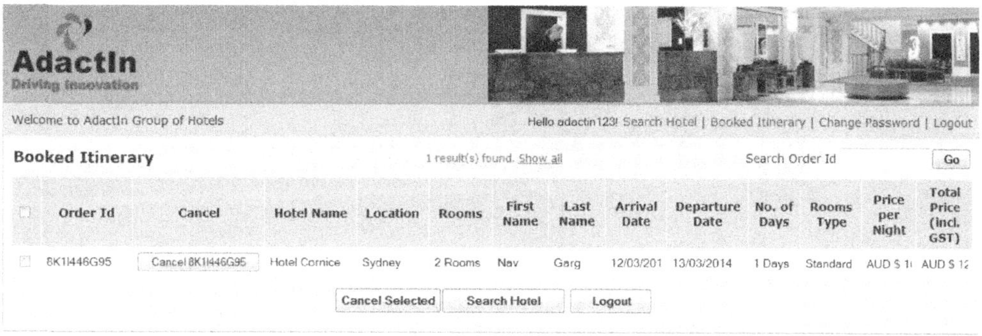

Figure 2-8 – Application Search Results Page

15. Click on **Logout** button or Logout link, on the top right corner to logout from the application. You will go to **"Click here to login again"** page.

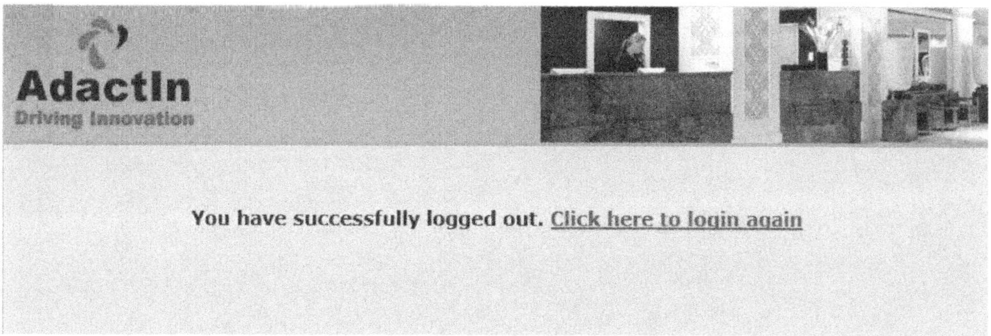

Figure 2-9 – Application Click here to login again Page

16. Click on "Click here to login again" link to go to login page

Note: Hotel Application has 2 builds:

- **Build 1** – Has been developed with known defects. Thus, few functional test cases and automation scripts will fail on this build.
- **Build 2** – Known defects have been fixed. Thus, functional test cases and automation test scripts should pass when executed on this build.

User can access either of the builds from the Login page of the Hotel Application

§

3

Planning before Automation

Introduction

Before you start recording your scripts and doing automation, it is important to plan the recording of your scripts. You need to plan:

- The test cases which need to be automated.
- Define priority of the test cases and automate key test cases first.
- The stability of your application.
- The data dependency of the tests.
- If there are tests which use the same data?
- If the automation tester knows the steps of the tests to be automated?
- Whether the automation testers have permission to access application components and systems?
- Who is going to automate which test within the team?
- When should the automation tasks be accomplished?

In this chapter, we will try to understand what we need to plan before we start our recording.

Key objectives:

- To understand the prerequisites before we record.
- To understand the Test Automation process.

3.1. Prerequisites Before you Start Recording

Let us try to understand some of the prerequisites before we start recording our scripts:

1. **Prepare your Test Environment - Check whether your environment and application are stable**

Determine that you have a test environment available in which you can record/replay and create your automation scripts.

Determine whether the application is stable from a development as well as functional point of view. Does the interface change very often?

As a recommendation, if the application interface is changing too often, or if the application is not stable from a functional point of view, or if the test environment is not stable, we should not start automation. It's important to understand the factual reason for that. Technically speaking, we can still perform automation, but it might increase maintenance and script modification effort later on, when functional issues or UI issues are fixed. So a better approach is to wait until the environment is stabilized.

2. Ensure that the automation tester has permission to access application components and systems

Ensure that the test suite and testers have permissions to access the database, host systems, and input/output data.

3. Execute the test case steps manually to determine the expected results

Execute the test case steps manually on the application, to verify that all steps are listed and ensure that you are able to understand the business process. One of the most important factors to consider in automated testing is to ensure that the test duplicates the test steps in the most straightforward manner possible. It is important to **capture the actions** stated in the test case **exactly as an end-user would perform them** in the business process.

Also, it helps you to understand if there is any **pre-data setup** required for the test. For example, if you need to automate login, you need to have a valid username and password. Rather than starting your recording first, you will realize that you need a valid login, when you first manually execute your tests.

example Let me give you another example from my personal experience:

Once we were implementing functional automation for a client and we had a test case which required us to verify that the login expires after two months. We thought of changing the system date of our PC to two months in the past and then verifying that the login expires. But the question was: How do we change the system date using OpenText UFT? So we did some research and tried a couple of examples, and were able to figure out a way. But that took us about two days.

Now once we had implemented the solution, we found that whenever we ran the UFT script, the login did not actually expire and the user could still login. We ran the test case steps manually and found the login does not expire even when manually executed which got us perplexed. After some more investigations, we realised that we were changing the system date of our local PC and not the server. But as the expiry date was linked to the server date and not to the local machine, the login did not expire and the user could login. Also, we did not have access to the server, and due to authorization issues it was not possible to change the date of the server machine.

The question is: Could we have foreseen this issue and saved our two days? If we would have manually tested this scenario, we would have realized our mistake and would not have spent two days trying to automate it.

4. Determine what data will be required to be used for test execution

Ensure that you understand what input data you would need for test creation. You need to understand valid and invalid input data. Also, a lot of times there are scenarios where you would need data in a specific format or type.

example Let me take an example here:

I used to work for a mortgage domain client. One of their applications required the input date to be greater than or equal to the current date. How do we design an automation script to take a date that's greater than the current date without the test case depending on the already defined data? It needs better planning. As a solution, instead of using hard-coded data, we used some VBScript function to generate a date greater than today's date.

Apart from this, there might be scenarios where after the test verification has finished; you would need to roll back specific data that was earlier set up as part of test execution. So make sure to understand data dependencies before you start automation.

5. Determine the start and end point of the test and follow it for all your automation scripts

Make sure that for all your automation scripts, you determine where your script will start from and where it will end.

Why is this important? This is important as you are going to run your scripts as a suite or a batch and not individually. So your current script should know where your previous script ended.

For instance, say you are working on a Web based application and you open your browser at the start of every test but fail to close the browser at the end of each test. If you are running 50 tests you will have 50 browser windows open at the end of your script run which will cause script execution issues.

The correct way is to determine the Start and End point of your test and follow it for all your scripts. This will ensure that any automation tester in your team would know which form or page of application will be open when they start creating their automation script and where they should finish their script.

A better solution for Web based applications will be to open the browser at the start of every test and close the browser at the end of every test.

6. Reset any master data, if data is modified as part of test

Another important thing to do is to reset any master data to the default value, after data is modified as part of your test. The reason is that future test scripts would be looking for default data and not the data which you have modified as part of your current test.

example Let me give an example here:

I was once working on a manufacturing based application which had units (centimetres, millimetres, and inches) defined as master data to measure the length of various manufacturing components. As a part of one test case we automated, we changed the master data of the Unit field from centimetres to inches, and verified that all the valid lengths are now in inches. Our automation script worked beautifully, and we integrated it with our automation suite and executed our overnight test run.

Next morning we found that all our scripts, following this script, failed. We realized that though we had changed the master data of the Unit field from centimetres to inches, we never changed it back to default (which was centimetres). Hence, all the sequential scripts failed as they expected the unit to be centimetres, but found the unit in inches. So we had to fix the script to reset the unit field back to default value at the end of the script.

So as a thumb rule, reset all the master data that you have modified at the end of your test to default values as it can impact other tests.

7. Standardize naming conventions

Create standards and conventions on how you are going to name your automation scripts, setup naming conventions for your temporary variables, functions, and other components of your automation framework.

This will help to ensure that the whole team is following standardized naming conventions and the complete framework can be easily maintained in the future.

8. Plan and prioritize your test cases. Identify your automation candidates

Plan and prioritize which test cases you should automate first. We use the term automation candidates for regression test cases, which we select for automation.

A few key criteria for selection of automation candidates include:

- Test cases which are high priority or linked to high priority requirements. Usually we automate Sanity or Build Acceptance test cases as a first step.
- Test cases which are data oriented or which need to be executed multiple times for different sets of data.

- Test cases which take a long time to execute and their automation will free up functional testers to perform more key tasks.
- Existing or fixed defects in the systems which are now converted to test cases.
- Frequency of execution of the test cases. Test cases which are executed very frequently are better candidates for automation giving more Return on Investment (ROI).
- Test cases for operating system compatibility or browser compatibility can be automated, as the same script can be executed for different operating systems or browsers.

9. Plan resources and schedule

Plan how many people will be automating the test cases, and what will be the delivery schedule.

3.2. Test Automation Process

This section describes automation processes usually followed as we automate the regression test cases.

1. Defining the scope for automation: Define the scope of test cases that should be automated, check feasibility, and confirm return on investment.
2. Selection of the Test Automation tool: Select right test automation tools, which will suit your application technology and fit into your budget. It can be an open source or a commercial tool.
3. Procurement of licenses: If a commercial tool is selected, procure the license for the commercial tool.
4. Training the testers to use the tool: If required, train the testing team on how to perform automation and how to use the automation tool.
5. Automation strategy and plan: Design the automation strategy and plan on how and when regression test cases will be automated. Also, define data dependencies, environmental needs and risks.
6. Identification and development of Automation Framework and Test Automation Lab: Automation framework is required to make sure automation scripts are maintainable. It involves setting up design and guidelines of automation components. This includes defining naming conventions, guideline documents, and structure of the automation scripts and setup of test machines in the test environment.
7. Creation of Automation Scripts: Actual recording or creation of automation scripts from regression test cases.
8. Peer Review and Testing: Review of Automation Scripts by peers to ensure that all conventions are followed and automation scripts are correctly mapped to

functional test cases. Test case verification points are also verified as part of the review.

9. Integration of scripts: This involves integration of automation scripts into a larger automation suite for overnight test execution, to be executed as a batch process.
10. Script maintenance: Regular script maintenance that is required when an application undergoes functional changes and needs fixes in automation scripts.

§

4

Introduction to Selenium

Introducing Selenium

Selenium is an Open Source tool for automating browser-based applications. Selenium is a set of different software tools, each with a different approach to support test automation. The tests can be written as HTML tables or coded in a number of popular programming languages and can be run directly in most modern Web browsers. Selenium can be deployed on Windows, Linux, and Macintosh and many OS for mobile applications like iOS, Windows Mobile, and Android.

Among all Open Source tools, Selenium functional testing tool is considered to be a highly portable software testing framework and one of the best tools available in the current market for automation of Web applications.

Key objectives:

1. Understand Selenium Tool Suite
2. Choosing right Selenium Tool for use
3. Requirements for Selenium Setup

4.1. Selenium's Tool Suite

Selenium is not just a single tool but a suite of software, each catering to different testing needs of an organization. It has four components.

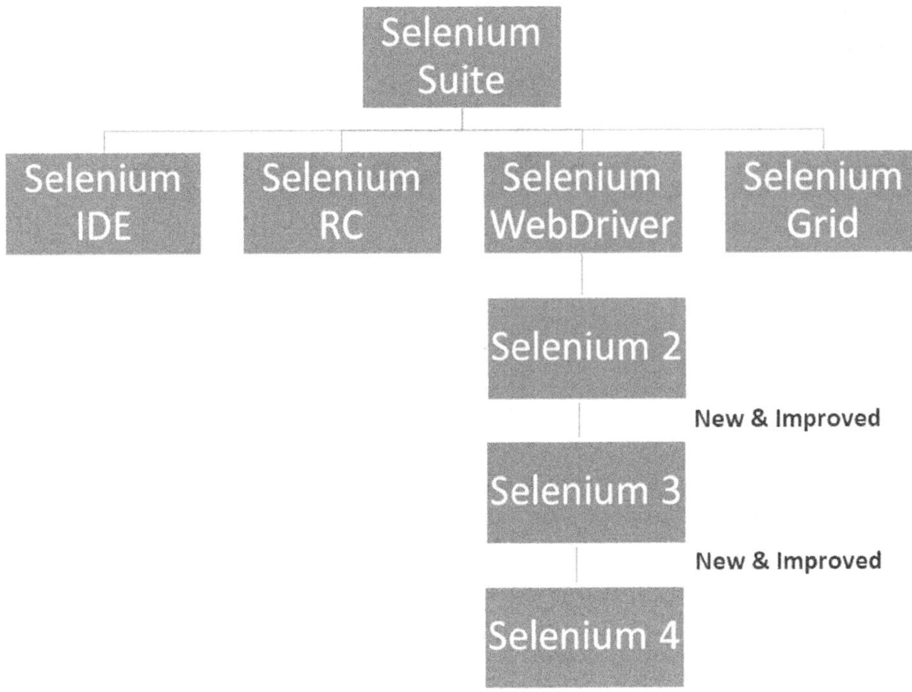

Figure 4-1 - Selenium Suite Structure

In the section below we will understand more about each of these components.

1. Selenium IDE

Selenium IDE (Integrated Development Environment) is a prototyping tool for building test scripts. It comes as a Firefox/Chrome/Edge plug-in and provides an easy-to-use interface for developing automated tests. Selenium IDE has a recording feature, which records user actions as they are performed and then exports them as a reusable script in one of many programming languages for execution later.

Selenium IDE is simply intended to be a rapid prototyping tool. Selenium IDE has a "Save" feature that allows users to keep the tests in a table-based format for later import and execution. Selenium IDE doesn't provide iteration or conditional statements for test scripts. Use Selenium IDE for basic automation. Selenium developers usually recommend Selenium 2 or Selenium 1 to be used for serious, robust test automation.

2. Selenium 1- Selenium RC or Remote Control

Selenium RC is the main Selenium project allowing user actions to be simulated in a browser like clicking a UI element, input data, etc. It executes the user commands in the browser by injecting Java script functions to the browser when the browser is loaded. As we know, Java Script has its own limitations and so does Selenium RC.

How Selenium RC Works

First, we will describe how the components of Selenium RC operate and the role each plays in running your test scripts.

RC Components: Selenium RC components are:

- The Selenium Server which launches and kills browsers, interprets and runs the Selenese commands passed from the test program, and acts as an HTTP proxy, intercepting and verifying HTTP messages passed between the browser and the AUT.
- Client libraries which provide the interface between each programming language and the Selenium RC Server.

Here is a simplified architecture

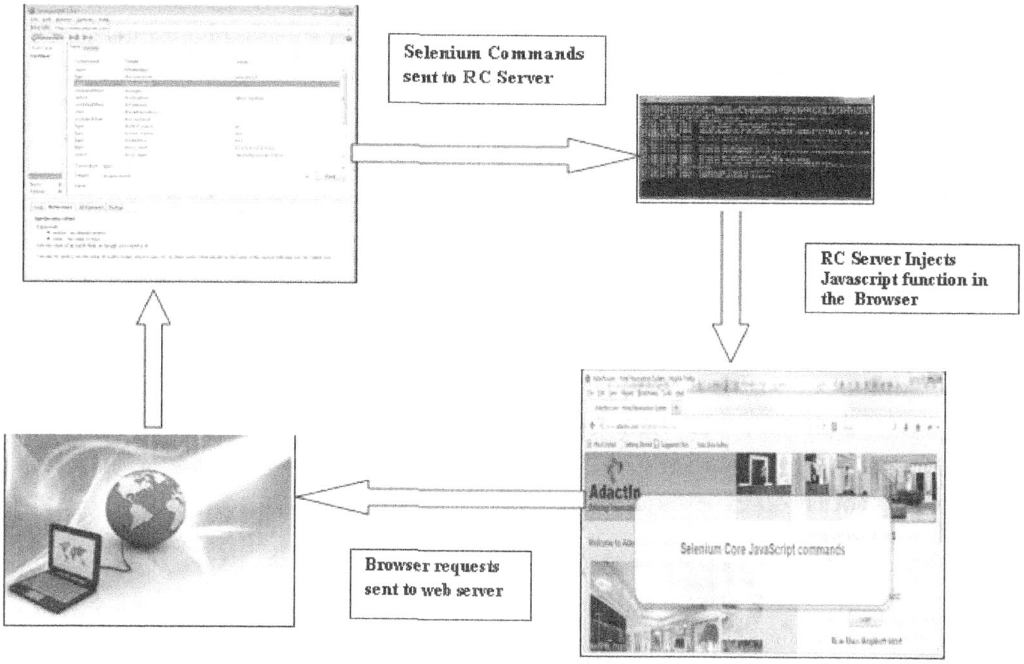

Figure 4-2 Selenium RC Architecture

The diagram shows how the client libraries communicate with the Server passing each Selenium command for execution. Then the server passes the Selenium command to the browser using Selenium-Core JavaScript commands. The browser, using its JavaScript interpreter, executes the Selenium command. This runs the Selenese action or verification you specified in your test script.

3. Selenium WebDriver

WebDriver was a new project developed due to inherent limitation of the Selenium RC. WebDriver interacted directly to the browser using the 'native' method for the browser and operating system, thus avoiding the restrictions of a sandboxed JavaScript environment. The WebDriver project began with the aim to solve Selenium's pain-points.

Developers of both Selenium RC and WebDriver decided to merge both the tools and create Selenium 2.0 aka Selenium WebDriver. It was a new addition to the Selenium toolkit which provided all sorts of awesome features, including a more cohesive and object-oriented API as well as an answer to the limitations of the old implementation (Selenium RC). Both tools had their own advantages and merging the two provides a much more robust automation tool.

Selenium WebDriver supports the WebDriver API and underlying technology, along with the Selenium 1 technology underneath the WebDriver API for maximum flexibility in porting your tests. In addition, Selenium 2 still runs Selenium 1's Selenium RC interface for backward compatibility.

In Selenium 3, the Selenium RC core has been removed and replaced with a backend WebDriver. In Selenium 3 instead of using the Selenium RC API's to interact with browser, the webdriver directly uses the implementations provided by the browser Vendor and hence the interactions are more robust and gives a better testing experience.

In Selenium 4, the major difference is that the communication between the selenium webdriver and the browsers have been moved from the JSON wire protocol to the W3C protocol. Along with this some code updates have been done as well to have a smoother automation experience.

Selenium WebDriver's architecture is simpler than Selenium RC's.

WebDriver uses a different underlying framework from Selenium's JavaScript Selenium-Core. It makes direct calls to the browser using each browser's native support for automation. How these direct calls are made, and the features they support depends on the browser you are using.

- It controls the browser from the OS level.
- All you need are your programming language's IDE (which contains your Selenium commands) and a browser.

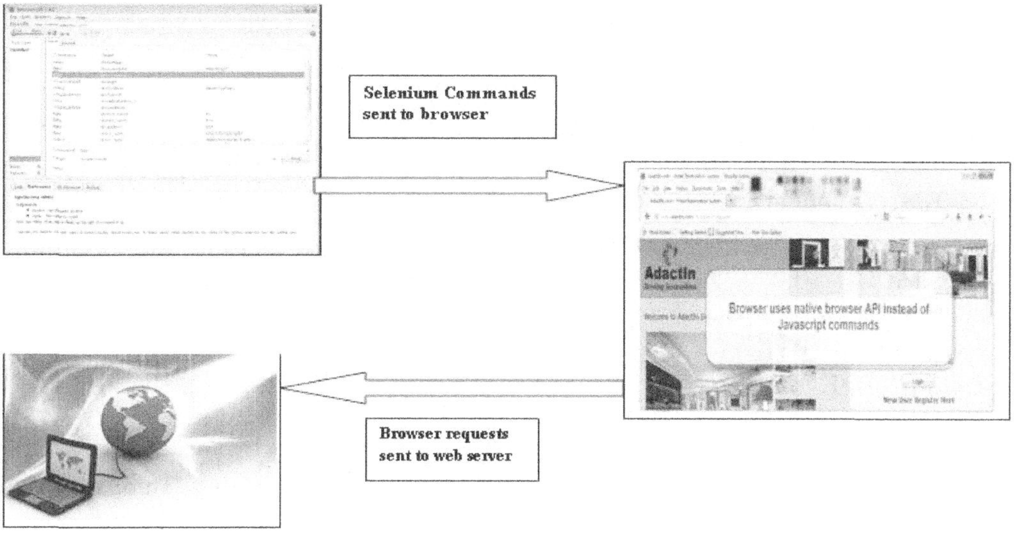

Figure 4-3 Selenium WebDriver Architecture

It provides the following:

- A well-designed standard programming interface for Web application testing.
- Improved consistency between browsers.
- Additional functionality addressing testing problems not well-supported in Selenium 1.0 like handling multiple frames, pop-up, and alerts, better page navigation, drag-and-drop and handling of AJAX-based UI elements.

4. Selenium-Grid

Selenium Grid is used to run parallel tests across different machines and different browsers all at the same time. Parallel execution means running multiple tests at once.

Features:

- Enables simultaneous running of tests in multiple browsers and environments.
- Utilizes the hub-and-nodes concept. The hub acts as a central source of Selenium commands to each node connected to it.

This has two advantages.

- First, if you have a large test suite, or a slow-running test suite, you can boost its performance substantially by using Selenium Grid to divide your test suite to run different tests at the same time which will result in significant time savings.
- Also, if you must run your test suite on multiple environments you can have different remote machines supporting and running your tests in them at the same time.

In each case Selenium Grid greatly improves the time it takes to run your suite by making use of parallel processing.

4.2. How to Choose the Right Selenium Tool for your need

Tool	Why Choose?
Selenium IDE	To learn about concepts on automated testing and Selenium, including: • Locators such as id, name, XPath, CSS selector, etc. • Exporting test cases in various programming languages. • To create tests with little or no prior knowledge in programming. • To create simple test cases and test suites that can be exported later and modified according to your need.
Selenium WebDriver	• To use a certain programming language in designing your test case. • To execute tests on the HTML Unit browser. • To create customized test results.
Selenium Grid	• To run your Selenium WebDriver scripts in multiple browsers and operating systems simultaneously. • To run a huge test suite that needs to complete as soon as possible.

Advantages/Limitations of Selenium Tools

You'll find that Selenium is highly flexible. There are many ways you can add functionality to both Selenium test scripts and Selenium's framework and customize your test automation. This is perhaps Selenium's greatest strength when compared with other automation tools. In addition, since Selenium is Open Source, the source code can always be downloaded and modified.

Advantages of Selenium

1. Open-Source tool.
2. Supports all browsers like, Edge, Chrome, Firefox, Safari .
3. Runs on all operating systems - Windows, Mac OS and Linux.
4. Supports various languages like Java, .NET, Ruby, Perl, PHP, Python etc.
5. Runs multiple tests at a time.
6. Provides the option of using wide range of IDEs such as Eclipse, IntelliJ IDEA, Netbeans, Visual Studio, etc., depending on the choice of development language.

Limitations of Selenium

1. Supports only browser based applications, not desktop/windows applications.
2. Does not support file upload from local machine.
3. Requires high technical skills to meet its full potential.
4. Being an open source, Selenium has no official technical support.

4.3. Installation Requirements for Selenium

In this section we will discuss some of the tools that we need to install for using Selenium. We will look into installing each of these tools and add-ons in upcoming chapters as and when needed.

Non-Selenium Installations:

- Windows 8 Or latest
- Install Chrome browser
- Get IDE according to the choice of your language like Eclipse or NetBeans or IntelliJ IDEA for Java programming
- Install WinANT
- Firefox / Chrome add-ons ChroPath to get XPath for UI elements
- Jenkins – Continuous Integration tool

Selenium related Installation:

- Selenium IDE Add-on - A Firefox /Chrome / Edge add-on to record and play scripts
- Latest Selenium binaries downloaded and unpacked into a directory from https://mvnrepository.com/artifact/org.seleniumhq.selenium/selenium-java. A safe choice is the latest version of selenium-server-standalone-x.y.z.jar x, y and z will be digits, e.g., 4.10.0 at the time of writing and save these in a folder, say, Lib for later use.
- TestNG Plugin

So let us get started!

§

5

Installing Selenium Components

Introduction

Before we can start using Selenium, there are a few Selenium and non-Selenium components that we need to install. In this chapter we will perform step by step installation and setup of the components which we will need to use over the scope of this book.

Note: You need access to the internet to download the required setup files.

Key objectives:

1. Setup Instructions for installing Selenium IDE
2. Setup Instructions to install add-on ChroPath
3. Setup Instructions to install Java Development toolkit
4. Setup Instructions to install and setup Eclipse
5. Setup Instructions to install WinANT

5.1 Installing Selenium IDE

Prerequisite – Chrome browser should be installed locally on the test machine.

1. Launch Chrome browser and open URL https://www.selenium.dev/downloads o download Selenium IDE from the SeleniumHQ download Page.
2. Click on the Selenium IDE link

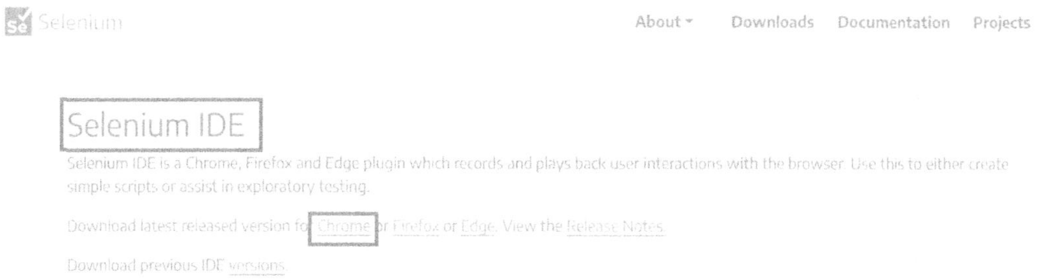

Figure 5-1 – Download Selenium IDE

3. Click on the Chrome link below Selenium IDE section.

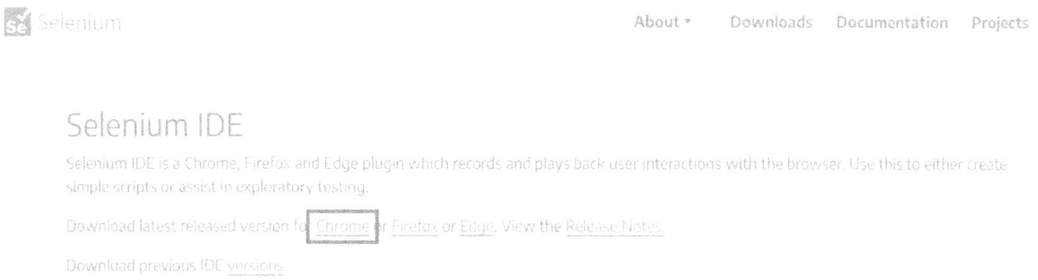

Figure 5-2 – Download Selenium IDE for Chrome

4. Click on 'Add to Chrome'

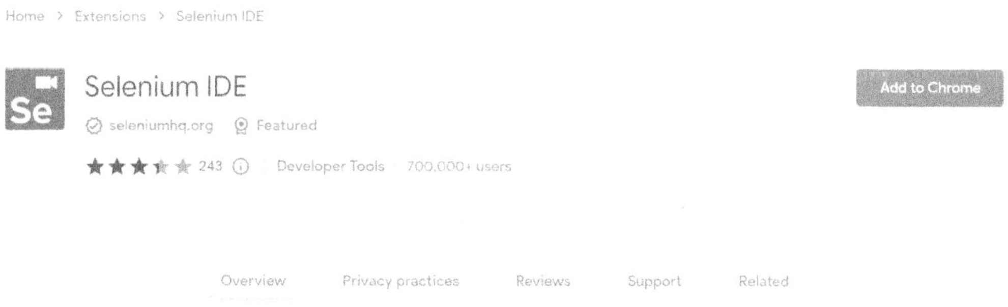

Figure 5-3 – Add to Chrome

5. Click on 'Add Extension'

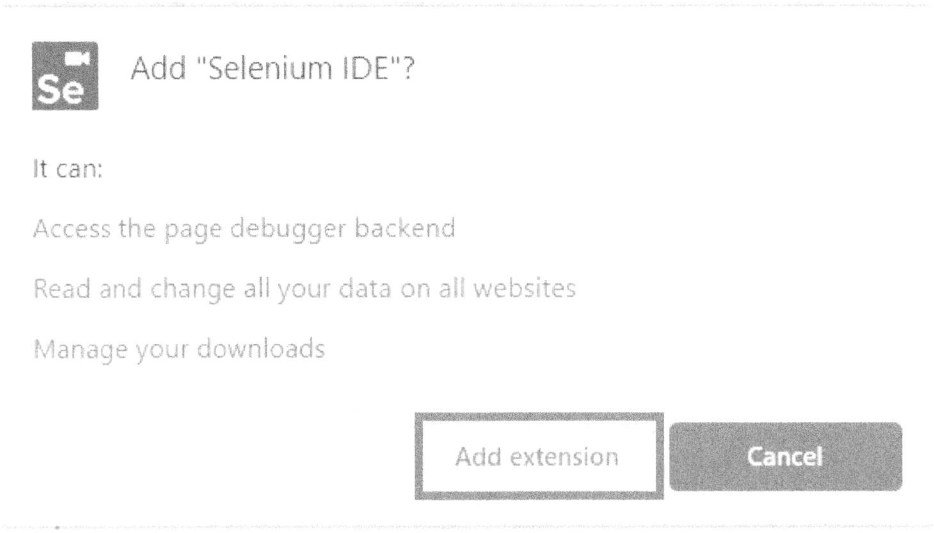

Figure 5-4 – Add Selenium IDE Extension

6. Check the tool bar to see if the Selenium IDE is present in Extensions

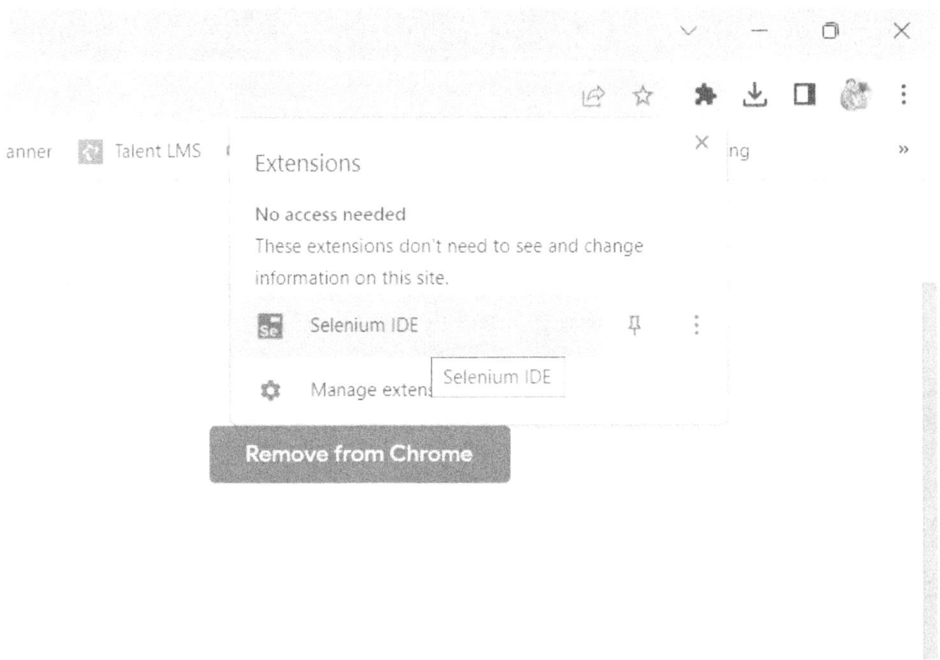

Figure 5-6 –Selenium IDE in Extensions

5.2. Installing ChroPath plug-in

Now we will install Chrome add-on **ChroPath** (if we haven't done that already).

ChroPath is a development tool to edit, inspect and generate XPath and CSS selectors. ChroPath makes it easy to write, edit, extract, and evaluate XPath and CSS queries on any webpage and saves atleast 40–50% manual effort in automation script writing

In Selenium, ChroPath helps in inspecting UI elements and finding its associated properties and values.

1. To install ChroPath add-ons, Open Chrome browser, and open URL '**https://chrome.google.com/webstore/category/extensions**' and search for '**ChroPath**'. Click on **ChroPath** section.

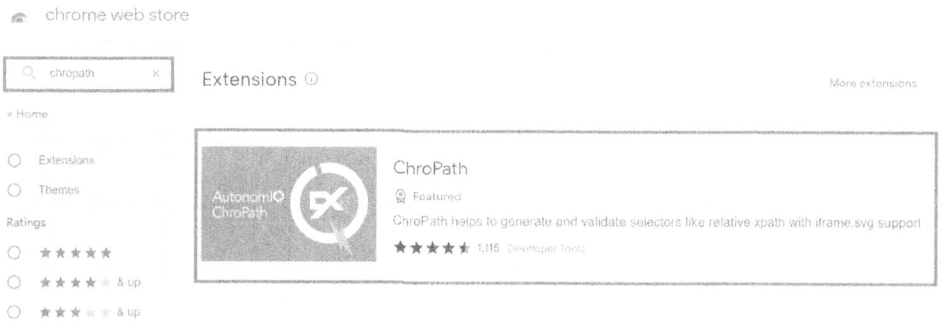

Figure 5-7 ChroPath link

2. Add-ons page appears. Click on + Add to Chrome button

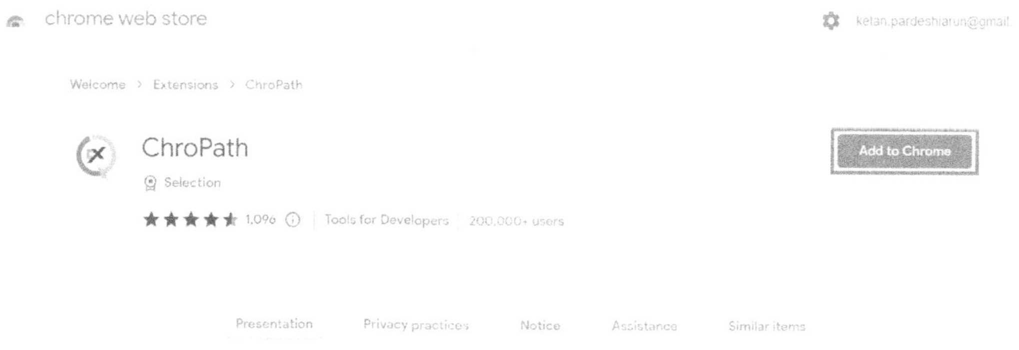

Figure 5-8 Add ChroPath to Chrome

3. Click on 'Add Extension' button on the popup

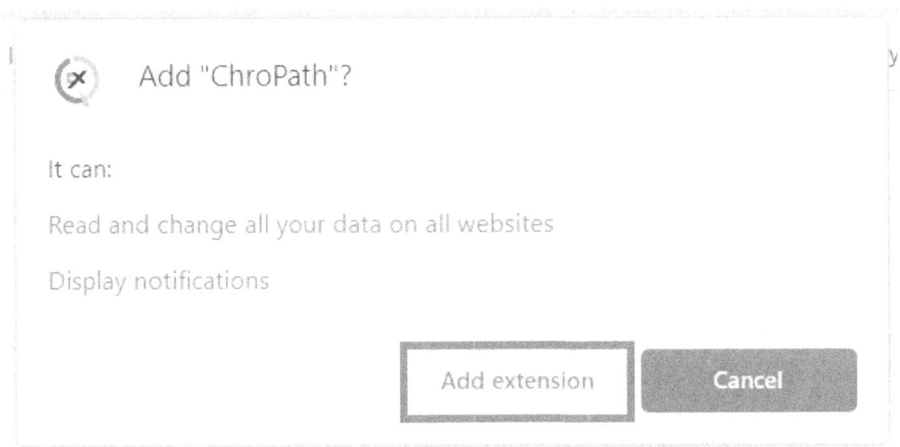

Figure 5-9 Add ChroPath Extension

4. Verify that Chropath Added confirmation is displayed.

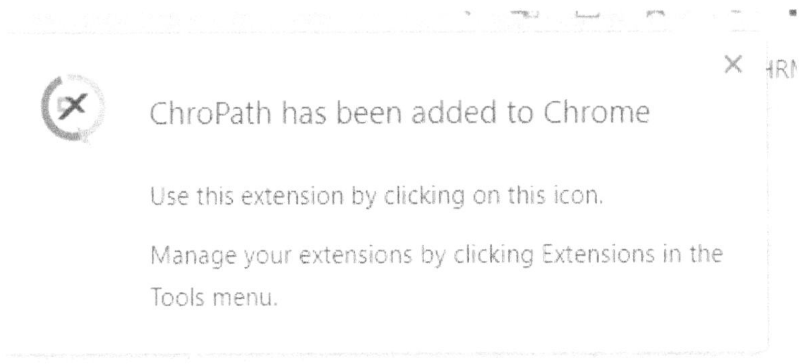

Figure 5-10 ChroPath Added confirmation

5. Restart the browser and load www.google.com

6. Right Click on 'Search' textbox on google.com webpage and select option as Inspect to see the developer options

Figure 5-11 Inspect Element

7. Click on '>>' icon to view the Dev Tools

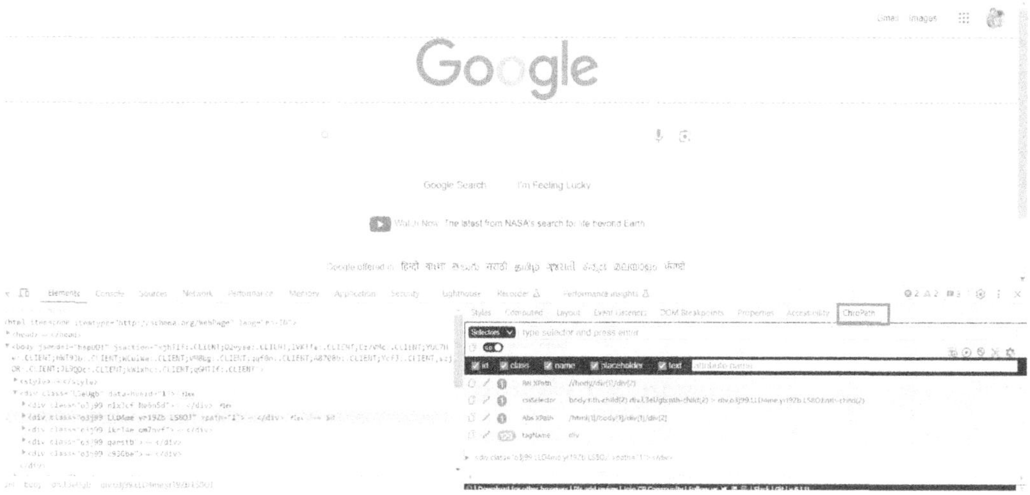

Figure 5-12 Navigation to ChroPath Interface

8. Click on 'ChroPath' Tab

Figure 5-13 Navigation to ChroPath Interface_Step 2

9. This will open ChroPath interface

Figure-5-14 ChroPath Interface

5.3. Installing Java Development Kit

Since we will be using Java as a base programming language for Selenium WebDriver we will download and install Java Development Kit.

Note: Apart from Java, Selenium also supports languages like .NET, Ruby, Python, Perl and PHP. Setup and using Selenium with other languages is not in the scope of this book. We will be focusing on using Selenium WebDriver with Java.

Downloading and Installing Java

A typical Java installation takes about 400 MB of disk space. The installation process reports the exact size.

1. Go to **https://www.oracle.com/java/technologies/downloads/** Select tab **Windows** and Click on URL present in front of **x64 Installer** .

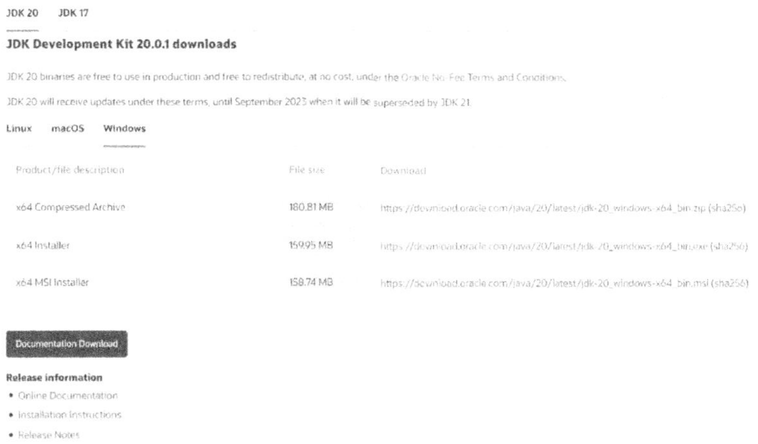

Figure 5-15 – Java Download

2. Wait till exe file gets downloaded. Store this exe file somewhere permanent on your disk drive.
3. Double-click the file that you just downloaded, to run it.
4. Click on 'Yes' to 'Do you want to allow this app to make changes to your device?'. You will see a 'Windows Installer' window temporarily then a 'Preparing to Install window' temporarily.
5. Clicking 'Next' on this wizard will guide you through the installation process for the Java SE Development Kit <version>'.

Figure 5-16 Java Installation Wizard

6. Click on 'Next'.
7. You will see the folder to which Java is saved. Click on 'Next'

Figure 5-17 Java Directory Path

8. You will see an Installing window (for Java SE Development Kit <version>), whose progress is shown by progress bars (on various panes). Be patient: Many megabytes of information are being downloaded and then used to perform the actual installation.
9. Click the Close button on the 'Installation Completed' window

5.4. Installing and Configuring Eclipse

About Eclipse: Eclipse is an Open-Source integrated development environment (IDE). Written mostly in Java, Eclipse can be used to develop Java applications. By means of various plug-ins, Eclipse may also be used to develop applications in other programming languages: Ada, C, C++, COBOL, Fortran, Haskell, JavaScript, Lasso, Perl, PHP, Python, R, Ruby (including Ruby on Rails framework), Scala, Clojure, Groovy, Scheme, and Erlang.

Alternatively, you could also use another popular IDE IntelliJ IDEA for Java development from JetBrains. Unlike Eclipse, IntelliJ IDEA is a commercial tool. However, JetBrains offers a community edition with limited features for free use.

In keeping with this book's scope, we will be using Eclipse IDE for creating our Junit and TestNG based Selenium WebDriver tests.

Downloading and Installing Eclipse

1. Go to http://www.eclipse.org/downloads/ . Click Eclipse Standard or Eclipse IDE for Java Developers. A page, with the label Eclipse Downloads, will be displayed in your browser.
2. Click on Download x86_64 button

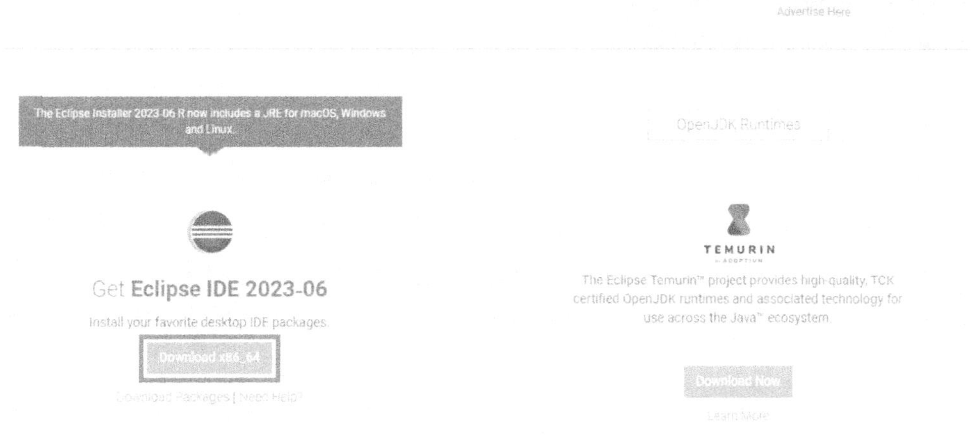

Figure 5-18 - Eclipse Download

2. Click on Download button. Wait till eclipse exe gets downloaded.

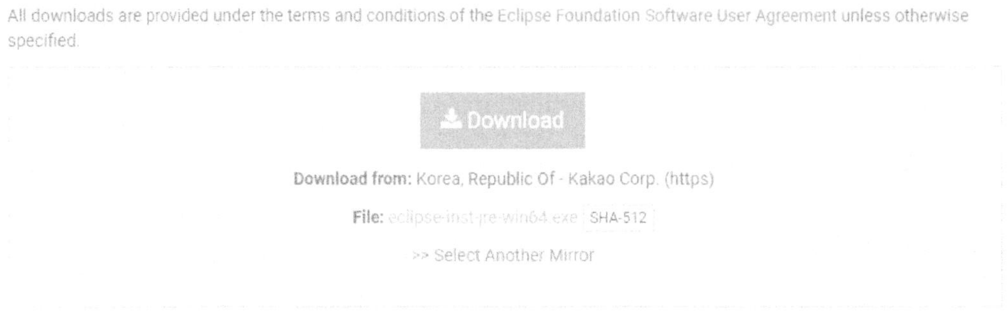

Figure 5-19 - Eclipse Download

3. Double click on the downloaded exe file and then click on 'Eclipse IDE for Java developers'.

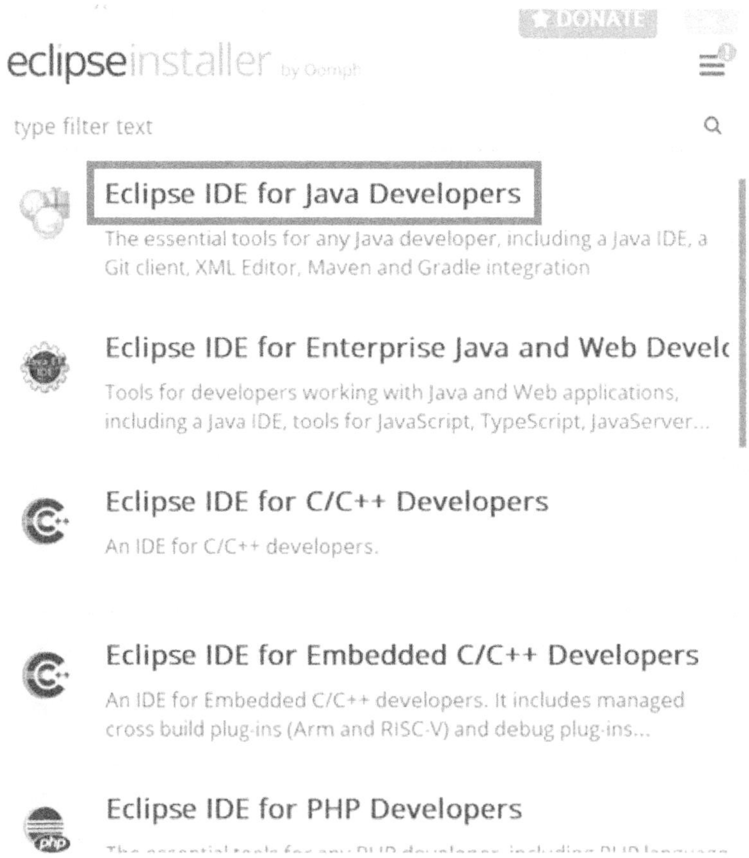

Figure 5-20 - Eclipse IDE for Java Developers

4. Click on 'Install' button after entering a valid location to save it. Wait till installation gets completed.

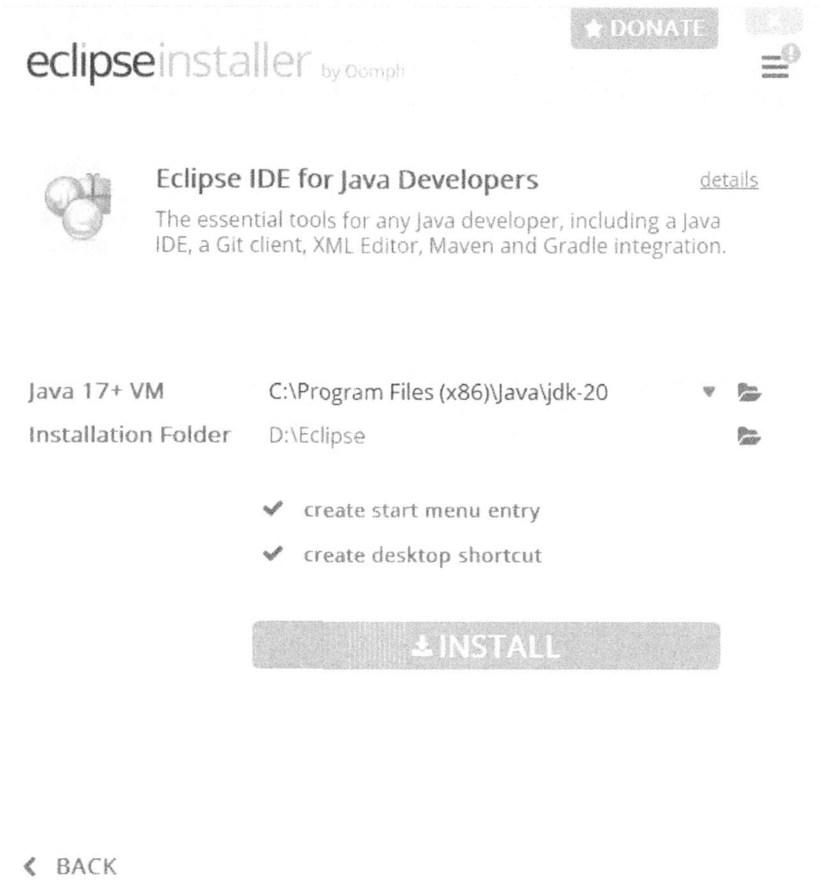

Figure 5-21 - Install Eclipse

Now you are ready to perform a one-time only setup of Eclipse.

1. Click on Launch Eclipse.

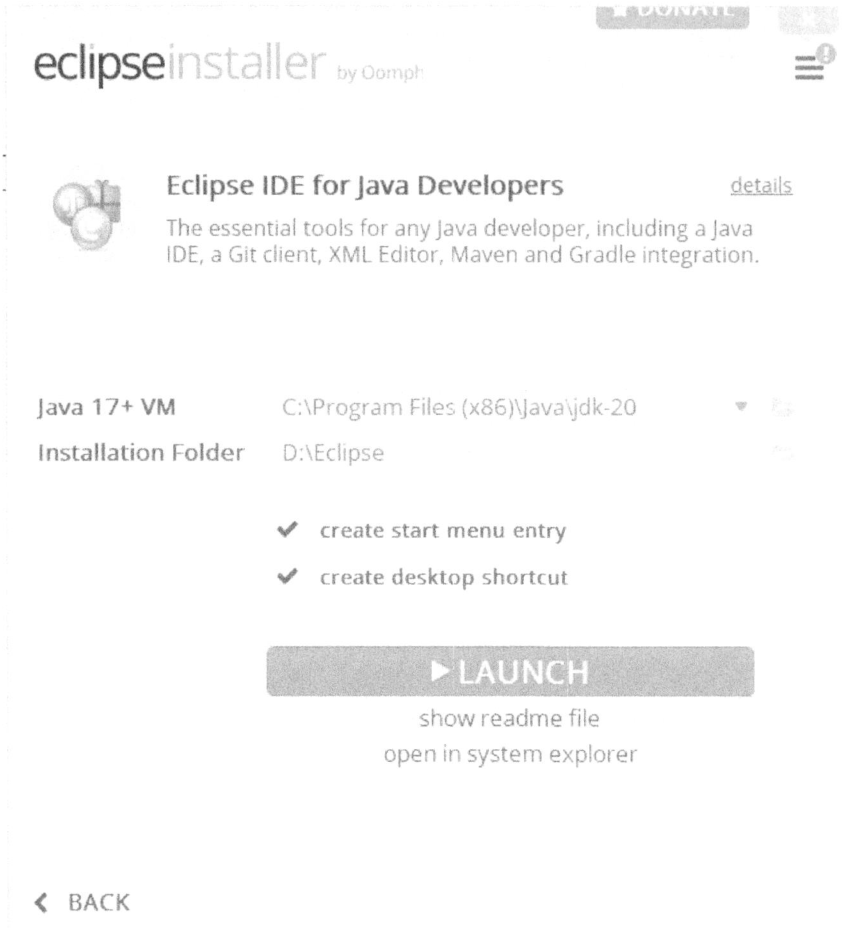

Figure 5-22- Eclipse location

2. In the Workspace Launcher window select the location for your workspace. By default it could be something like C:\Documents and Settings\username\workspace (where username is your login on the machine). We can also browse to another location for the workspace file to be created.
3. Create a folder in C:\ drive named Selenium and save the workspace in that folder. You can click on the Browse button to Browse to Selenium folder.

So use Path **C:\Selenium\workspace** for eclipse

Note: Workspace is the location where all the Java/Selenium project files will reside.

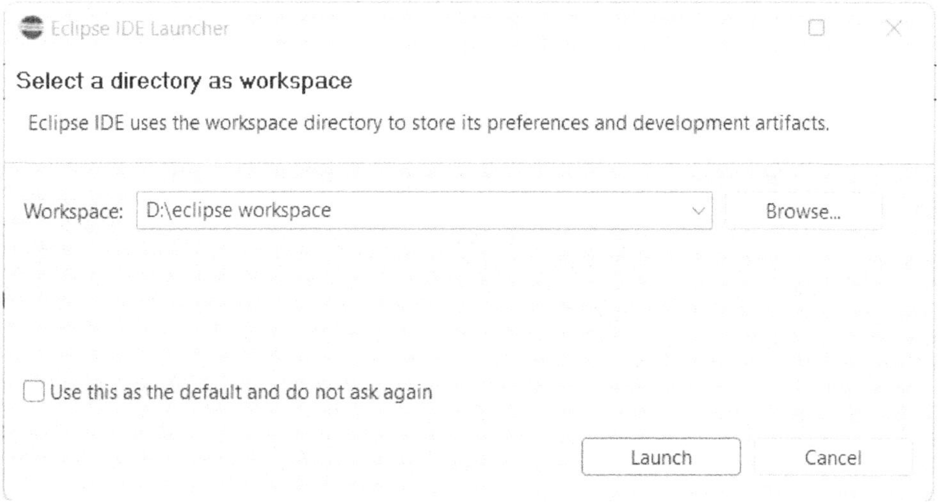

Figure 5-23 - Eclipse Workspace location

4. Click on 'Launch' button and you will see the Eclipse interface

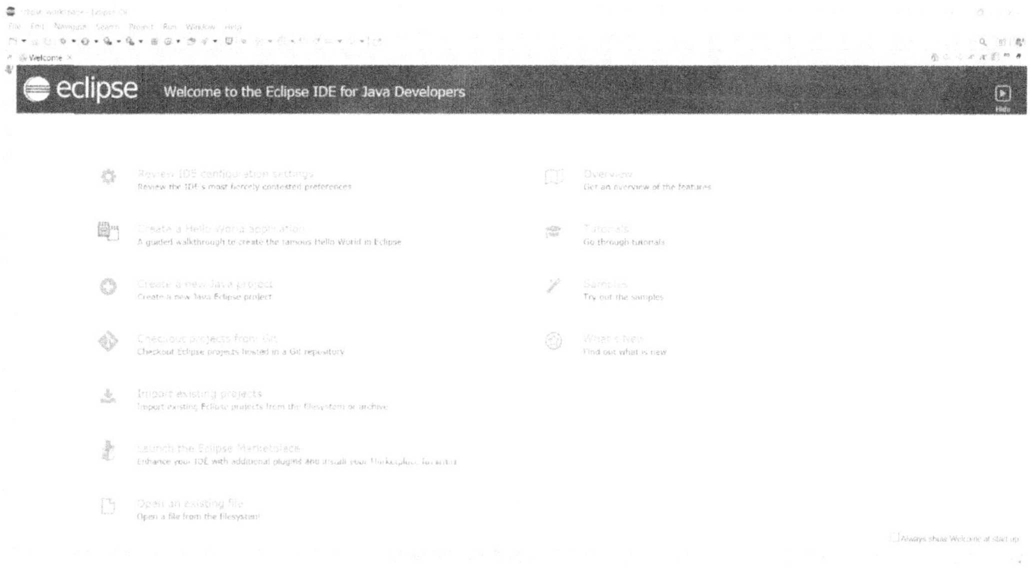

Figure 5-24 - Eclipse Interface

5. Eclipse, select Windows → Preferences→ Java → Installed JREs

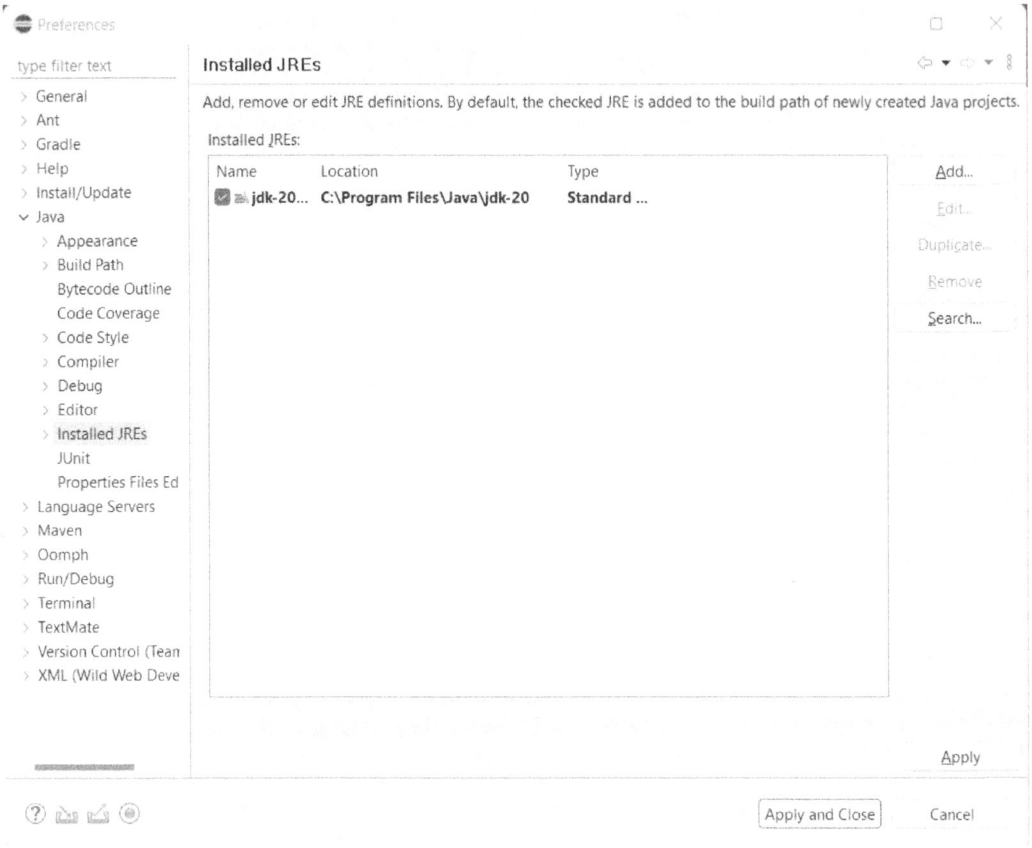

Figure 5-25 – Eclipse Preferences6.

6. Under the Name column, make sure jdk<version> checkbox is selected. Or else Click on Add button to browse to location of JDK<version> and add it to list.

7. Close the Preferences dialog.

8. We are done with the one-time setup. We can keep the eclipse open or close it for now to be used later with Selenium WebDriver.

5.5. Installing WinANT

Apache Ant is a Java library and command-line tool for automating software build processes. The main known usage of Ant is the build of Java applications. It supplies several built-in tasks allowing to compile, assemble, test and run Java applications.

WinANT - WinAnt is a Windows installer for Apache Ant.

Since we will be working on Windows Environment, we will install WinANT. Libraries to be used in other operating systems can be found at http://ant.apache.org/

Downloading and Installing WinANT

1. Go to https://code.google.com/archive/p/winant/downloads and download the latest version.

Figure 5-26– Download WinANT

2. Double click on the downloaded exe file. Click on 'Yes' when a popup which displays 'Do you want to allow this app to make changes?' is displayed

3. Click on the **Save File** to save the install on your local machine

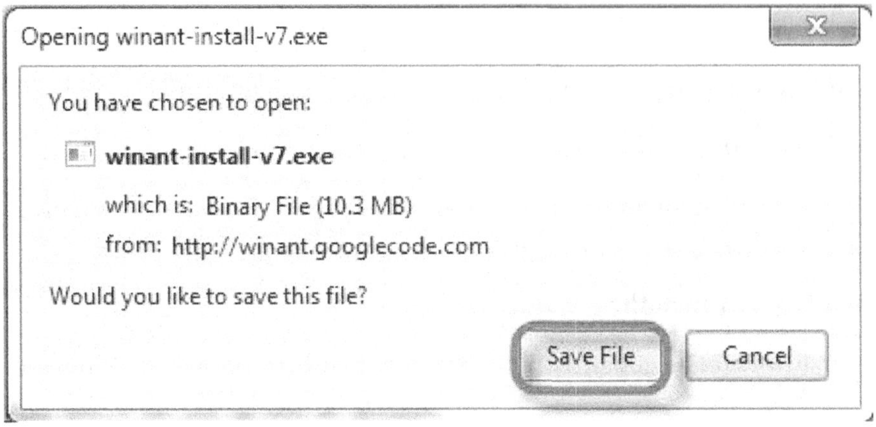

Figure 5-27 – Save WinANT installer

4. Once download is completed, **run** the **winant-install-v*.exe** file
5. Read the details of license agreement dialog and click **Next**

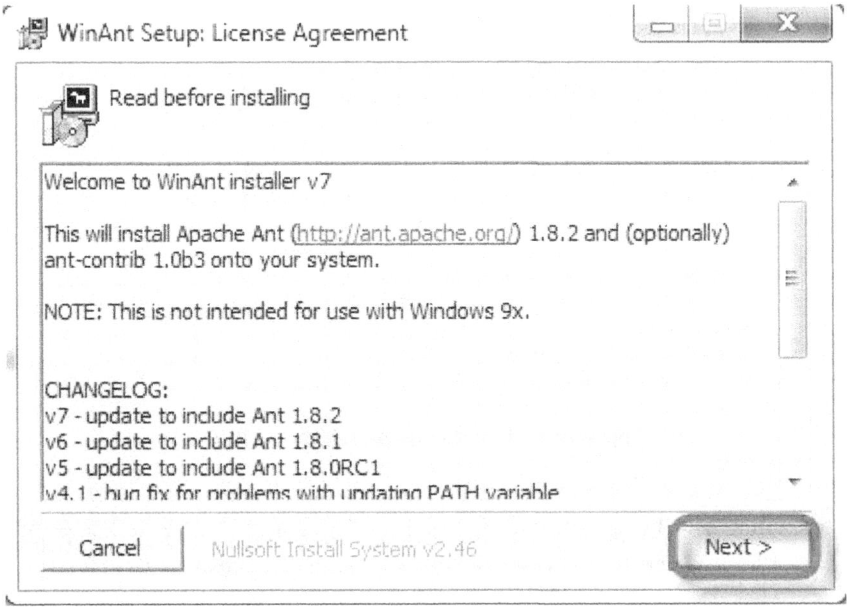

Figure 5-28 – License Agreement

6. Verify installation options dialog and click **Next**

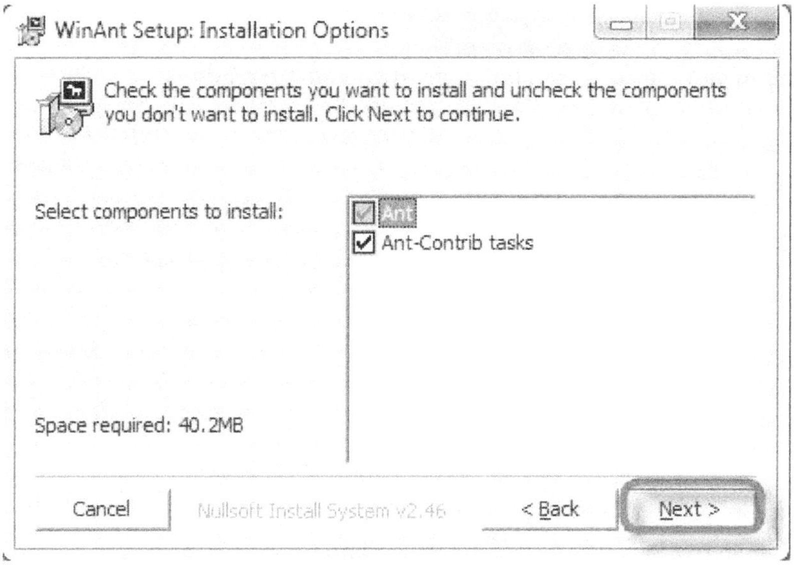

Figure 5-29 – WinANT Installation Options

7. ANT requires a path to your Java Directory. By default it should be C:\Program Files\Java\jdk**.*.* (it is the location at which you had installed the Java development toolkit). Browse to your JDK directory and Click **Next**
8. Confirm the default WinANT installation location and click **Install**

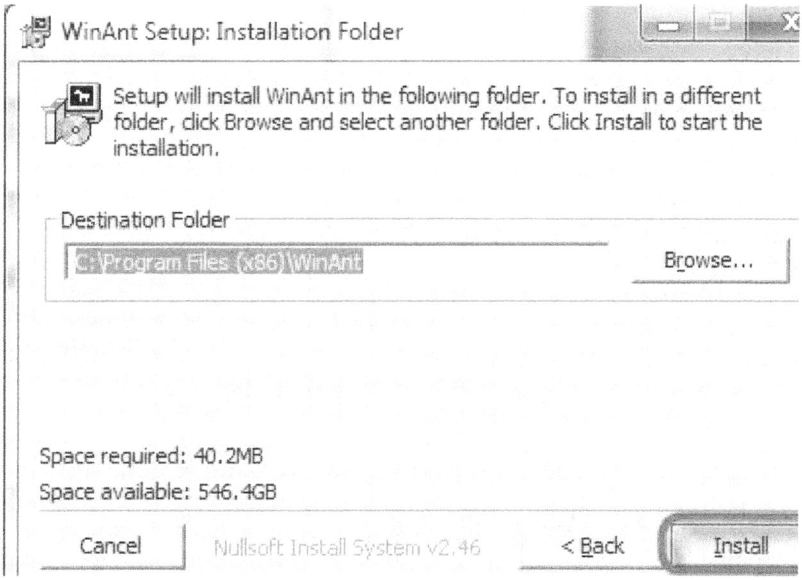

Figure 5-30 – WinANT Installation Folder

9. Once you have completed the installation, installer will ask to Log off and log back in to complete the installation process. Click **OK**

Note: You can log off at a later stage once you are ready.

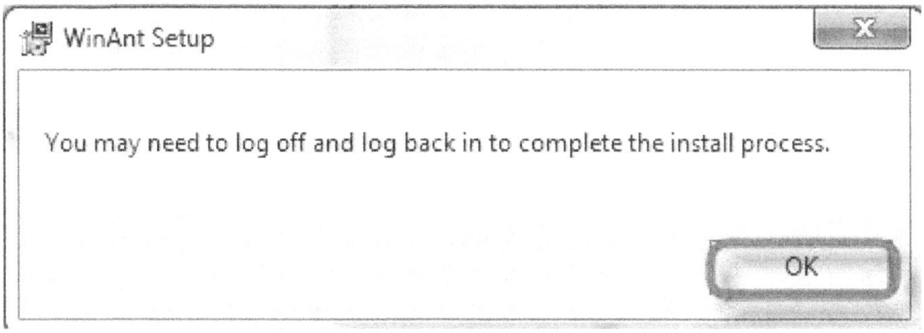

Figure 5-31 – Log-off Dialog

10. Click **Close** to close the WinAnt Setup dialog

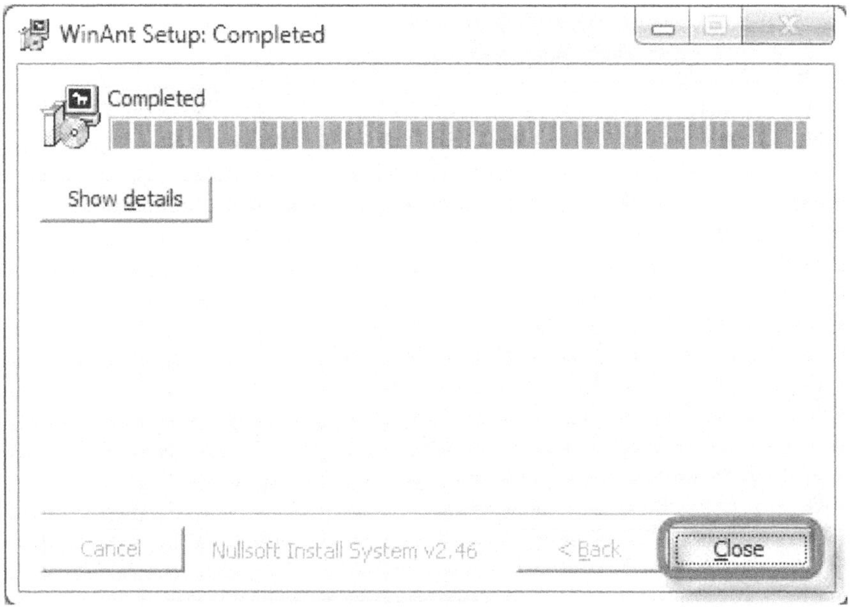

Figure 5-32 – Close Installer

11. You can log off and log back into to the system to complete the installation.
12. We are done with WINAnt installation!

§

6

Using Selenium IDE

Introduction

Many times, functional testers believe that we need to write programs to automate applications. This is not entirely true. Most of the automation tools come with record/replay features which allow you to record user actions and replay those actions back without writing a single line of program. Yes, you might need to make some enhancements to your script, which again can be accomplished without any programming.

In Selenium we can perform record/replay and automate test cases using Selenium IDE. Selenium IDE is an easy-to-use Firefox/Chrome/Edge plug-in and is generally a quick and efficient way to develop test cases. It also contains a context menu that allows you to first select a UI element from the browser's currently displayed page and then select from a list of Selenium commands with parameters pre-defined according to the context of the selected UI element. This chapter is all about understanding features of Selenium IDE and how to use it effectively.

example - As a consultant we have to perform automation feasibility studies across various client applications. The simplest way to do the feasibility study using Selenium is by using Selenium IDE. By recording and replaying back a basic script we can figure out if application can be automated by Selenium. If yes, we can delve into the bigger effort of configuring Selenium WebDriver.

In this chapter we will define how to record a basic script, replay the script, and save the script using Selenium IDE.

Key objectives:

- Understand Selenium interface
- Record a basic script
- Replay the script
- Save the script

6.1. Selenium IDE Interface

1. To launch the Selenium-IDE, open Chrome browser and click on Selenium IDE extension on the Toolbar

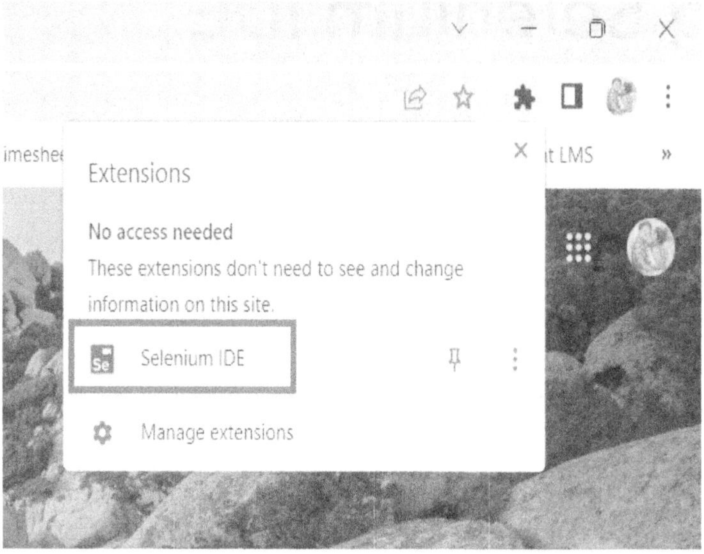

Figure 6-1 - Open Selenium IDE

2. It opens with a welcome message with options to create a new project or open an existing project

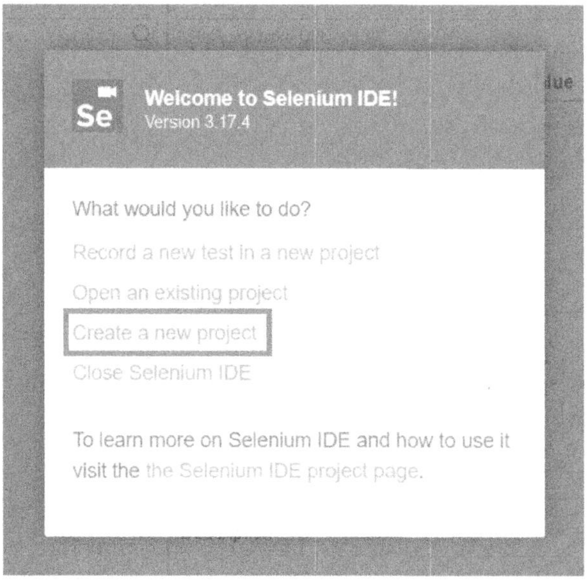

Figure 6-2 - Selenium IDE Welcome Screen

3. Click on **'Create a new project'**. Enter any project name as **HotelAppApplication** and click on **'Ok'**

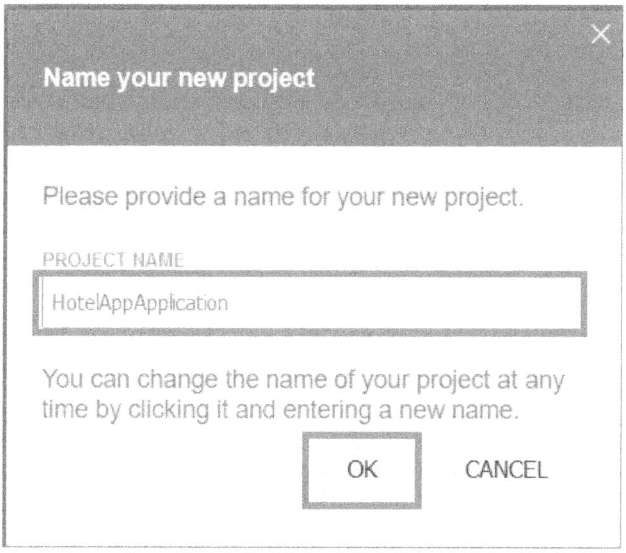

Figure 6-3 - Creating New Project

4. Click on **record** button.

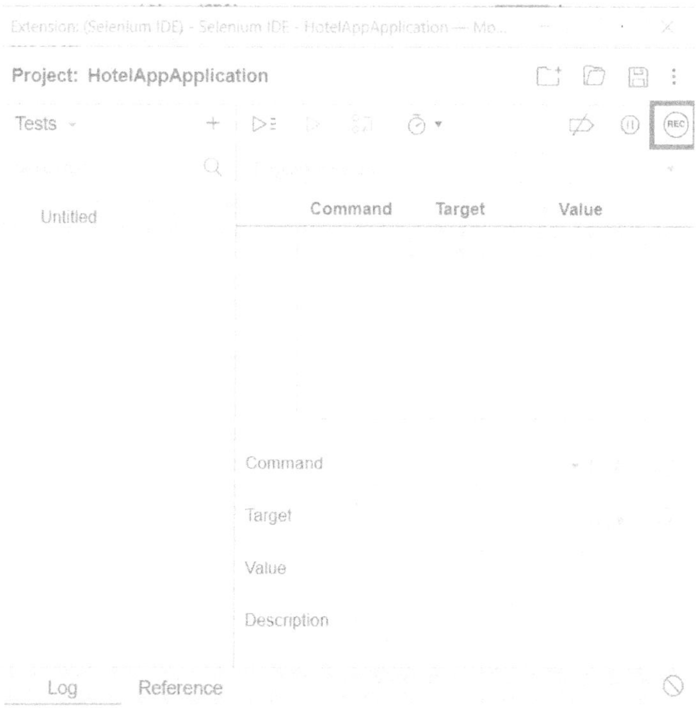

Figure 6.4 Record Session

5. Enter Project Base URL: http://adactinhotelapp.com and click on **Start Recording**

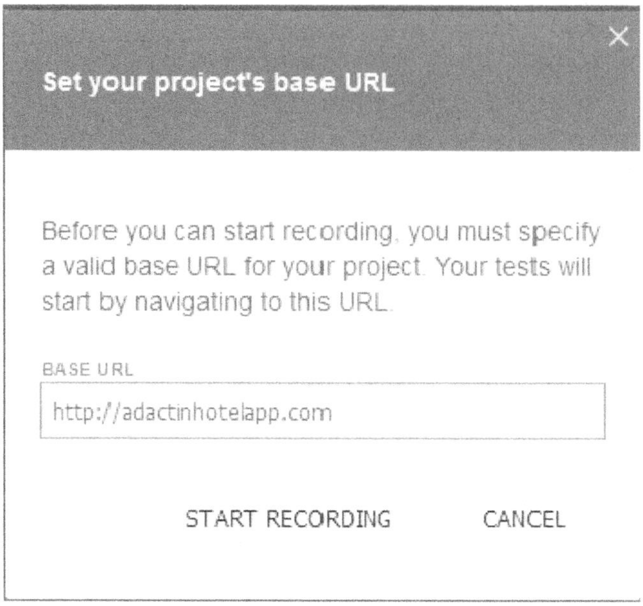

Figure 6-5 Set Project's base URL

6. The Selenium IDE window is now opened

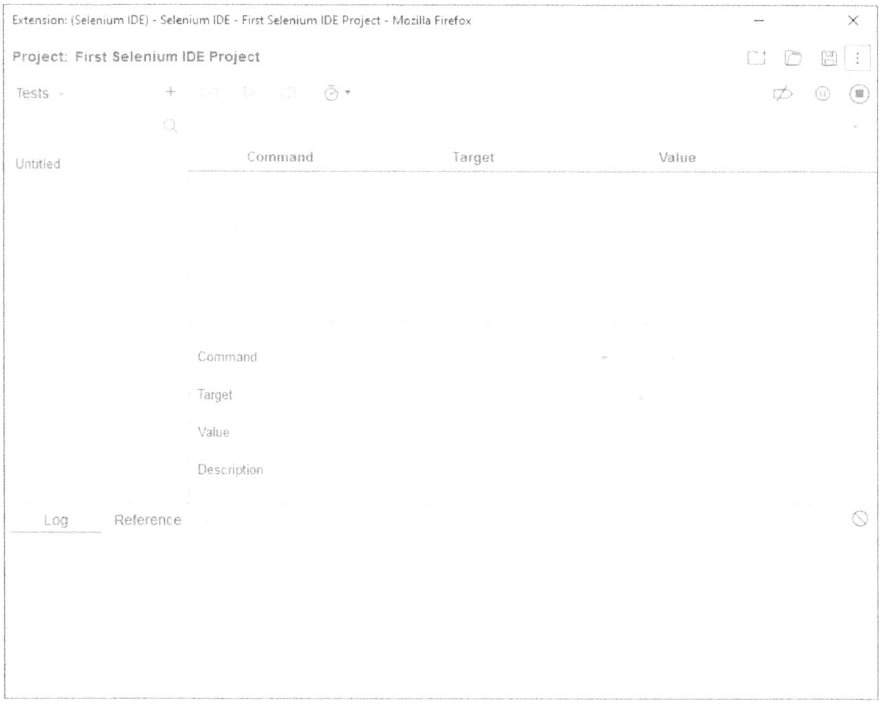

Figure 6-6 Selenium IDE

Menu Bar

The Menu Bar has options to:

1) Create New Project
2) Open Project
    a) Save Project
    b) Running in CI option which opens a page in Chrome with information about Command Line Runner
    c) What's New option which provides information about the latest release of Selenium IDE
    d) Help option which navigates to a Chrome page with information about the installation, launching and troubleshooting of IDE. It also has information about the Welcome Screen, , saving and playback of tests.

Toolbar

The toolbar contains buttons for controlling the execution of your test cases, including a step feature for debugging your test cases. The extreme right button, the one with 'REC' text present inside a circle is the record button.

Figure 6-7 - Selenium IDE Toolbar

<Run all tests>	Run all tests: Runs the entire test suite when a test suite with multiple test cases is loaded.
<Run current test>	Run current test: Runs the currently selected test. When only a single test is loaded this button and the Run All button have the same effect.
<Stop test execution>	Stop test execution: Allows stopping of a running test case
<Pause_Resume test execution>	Pause/Resume test execution: Allows pausing/resuming of a running test case.
<Step over current command>	Step over current command: Allows you to "step" through a test case by running it one command at a time. Use for debugging test cases.
<Test execution speed>	Test execution speed: Controls how fast your test case runs.

<Disable breakpoints>	Disable breakpoints: when enabled,it ignores the breakpoints defined for the test.
<Pause on exceptions>	Pause on exceptions: If this button is enabled, it pauses the execution in case any exception is encountered.
<Start recording>	Start recording: Records the user's browser actions.

6.2. Recording Using Selenium IDE

Alright! Let us record our first script now. **Recording your First Test Case**

1. Open Chrome browser.
2. Enter URL for the site you want test. Let's take our sample application http://adactinhotelapp.com/
3. Click on Selenium IDE extension on the toolbar
4. Click on 'Create a new project'. Enter any project name and click on 'Ok'
5. Click on Record button
6. IDE is already in a recording mode. If not, press REC button to start recording.

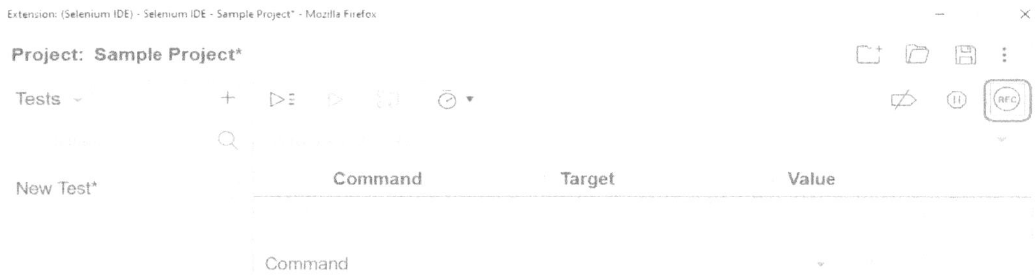

Figure 6-8 Recording Button

7. Enter Project Base URL: http://adactinhotelapp.com and click on Start Recording.
8. The Selenium IDE will open chrome browser with given URL.
9. Assuming that application is already open in Chrome browser with login page visible, perform the following steps:
   a. Login (Use the username/password with which you have registered earlier).
   b. Search for Hotel.
      i. Select a location, e.g., Sydney
      ii. Select number of rooms, e.g., 2-Two
      iii. Select adults per rooms, e.g., 2-Two
      iv. Click on Search button

c. Select a Hotel and click on continue button.
    i. Select one of the Hotel Radio buttons, e.g., select radio button next to Hotel Creek.
d. Book a Hotel.
    i. Enter First Name
    ii. Enter Last Name
    iii. Enter Address
    iv. Enter 16-digit Credit Card no:
    v. Enter Credit Card type
    vi. Enter Expiry Month
    vii. Enter Expiry Year
    viii. Enter CVV number
    ix. Click on Book Now
e. After you see the Booking confirmation page, click on Logout link in the top right corner
f. Click on "Click here to Login again" link to go back to Home page.

10. Stop recording by clicking on Stop Recording button in record toolbar.

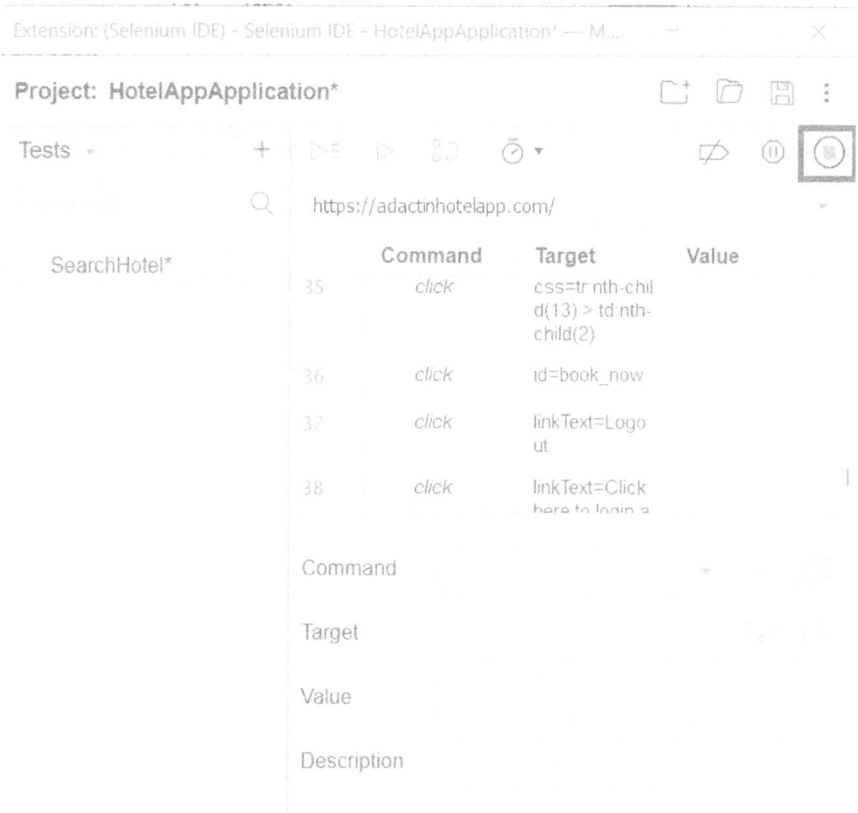

Figure 6-9 Stop Recording Button

11. Enter Test case Name and click on Save button.

12. Verify the steps below that are recorded in Selenium IDE.

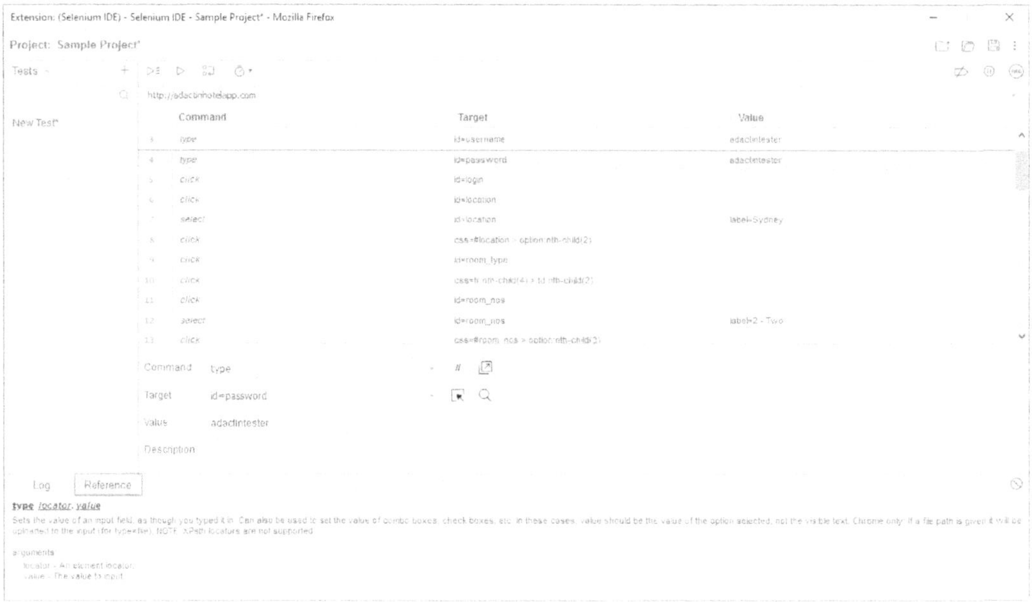

Figure 6-10 IDE Steps

If you look at the above steps, first column "Command" represents the operation or method performed on user controls. For example, type method to enter value in username field.

Second column- "Target" represents the user controls of the application on which actions are done.

Third column "Value" represents all the input data that has been entered into the application for testing.

Note: Note that all the actions which we performed are captured as separate steps in the IDE script.

During recording, Selenium-IDE will automatically insert commands into your test case based on your actions. Typically, this will include:

clicking a Web element like a button or link - click or clickAndWait commands
entering values - type command
selecting options from a drop-down list box - select command
clicking checkboxes or radio buttons - click command
Select step of where location selection is Sydney

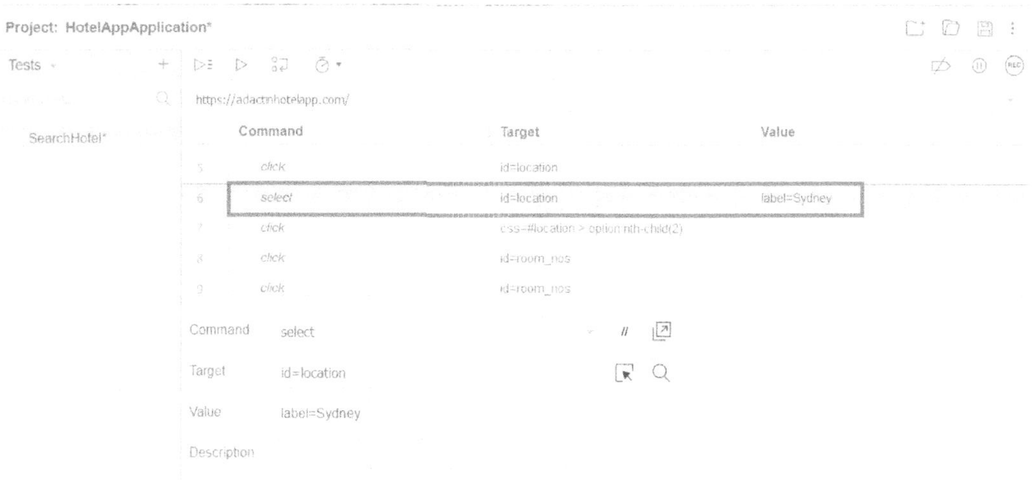

Figure 6-11 IDE Commands, Target and Values

Note that in the bottom pane you can select values for Command, change Target and Data values.

6.3. Save and Replay the Script using IDE

Saving the Project

Now you can save your script. Like most programs, there are Save and Open commands under the File menu. However, Selenium distinguishes between test cases and test suites. To save your Selenium-IDE tests for later use, you can either save the individual test cases or save the test suite. If the test cases of your test suite have not been saved, you'll be prompted to save them before saving the test suite.

1. Select 'Save Project' option in the Menu bar

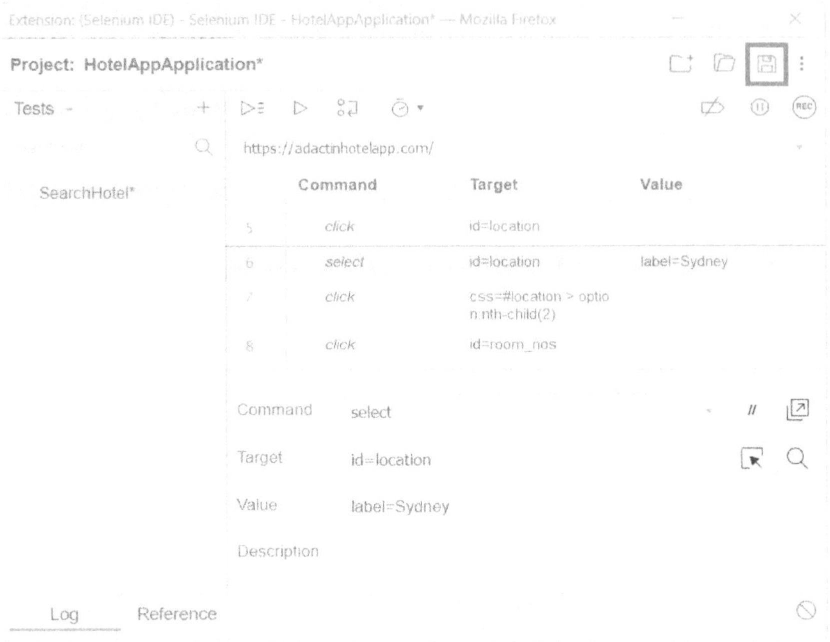

Figure 6-12 Option to Save Project

2. Rename the Test in the Test panel as 'MyIDEScript'

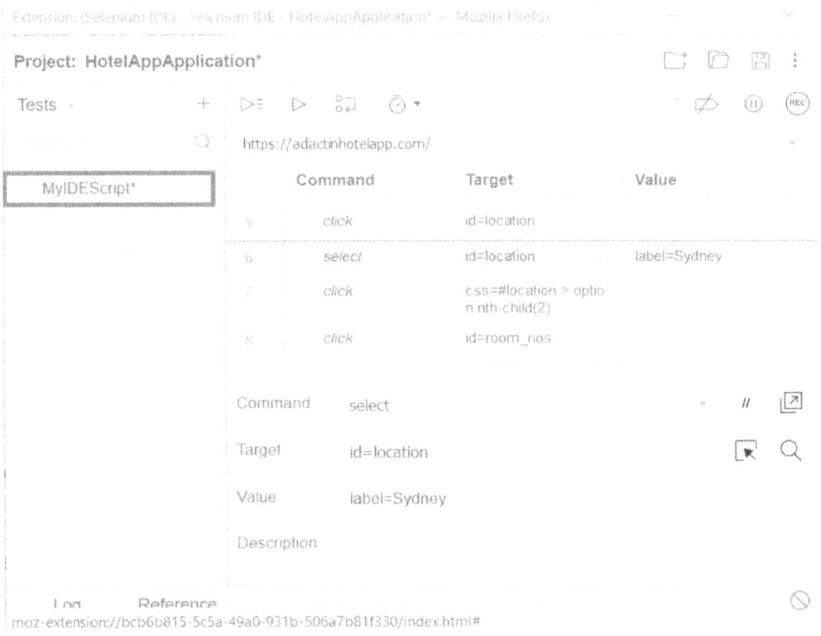

Figure 6-13 Renamed IDE Test case

3. Running the Test cases

The IDE allows many options for running your test case. You can run a test case all at once, stop and start it, run it one line at a time, run a single command you are currently developing, and you can do a batch run of an entire test suite. Execution of test cases is very flexible in the IDE.

Run a Test Case - Click the Run button to run the currently displayed test case.

Run a Test Suite - Click the Run All button to run all the test cases in the currently loaded test suite.

Stop and Start - The Pause button can be used to stop the test case while it is running. The icon of this button then changes to indicate the Resume button. To continue click Resume.

Stop in the Middle - You can set a breakpoint in the test case to cause it to stop on a particular command. This is useful for debugging your test case. To set a breakpoint, select a command, right-click, and from the context menu select Toggle Breakpoint. 'Play to this point' option in the context menu can be used to execute till a particular step.

Start from the Middle - You can tell the IDE to begin running from a specific command in the middle of the test case. This also is used for debugging. To set a start point, select a command, right-click, and from the context menu select Play from here. Inorder to start recording from the middle, click on 'Record from here'

Run Any Single Command - Double-click any single command to run it by itself. This is useful when writing a single command. It lets you immediately test a command you are constructing, when you are not sure if it is correct. You can double-click it to see if it runs correctly.

1. Click on Run button and Run and verify test case run until completion

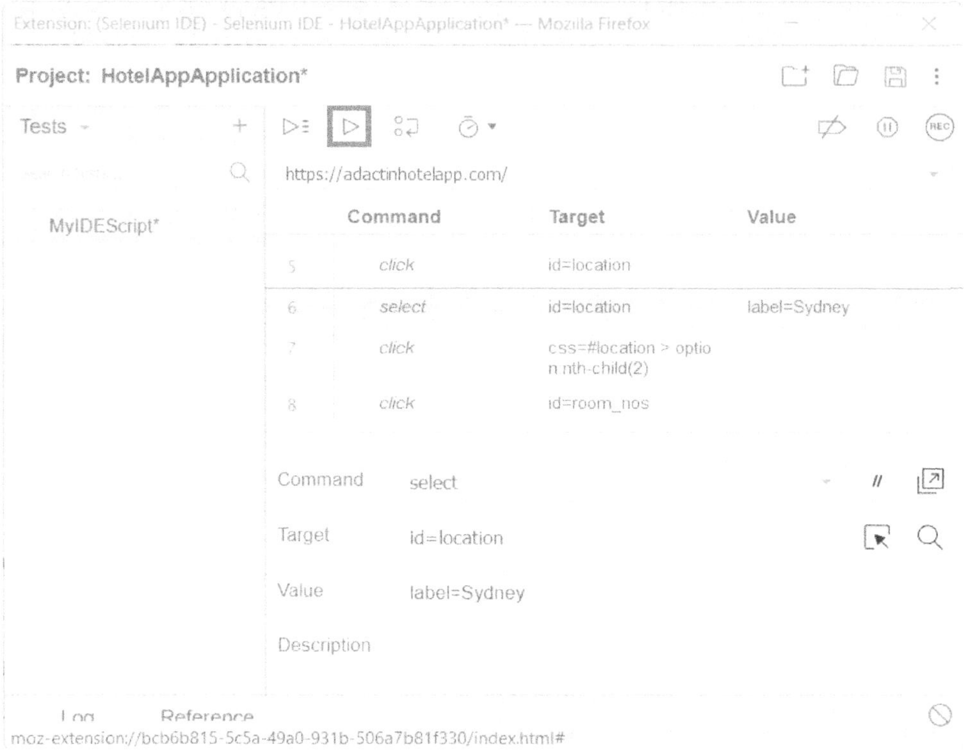

Figure 6-14 Run Test Case

2. In the Test Case pane, click on the drop-down next to Tests and select the context-menu 'Executing'

Note: In case you have clicked on the Logout button (at the bottom of Booking Page) instead of Logout link (on the top right corner of Booking page) your script might fail due to synchronization issue. Re-record your script and click on Logout link on top right corner of application.

In coming chapters, we will see how to deal with synchronization issues.

3. Verify the Results in the Log tab

Figure 6-15 Result Logs

Log/Reference

The bottom pane is used for four different functions – Log, Reference depending on which tab is selected.

Log

When you run your test case, error messages and information messages showing the progress are displayed in this pane automatically, even if you do not first select the Log tab. These messages are often useful for test case debugging. Notice the Clear button for clearing the Log. Also notice the Info button is a drop-down allowing selection of different levels of information to log.

Reference Tab

The Reference tab is the default selection whenever you are entering or modifying Selenese commands and parameters in Table mode. In Table mode, the Reference pane will display documentation on the current command. When entering or modifying commands, whether from Table or Source mode, it is critically important to ensure that the parameters specified in the Target and Value fields match those specified in the parameter list in the Reference pane. The number of parameters provided must match the number specified, the order of parameters provided must match the order specified, and the type of parameters provided must match the type specified. If there is a mismatch in any of these three areas, the command will not run correctly.

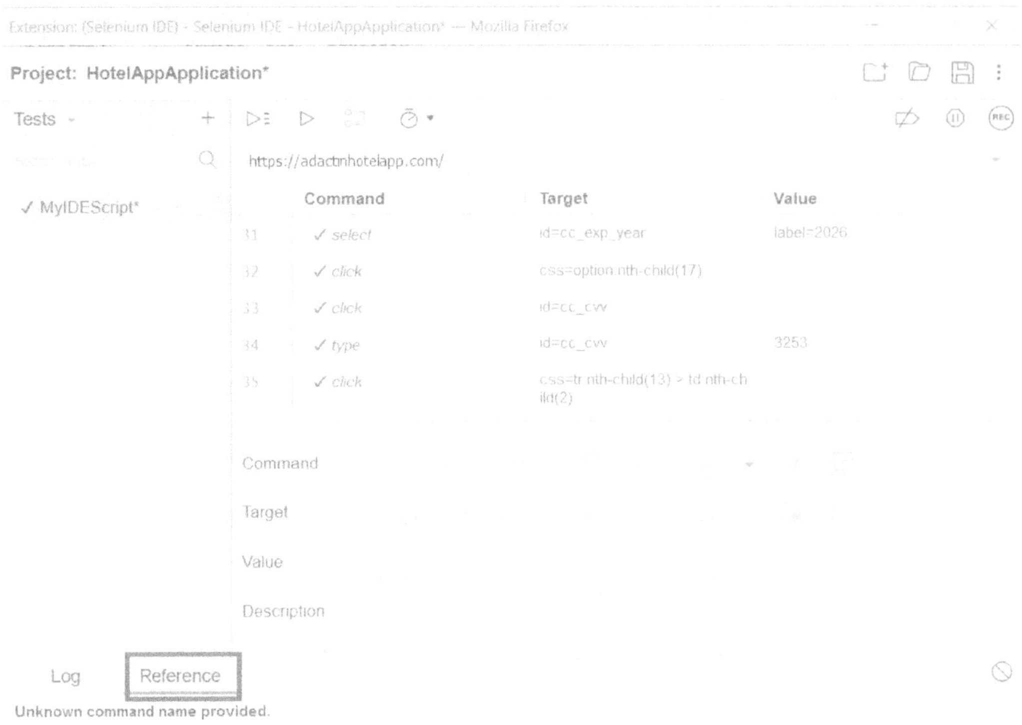

Figure 6-16 Reference Info

While the Reference tab is invaluable as a quick reference, it is still often necessary to consult the Selenium Reference document.

6.4. Inserting/Editing Test Steps Manually

You can not only edit the steps inside the selenium IDE but can also add new steps or change the sequence of steps inside the IDE itself. To do that we need to understand the following:

Test Script Editor Box: Your script is displayed in the test script editor box. It displays the command and their parameters in a readable "table" format.

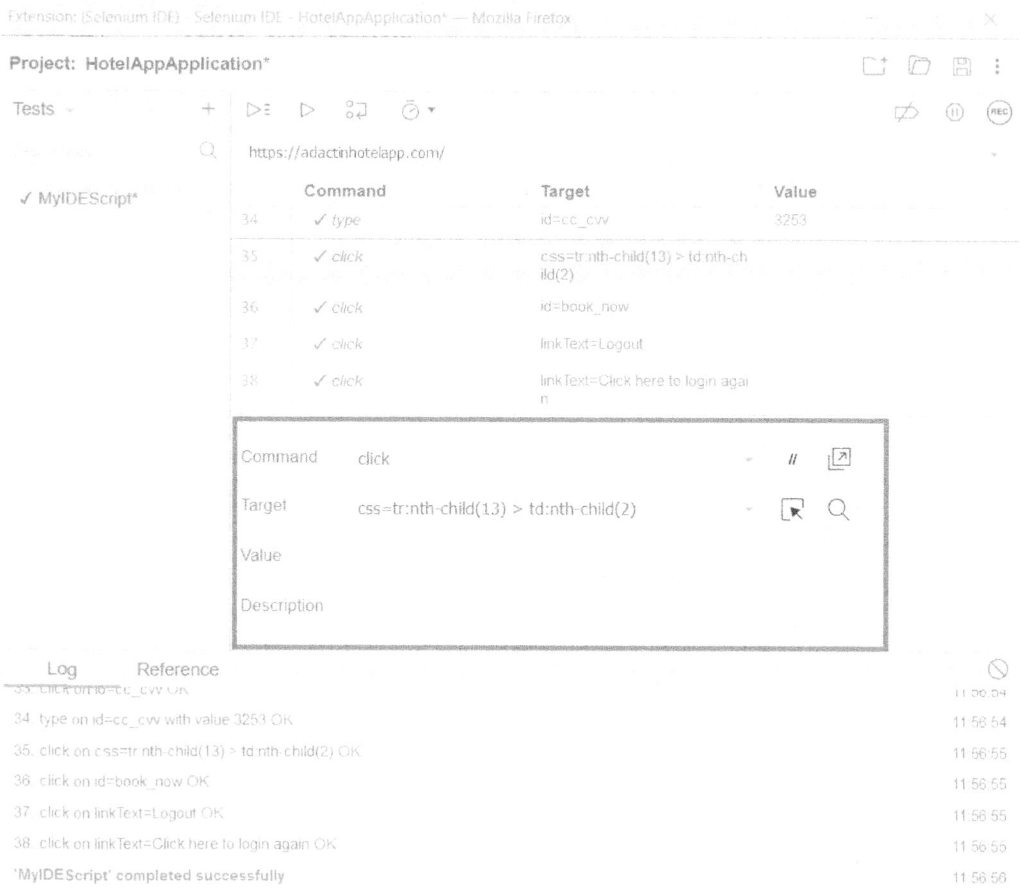

Figure 6-17 Test Script Editor Box

Operations Pane - The Command, Target, and Value entry fields display the currently selected command along with its parameters. User can modify Command, Target or Value in this pane.

If you select the step where we select location as Sydney you would see Command, Target and Value as in below snapshot.

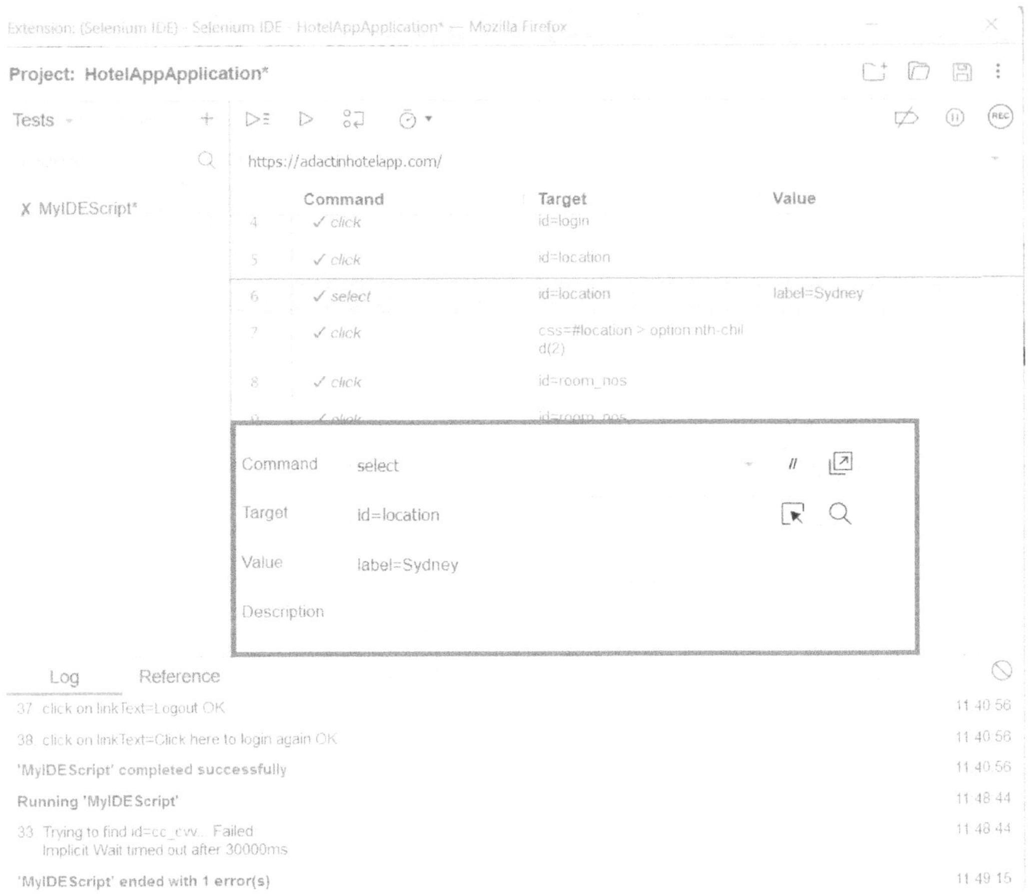

Figure 6-18 - Operations Pane

Additional operations such as Enable/Disable command, Add new window configuration, find and select target in page can be performed in this panel.

Also, if you select Reference Tab, you will see details on the command and arguments that should be entered for that command (in this case select command),

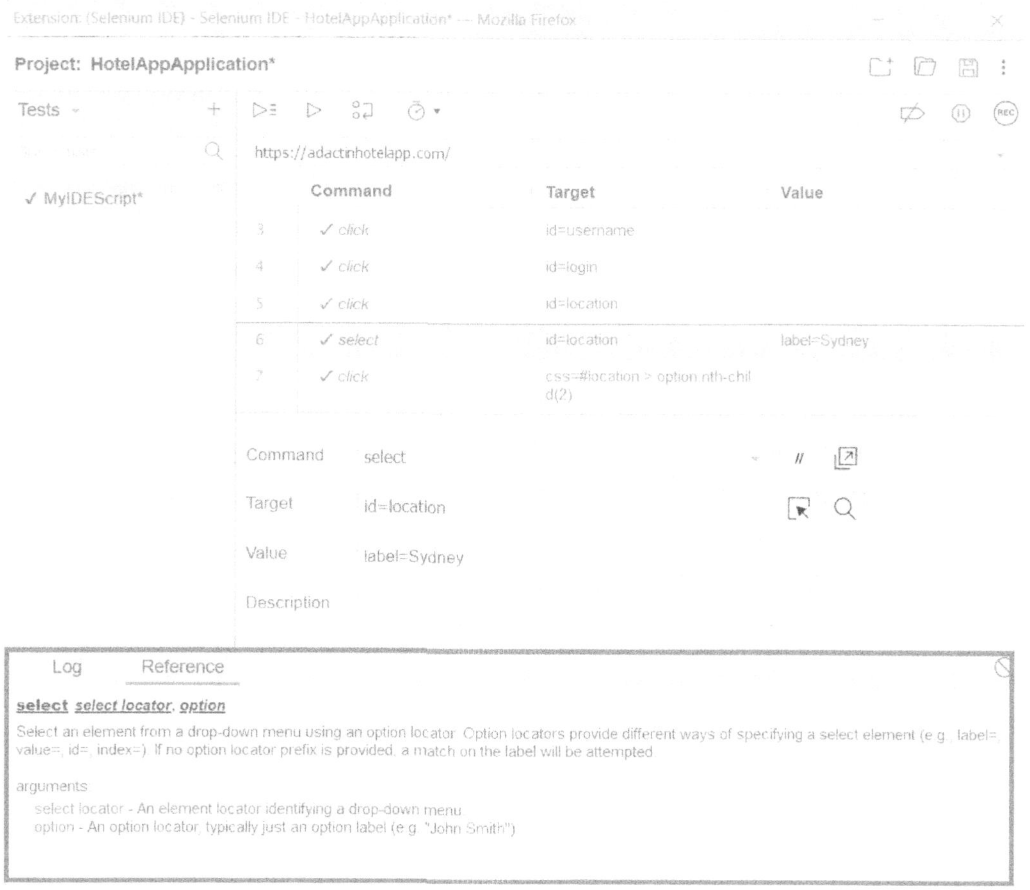

Figure 6-19 – Reference Tab

If you start typing in the Command field, a drop-down list will be populated based on the first characters you type; you can then select your desired command from the drop-down.

6.5. Adding Verifications and Asserts with the Context Menu

One crucial part of any test case is the verification of your results against expected results. Your test cases will also need to check the properties of a Web page. To do the same in your script you can insert assert or verify commands in Selenium IDE. Both the commands work to compare results, with different outcome to the failure of a test condition.

Assert	"assert" will fail the test and abort the current test case
Verify	"verify" will fail the test and continue to run the test case

Choosing between "assert" and "verify" comes down to convenience and management of failures. There's little point checking that the first paragraph on the page is the correct one if your test has already failed when checking that the browser is not displaying the expected page. If you're not on the correct page, you'll probably want to abort your test case so that you can investigate the cause and fix the issue(s) promptly. So it is better to use the "assert" command.

On the other hand, you may want to check many attributes of a page without aborting the test case on the first failure as this will allow you to review all failures on the page and take appropriate action. You can use the "verify" command in this case.

Problem: As an example, let us say we want to verify that our login is successful. We can verify this by checking that the Logout link exists once a user logs in.

To add a verification point:

1. Open your IDE test case **MyIDEScript**
2. Right Click on **'MyIDEScript'** in the Test Case pane and select **Rename**

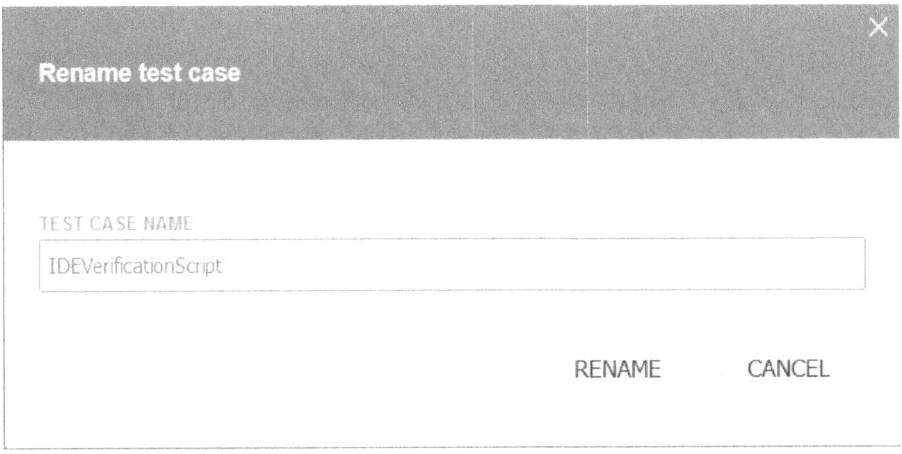

Figure 6-20 – Rename Test Case

3. Rename the Test script as **IDEVerificationScript**

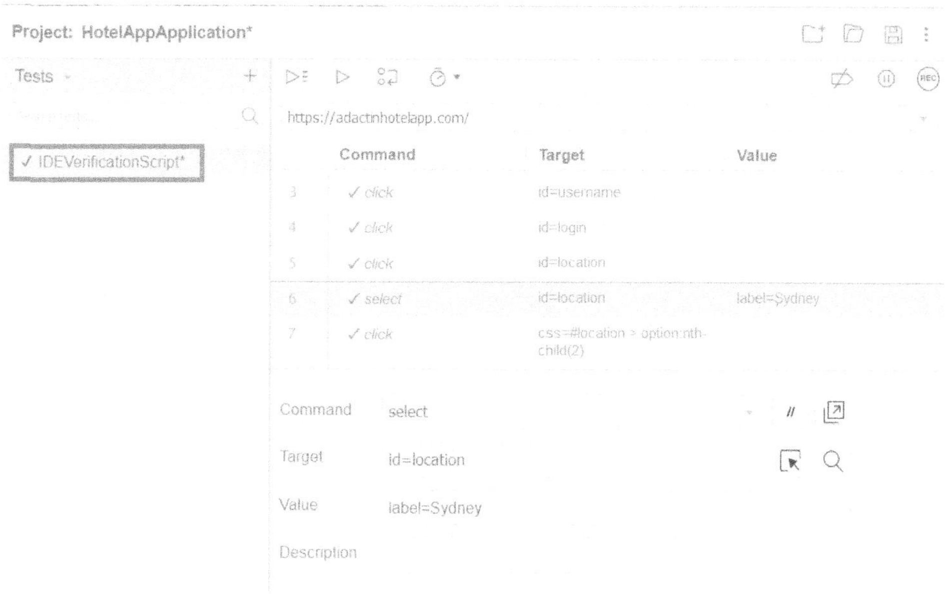

Figure 6-21 – Renamed Test Case

4. Select step where user selects location Sydney

Figure 6-22 – Select Location Step

5. Right-click on the step and select Toggle Breakpoint to insert a Breakpoint at location step

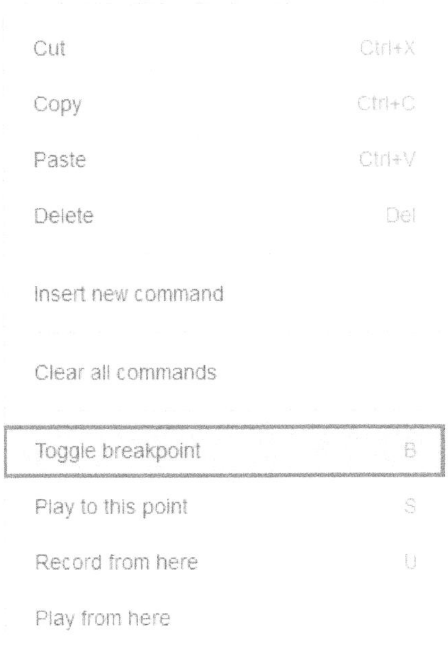

Figure 6-23 – Insert Breakpoint

6. Run a test via Selenium-IDE to go to the browser displaying your test application by selecting **Run current test case**. You will notice that script runs until login and pauses.

Figure 6-24 – Script Pause at breakpoint

7. To add an assertion statement after this step, right-click where the command is to be added and click on **'Insert new command'**

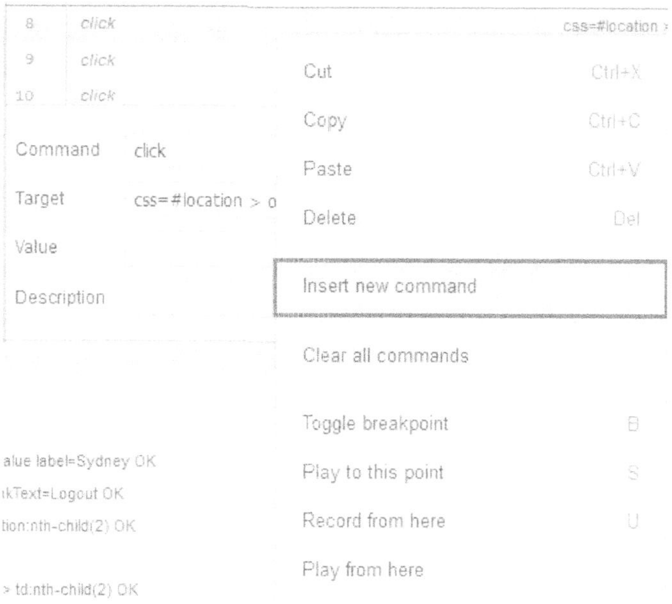

Figure 6-25 Insert new command

8. Select the empty command: enter 'assert' in Command dropdown, and select assert element present option

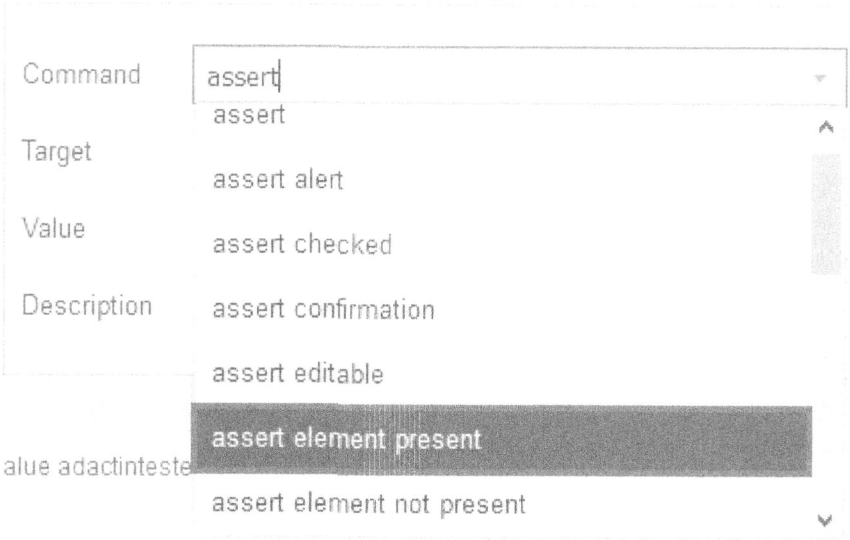

Figure 6-26 Assert element present

9. Click on **'Select Target in page'** button to the right of Target textbox and select **'Logout'** link.

Figure 6-27 Select Target in page

10. The Target field will be populated as below

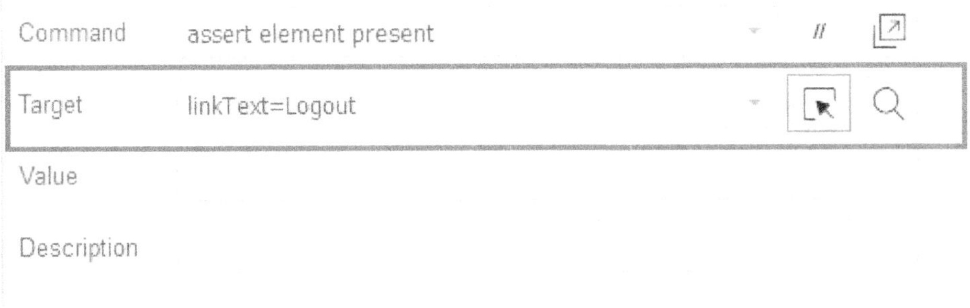

Figure 6-28 Target

11. Run the test case again. Once it reaches Select command, click on 'Step over current command'

12. Again click on 'Step Over' command to see the result of the assertion statements

§

7

Managing User Interface Controls

Introduction

In the last chapter we recorded a sample script and replayed it. You would be very curious to understand how Selenium IDE actually replayed the whole script. How could it identify the location field and enter the value that we had earlier recorded in the previous chapter? Was it like a video recording that got replayed? In this chapter we will understand test automation fundamentals of how a recorded script is replayed and how automation tools recognize objects on the application.

Key objectives:

- Understanding Object Recognition fundamentals and how Selenium replays
- Understand various locators to identify objects
- Using Chropath plugin to get object properties

7.1. How Does Selenium IDE Replay Scripts?

Let us take a simple example:

Assume that you parked your car on some level of a big shopping mall before going to a party. For the sake of this example, say you had too many drinks at the party and so took a cab to get home. Next morning, you come back to the mall to pick up your car but you do not remember the location of the car apart from knowing the level on which your car was parked. How will you find your car? Assume that you do not have a remote control for the car!

Figure 7-1 – Sample Car

If I were to find my car, I will go looking for my car in the first row and look for the Make and Colour of my car. If I find a car with the same make and colour, I will go closer to the car and try to identify my car based on the registration number. If I can match all these three properties, I am sure I will find my car. So the three properties I will look for will be:

- Make of the car
- Colour of the car
- Registration No. of the car

I do not really need to care about height, width or any other details about my car.

This is what Selenium as a tool does and as a matter of fact, this principle is followed by all other automation tools available in the market. They use some key properties of the objects to identify the user interface controls and then use those properties to identify the objects. For instance, if the user clicks on a button, Selenium uses the label of the button to identify the object.

So this is the process of how Selenium IDE replays a script:

- While recording, Selenium stores object property information somewhere in the script
- When we replay the script, Selenium will pick up the Object properties and try to find the object in the application by matching properties
- Once it finds the object, it will perform the operation (click, select, etc.) on that object

This is the basic automation fundamental required to understand how functional automation tools work. The key point to remember is that the Selenium script is not a video recording of functionality, but a step by step execution of actions recorded in the script.

7.2. Locate the elements on a Web page

Every Web page is nothing but a set of different UI Web elements or objects. Before we work with an element on a page, we need Selenium to locate that element. The element on the Web page can be located by various locator types.

To understand the various locators one needs to have a basic understanding of HTML. Id, name, input, type, etc., are the HTML tags/attributes. Using these HTML tags, attributes like "xpath" can be constructed. We can use these tags or attributes to identify elements.

Let us follow these steps to understand tags and locators.

1. Launch Sample Application URL www.adactinhotelapp.com
2. Right click on the home page and select View Page Source

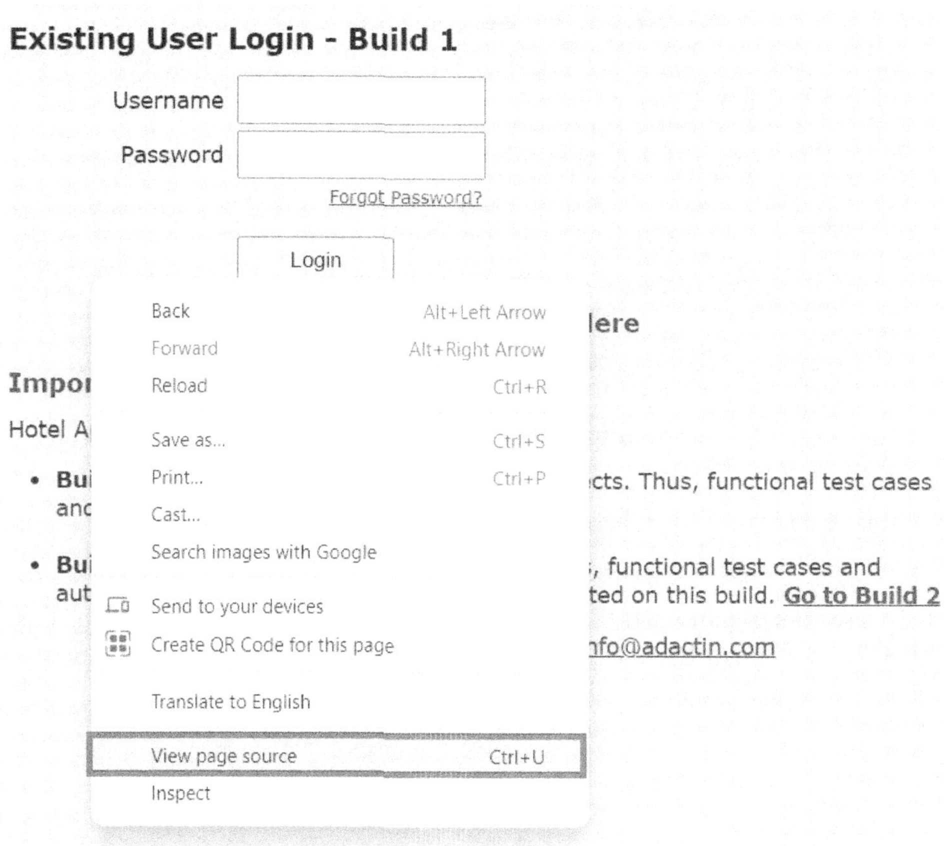

Figure 7-2 – View Page Source Option

3. You will see below source code

```
<tr>
  <td colspan="2" class="login_title">Existing User Login - Build 1</td>
  </tr>
<tr>
  <td width="150" align="right">Username </td>
  <td><input type="text" name="username" id="username" class="login_input" value="" />&nb
  </tr>
<tr>
  <td align="right">Password </td>
  <td><input type="password" name="password" id="password" class="login_input" value="" /:
</tr>
<tr>
  <td> </td>
  <td><div class="login_forgot"><a href="ForgotPassword.php">Forgot Password?</a></div></
</tr>
<tr>
  <td></td>
  <td><div class="auth_error"></div></td>
</tr>
<tr>
  <td> </td>
  <td><input type="Submit" name="login" id="login" class="login_button" value="Login"/></
</tr>
```

Figure 7-3 – Page Source

Above source code is HTML based with tags: type, name, class, href which are used to define and identify elements.

4. Let us see our Selenium IDE script again. Notice the Target field which is used to identify an element based on locator.

Figure 7-4 – Target field in IDE script

It identifies the Web element password using locator id as id=password

Similarly, we can use other tags to identify other Web elements on the Web page.

Given below is the list of locators which are supported by Selenium:

- id
- name
- xpath
- tag name
- class name
- link text
- DOM
- CSS

We will understand each of these locators in details in the next point.

5. Based on the following source code let us see how Selenium can recognize a few of the elements in our application

```
<tr>
  <td colspan="2" class="login_title">Existing User Login - Build 1</td>
  </tr>
<tr>
  <td width="150" align="right">Username </td>
  <td><input type="text" name="username" id="username" class="login_input" value="" />&nb
  </tr>
<tr>
  <td align="right">Password </td>
  <td><input type="password" name="password" id="password" class="login_input" value="" /
</tr>
<tr>
  <td> </td>
  <td><div class="login_forgot"><a href="ForgotPassword.php">Forgot Password?</a></div></
</tr>
<tr>
  <td></td>
  <td><div class="auth_error"></div></td>
</tr>
<tr>
  <td> </td>
  <td><input type="Submit" name="login" id="login" class="login_button" value="Login"/></
</tr>
```

Figure 7-5 – Page Source

Locating by Name

The name locator type will locate the first element with a matching name attribute. For instance, to recognize username Web element we can use:

name=username

If multiple elements have the same value for a name attribute, then you can use filters to further refine your location strategy. The default filter type is value (matching the value attribute).

Locating by ID

The ID locator type will locate the first element with a matching id attribute. Use this when you know an element's id attribute.

id=username

There is one big difference between the 'id' and 'name' attributes though. The name attributes do not have to be unique in a page. If there are multiple elements with the same name, then the first element in the page is selected. In such a case you can use filters to further refine your location strategy.

Unlike some other types of locators like XPath and DOM locators, ID and Name allow Selenium to test a UI element independent of its position on the page. So if the page structure and organization is altered, the test will still run. In cases where Web designers frequently alter the page display and if its functionality must be regression tested, testing via id and name attributes, or really via any HTML property, becomes very important.

Note: Locating by ID is the most preferable way to locate elements if the element in your application has an id attribute assigned.

Even though id is a great locator, it is not realistic for all elements on a page to have ids. The developers add ids to key elements on the page to better control the look and feel or provide dynamic user interaction.

Locating by XPath

XPath is the language used for locating nodes in an XML document. As HTML can be an implementation of XML (XHTML), Selenium users can leverage this powerful language to target elements in their Web applications. XPath extends beyond (as well as supporting) the simple methods of locating by id or name attributes, and opens up all sorts of new possibilities such as locating the third checkbox on the page.

One of the main reasons for using XPath is to locate an element when you don't have a suitable id or name attribute for the element. You can use XPath to either locate the element in absolute terms (not advised), or relative to an element that does have an id or name attribute. XPath locators can also be used to specify elements via attributes other than id and name.

Absolute XPaths contain the location of all elements from the root (html tag) and are likely to change with only the slightest adjustment to the user interface of the web application. By finding a nearby element with an id or name attribute (ideally a parent element) you can locate your target element based on that relationship. This is much less likely to change and can make your element search more robust.

Note: In the next section, we will see how to get XPath of any Web element using Chropath add-on.

Locating Hyperlinks by Link Text

This is a simple method of locating a hyperlink in your Web page by using the text of the link. If two links with the same text are present, then the first match will be used.

For e.g., we can identify Logout link in our application by using

Link=Logout

See below snapshot for IDE command using link locator

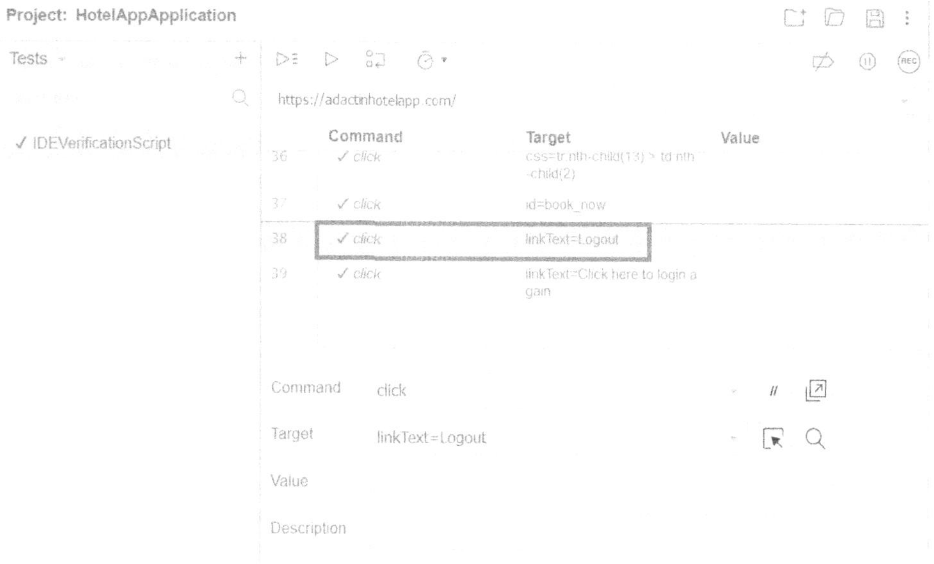

Figure 7-6 – Locating by link attribute in IDE

See below snapshot for IDE command using link locator

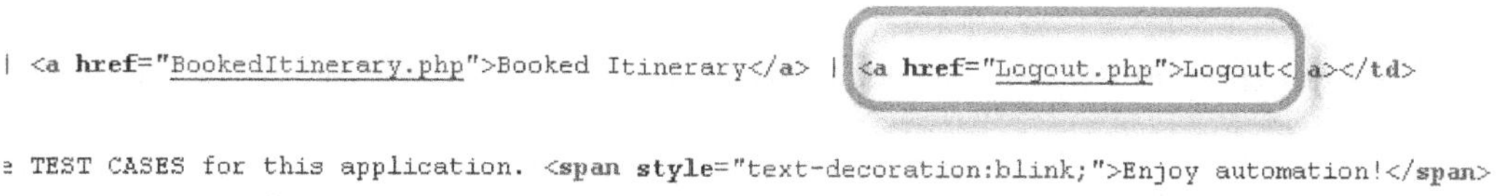
```
| <a href="BookedItinerary.php">Booked Itinerary</a> | <a href="Logout.php">Logout</a></td>
e TEST CASES for this application. <span style="text-decoration:blink;">Enjoy automation!</span>
```

Figure 7-7 – View source for link

Locating elements by class

The class locator type will locate the first element with a matching class attribute. For instance, to recognize class name attribute

Class=login_forgot

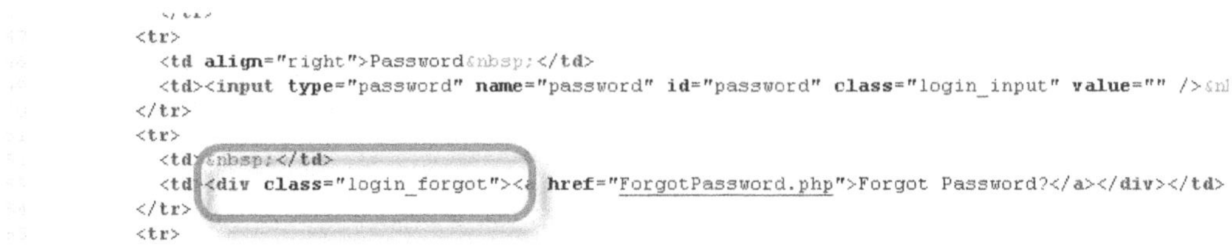

```
<tr>
  <td align="right">Password </td>
  <td><input type="password" name="password" id="password" class="login_input" value="" />&nl
</tr>
<tr>
  <td> </td>
  <td><div class="login_forgot"><a href="ForgotPassword.php">Forgot Password?</a></div></td>
</tr>
<tr>
```

Figure 7-8 – Locating by class attribute

Locating by DOM

The Document Object Model represents an HTML document and can be accessed using JavaScript. This location strategy uses a JavaScript that evaluates the location of an element on the page using the hierarchical dotted notation.

Since DOM locators start with "document", it is not necessary to include the dom= label when specifying a DOM locator. Let us see how we can locate username field in the following page source using DOM locators.

```
<html>
 <body>
  <form id="HotelAppLogin">
   <input name="username" type="text" />
   <input name="password" type="password" />
   <input name="login" type="submit" value="Login" />
  </form>
</body>
<html>
```

Highlighted element (username field) can be located by any of the following DOM locators

Possible DOM locators

- document.forms[0].username
- document.forms[0].elements['username']
- document.forms[0].elements[0]

Locating by CSS

CSS (Cascading Style Sheets) is a language for describing the rendering of HTML and XML documents. CSS uses Selectors for binding style properties to elements in the document. These Selectors can be used by Selenium as another locating strategy. Let us see how we can locate the input field element in the following page source.

```
<table name="Booking Itinerary ">
   <tr id="item1">
      <td class="label">Order RE111</td>
      <td class="item"><input name="qty" class="formfield disabled" /></td>
   </tr>
   <tr id="item2">
      <td class="label"> Order RE112</td>
      <td class="item">
<input id="item2_quantity" name="qty" class="Select_text" type="text"/></td>
   </tr>
   ...
</table>
```

Possible CSS locators

- css= input [class= 'Select_text']
- css=input.Select_text [type='text']
- css=#item2_quantity
- css=input#item2_quantity

In this case, any of the above CSS locator variations will help us locate the highlighted input field.

7.3. Find XPath using Chrome Add-on

In this section we will see how to use ChroPath plugin (installed in previous chapters) to find the XPath of the element.

1. Open Chrome and open our AUT- **http://www.adactinhotelapp.com/** home page
2. Right click and click on Inspect
3. Select ChroPath tab and click on the Inspect Element icon (Arrow) from the window

Figure 7-9 Inspect Element icon

4. Now click on the **username edit field**. Note the XPath value generated under Rel Xpath field.

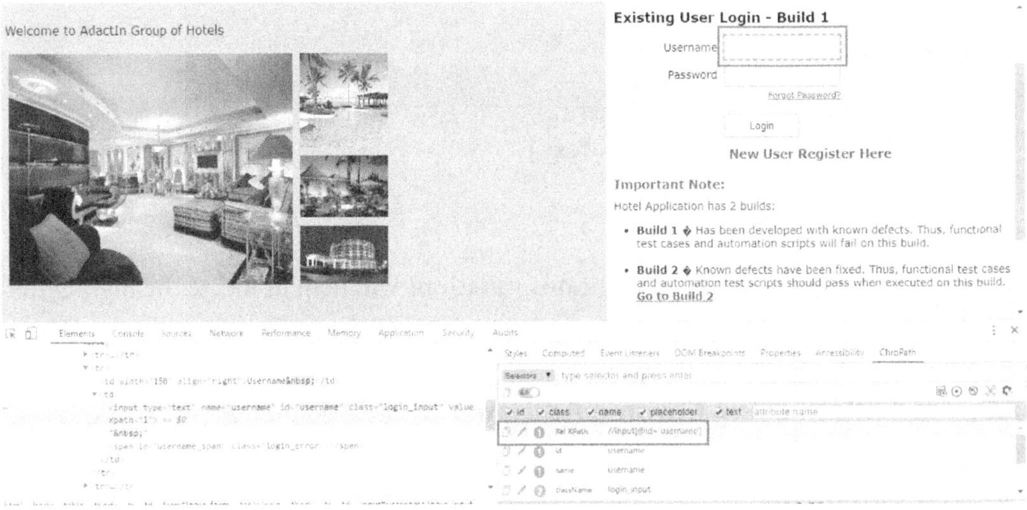

Figure 7-10 – XPath Value

5. Verify that it highlights the Username field.

7.4. Relative Locators

In this section we will see what relative locators are. Selenium introduces Relative Locators which are helpful when it is not easy to construct a locator for the desired element, but easy to describe spatially where the element is in relation to an element that does have an easily constructed locator. Selenium supports five type of relative locator.

a. **Left** – This specifies the locator on which we want to perform action is present on left side of the given element.

```
By submitBtn = RelativeLocator.with(By.tagName("input")).toLeftOf(By.id("Reset"));
driver.findElement(submitBtn).click();
```

b. **Right** - This specifies the locator on which we want to perform action is present on right side of the given element.

```
By resetBtn = RelativeLocator.with(By.tagName("input")).toRightOf(By.id("Submit"));
driver.findElement(resetBtn).click();
```

c. **Above** - This specifies the locator on which we want to perform action is present above of the given element.

```
By emailIdLocator = RelativeLocator.with(By.tagName("input")).above(By.id("password"));
driver.findElement(emailIdLocator).sendKeys("adactin");
```

d. **Below** - This specifies the locator on which we want to perform action is present below of the given element.

```
By passwordLocator = RelativeLocator.with(By.tagName("input")).below(By.id("username"));
driver.findElement(passwordLocator).sendKeys("Admin1");
```

e. **Near** - This specifies the locator on which we want to perform action is present near of the given element.

```
By resetBtn = RelativeLocator.with(By.tagName("input")).near(By.id("Submit"));
driver.findElement(resetBtn).click();
```

EXERCISE

1. Go back to our Login Script in IDE, perform Save As and replace all the locators by XPath. Re-run the script and see if it works.

Hint - Use Chrome add-on: ChroPath to find the XPath of the elements.Then replace the Target field for each step with XPath value; Or you could select the xpath from the drop down in Target for each step.

Figure 7-11 – XPath Value Replaced in script

§

8

Basics of Java

Since we will be focusing on Java as a programming language in Selenium WebDriver we need to be comfortable with the basics of Java. As part of this chapter we will discuss about basic concepts around Java language, syntax for basic Java statements, exception handling and concept of classes and objects.

We do not expect you to be a perfect programmer and Java expert to work with Selenium as we will be using core and basic Java. So be confident and let us understand these concepts!

Key objectives

- Object-oriented Programming Concepts
- Language and Syntax Basics
- Object, Classes and methods
- Exception Handling

8.1. Object-oriented Programming Concepts

If you've never used an object-oriented programming language before, you'll need to learn a few basic concepts before you can begin writing any code. This chapter will introduce you to terms like objects, classes, inheritance, interfaces, and packages. Each topic focuses on how these concepts relate to the real world, while simultaneously providing an introduction to the syntax of the Java programming language.

Note – Since part of the scope of this book focuses on WebDriver using Java, we will be working on Java syntax only.

What Is an Object?

Objects are a key to understanding 'object-oriented' technology. Look around right now and you'll find many examples of real-world objects: Your dog, your desk, your television set, your bicycle.

Real-world objects share two characteristics: They all have 'state' and 'behavior'. Dogs have state (name, color, breed, hungry) and behavior (barking, fetching, wagging tail). Bicycles also have state (current gear, current pedal cadence, current speed) and behavior (changing gear, changing pedal cadence, applying brakes). Identifying the state and behavior for real-world objects is a great way to begin thinking in terms of object-oriented programming.

Take a minute right now to observe the real-world objects that are in your immediate area. For each object that you see, ask yourself two questions: "What possible states can this object be in?" and "What possible behavior can this object perform?" Make sure to write down your observations. As you do, you'll notice that real-world objects vary in complexity; your desktop lamp may have only two possible states (on and off) and two possible behaviors (turn on, turn off), but your desktop radio might have additional states (on, off, current volume, current station) and behavior (turn on, turn off, increase volume, decrease volume, seek, scan, and tune). You may also notice that some objects, in turn, will also contain other objects. These real-world observations all translate into the world of object-oriented programming.

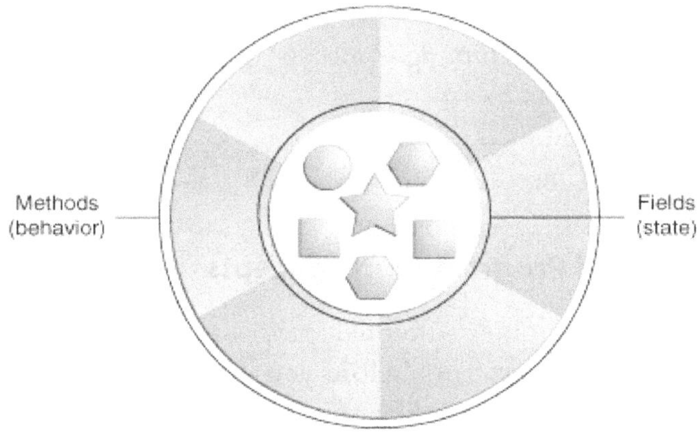

Figure 8-1 - A software object

Software objects are conceptually similar to real-world objects: they too consist of state and related behavior. An object stores its state in 'fields' (variables in some programming languages) and exposes its behavior through 'methods'(functions in some programming languages). Methods operate on an object's internal state and serve as the primary mechanism for object-to-object communication. Hiding internal state and requiring all interaction to be performed through an object's methods is known as '**data encapsulation'**— a fundamental principle of object-oriented programming.

Consider a bicycle, for example:

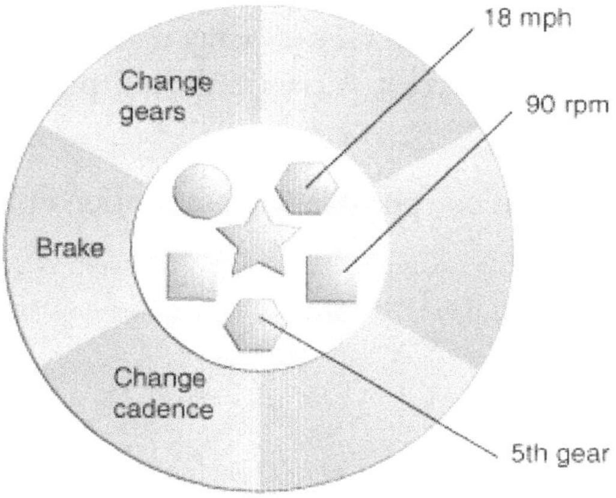

Figure 8-2 - A bicycle modelled as a software object

By attributing state (current speed, current pedal cadence, and current gear) and providing methods for changing that state, the object remains in control of how the outside world is allowed to use it. For example, if the bicycle only has 6 gears, a method to change gears could reject any value that is less than 1 or greater than 6.

Bundling code into individual software objects provides a number of benefits, including:

1. **Modularity**: The Source code for an object can be written and maintained independently of the Source code for other objects. Once created, an object can be easily passed around inside the system.
2. **Information-hiding**: By interacting only with an object's methods, the details of its internal implementation remain hidden from the outside world.
3. **Code re-use**: If an object already exists (perhaps written by another software developer), you can use that object in your program. This allows specialists to implement/test/debug complex, task-specific objects, which you can then trust to run in your own code.
4. **Plug-ability and debugging ease**: If a particular object turns out to be problematic, you can simply remove it from your application and plug in a different object as its replacement. This is analogous to fixing mechanical problems in the real world. If a bolt breaks, you replace it, not the entire machine.

What Is a Class?

A class is a blueprint or prototype from which objects are created. This section defines a class that models the state and behavior of a real-world object. It intentionally focuses on the basics and shows how a simple class can model state and behavior.

In the real world, you'll often find many individual objects all of the same kind. There may be thousands of other bicycles in existence, all of the same make and model. Each bicycle was built from the same set of blueprints and therefore contains the same components. In object-oriented terms, we say that your bicycle is an instance of the class of objects known as bicycles. A class is the blueprint from which individual objects are created.

The following 'Bicycle' class is one possible implementation of a bicycle:

```
class Bicycle {

    int cadence = 0;
    int speed = 0;
    int gear = 1;

    void changeCadence(int newValue) {
        cadence = newValue;
    }

    void changeGear(int newValue) {
        gear = newValue;
    }

    void speedUp(int increment) {
        speed = speed + increment;
    }

    void applyBrakes(int decrement) {
        speed = speed - decrement;
    }

    void printStates() {
        System.out.println("cadence:" +
            cadence + " speed:" +
            speed + " gear:" + gear);
    }
}
```

The syntax of the Java programming language will look new to you, but the design of this class is based on the previous discussion of bicycle objects. The field's cadence, speed, and gear represent the object's state, and the methods (changeCadence, changeGear, speedup,.etc.) define its interaction with the outside world.

You may have noticed that the 'Bicycle' class does not contain a 'main' method. That's because it's not a complete application; it's just the blueprint for bicycles that might be 'used' in an application. The responsibility of creating and using new 'Bicycle' objects belongs to some other class in your application.

Here's a BicycleDemo class that creates two separate 'Bicycle' objects and invokes their methods:

```
class BicycleDemo {
  public static void main(String[] args) {

    // Create two different
    // Bicycle objects
    Bicycle bike1 = new Bicycle();
    Bicycle bike2 = new Bicycle();
    // Invoke methods on
    // those objects
    bike1.changeCadence(50);
    bike1.speedUp(10);
    bike1.changeGear(2);
    bike1.printStates();

    bike2.changeCadence(50);
    bike2.speedUp(10);
    bike2.changeGear(2);
    bike2.changeCadence(40);
    bike2.speedUp(10);
    bike2.changeGear(3);
    bike2.printStates();
  }
}
```

The output of this test prints the ending pedal cadence, speed, and gear for the two bicycles:

```
cadence:50 speed:10 gear:2
cadence:40 speed:20 gear:3
```

What Is Inheritance?

Inheritance provides a powerful and natural mechanism for organizing and structuring your software. This section explains how classes inherit state and behavior from their superclasses, and explains how to derive one class from another using the simple syntax provided by the Java programming language.

Different kinds of objects often have a certain amount in common with each other. Mountain bikes, road bikes, and tandem bikes, for example, all share the characteristics of bicycles (current speed, current pedal cadence, current gear). Yet each also defines additional features that make them different: tandem bicycles have two seats and two sets of handlebars; road bikes have drop handlebars; some mountain bikes have an additional chain ring, giving them a lower gear ratio.

Object-oriented programming allows classes to inherit commonly used state and behavior from other classes. In this example, 'Bicycle' now becomes the superclass of MountainBike, RoadBike, and TandemBike. In the Java programming language, each class is allowed to have one direct superclass, and each superclass has the potential for an unlimited number of subclasses:

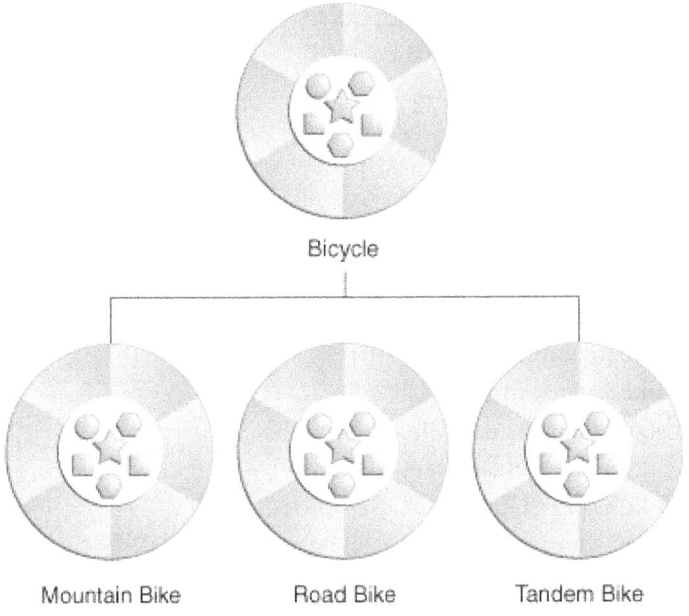

Figure 8-3 - A hierarchy of bicycle classes

The syntax for creating a subclass is simple. At the beginning of your class declaration, use the '**extends**' keyword, followed by the name of the class to inherit from:

```
class MountainBike extends Bicycle {

    // new fields and methods defining
    // a mountain bike would go here

}
```

This gives 'MountainBike' the same fields and methods as 'Bicycle', yet allows its code to focus exclusively on the features that make it unique. This makes code for your subclasses easy to read. However, you must take care to properly document the state and behavior that each superclass defines, since that code will not appear in the source file of each subclass.

What Is an Interface?

An interface is a contract between a class and the outside world. When a class implements an interface, it promises to provide the behavior published by that interface. This section defines a simple interface and explains the necessary changes for any class that implements it.

As you've already learned, objects define their interaction with the outside world through the methods that they expose. Methods form the object's interface with the outside world; the buttons on the front of your television set, for example, are the interface between you and the electrical wiring on the other side of its plastic casing. You press the "power" button to turn the television on and off.

In its most common form, an interface is a group of related methods with empty bodies. A bicycle's behavior, if specified as an interface, might appear as follows:

```
interface Bicycle {

   // wheel revolutions per minute
   void changeCadence(int newValue);

   void changeGear(int newValue);

   void speedUp(int increment);

   void applyBrakes(int decrement);
}
```

To implement this interface, the name of your class would change (to a particular brand of bicycle, for example, such as ACMEBicycle), and you would use the **implements** keyword in the class declaration:

```
class ACMEBicycle implements Bicycle {

   // remainder of this class
   // implemented as before
}
```

Implementing an interface allows a class to become more formal about the behavior it promises to provide. Interfaces form a contract between the class and the outside world, and this contract is enforced at build time by the compiler. If your class claims to implement an interface, all methods defined by that interface must appear in its source code before the class will successfully compile.

What Is a Package?

A package is a namespace for organizing classes and interfaces in a logical manner. Placing your code into packages makes large software projects easier to manage.

Conceptually you can think of packages as being similar to different folders on your computer. You might keep HTML pages in one folder, images in another, and scripts or applications in yet another. Because software written in the Java programming language can be composed of hundreds or thousands of individual classes, it makes sense to keep things organized by placing related classes and interfaces into packages.

What Is a Java API?

The Java platform provides an enormous class library (a set of packages) suitable for use in your own applications. This library is known as the "Application Programming Interface" or "API" for short. Its packages represent the tasks most commonly associated with general-purpose programming. For example, a 'String' object contains state and behavior for character strings; a 'File' object allows a programmer to easily create, delete, inspect, compare, or modify a file on the file system; a 'Socket' object allows for the creation and use of network sockets; various GUI objects control buttons and checkboxes and anything else related to graphical user interfaces.

There are literally thousands of classes to choose from. This allows you, the programmer, to focus on the design of your particular application, rather than the infrastructure required to make it work.

The Java Platform API Specification (http://docs.oracle.com/javase/7/docs/api/index.html) contains the complete listing for all packages, interfaces, classes, fields, and methods supplied by the Java SE platform. Load the page in your browser and bookmark it. As a programmer, it will become your single most important piece of reference documentation.

8.2. Language and Syntax Basics

Variables

You've already learned that objects store their state in fields. However, the Java programming language also uses the term "variable". Let's discuss this relationship, plus variable naming rules and conventions, basic data types (primitive types, character strings, and arrays), default values, and literals.

```
int cadence = 0;
int speed = 0;
int gear = 1;
```

Primitive Data Types

The Java programming language is statically-typed, which means that all variables must first be declared before they can be used. This involves stating the variable's type and name, as you've already seen:

```
int gear = 1;
```

The following chart summarizes the primitive data types and their default values.

Data Type	Default Value (for fields)
Byte	0
Short	0
Int	0
Long	0L
Float	0.0f
Double	0.0d
Char	'\u0000'
String (or any object)	Null
Boolean	False

The Java programming language defines the following kinds of variables:

- **Instance Variables (Non-Static Fields)** Technically speaking, objects store their individual states in "non-static fields", that is, fields declared without the **static** keyword. Non-static fields are also known as **instance variables** because their values are unique to each instance of a class (to each object, in other words); the currentSpeed of one bicycle is independent from the currentSpeed of another.
- **Class Variables (Static Fields)** A class variable is any field declared with the static modifier; this tells the compiler that there is exactly one copy of this variable in existence, regardless of how many times the class has been instantiated. A field defining the number of gears for a particular kind of bicycle could be marked as static since conceptually the same number of gears will apply to all instances. The code 'static int numGears = 6'; would create such a static field. Additionally, the keyword 'final' could be added to indicate that the number of gears will never change.

- **Local Variables** Similar to how an object stores its state in fields, a method will often store its temporary state in 'local variables'. The syntax for declaring a local variable is similar to declaring a field (for example, 'int count = 0';). There is no special keyword designating a variable as local; that determination comes entirely from the location in which the variable is declared — which is between the opening and closing braces of a method. As such, local variables are only visible to the methods in which they are declared; they are not accessible from the rest of the class.
- **Parameters** You've already seen examples of parameters in the Bicycle class. Recall that the signature for the main method is public static void main(String[] args). Here, the args variable is the parameter to this method. The important thing to remember is that parameters are always classified as "variables" not "fields". This applies to other parameter-accepting constructs as well (such as constructors and exception handlers) that you'll learn about later in this book.

Naming Conventions

Every programming language has its own set of rules and conventions for the kinds of names that you're allowed to use, and the Java programming language is no different. The rules and conventions for naming your variables can be summarized as follows:

- Variable names are case sensitive. A variable's name can be any legal identifier — an unlimited-length sequence of Unicode letters and digits, beginning with a letter, the dollar sign "$", or the underscore character "_". The convention, however, is to always begin your variable names with a letter, not "$" or "_". Additionally, the dollar sign character, by convention, is never used at all. You may find some situations where auto-generated names will contain the dollar sign, but your variable names should always avoid using it. A similar convention exists for the underscore character; while it's technically legal to begin your variable's name with "_", this practice is discouraged. White space is not permitted.
- Subsequent characters may be letters, digits, dollar signs, or underscore characters. Conventions (and common sense) apply to this rule as well. When choosing a name for your variables, use full words instead of cryptic abbreviations. Doing so will make your code easier to read and understand. In many cases it will also make your code self-documenting; fields named cadence, speed, and gear, for example, are much more intuitive than abbreviated versions, such as c, s, and g. Also, keep in mind that the name you choose must not be a keyword or reserved word like - this, return, super, etc.

- If the name you choose consists of only one word, spell that word in lowercase letters. If it consists of more than one word, capitalize the first letter of each subsequent word. The names 'gearRatio' and 'currentGear' are prime examples of this convention. If your variable stores a constant value, such as 'static final int **NUM_GEARS** = 6', the convention changes slightly, capitalizing every letter and separating subsequent words with the underscore character. By convention, the underscore character is never used elsewhere.

Arrays

An array is a container object that holds a fixed number of values of a single type. The length of an array is established when the array is created. After creation, its length is fixed. This section discusses arrays in greater detail.

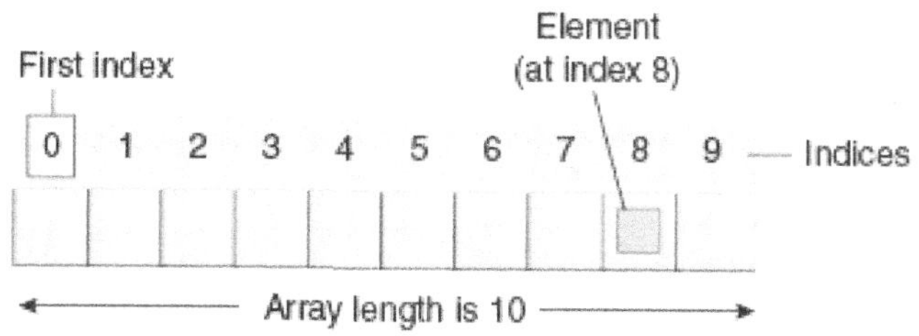

Figure 8-4 - An array of ten elements

Each item in an array is called an element, and each element is accessed by its numerical index. As shown in the above illustration, numbering begins with 0. The 9th element, for example, would therefore be accessed at index 8.The following program, ArrayDemo, creates an array of integers, puts some values in it, and prints each value to standard output.

```
class ArrayDemo {
    public static void main(String[] args) {
        // declares an array of integers
        int[] anArray;

        // allocates memory for 10 integers
        anArray = new int[10];

        // initialize first element
        anArray[0] = 100;
        // initialize second element
        anArray[1] = 200;
        // etc.
        anArray[2] = 300;
        anArray[3] = 400;
        anArray[4] = 500;
        anArray[5] = 600;
        anArray[6] = 700;
        anArray[7] = 800;
        anArray[8] = 900;
        anArray[9] = 1000;

        System.out.println("Element at index 0: "
                + anArray[0]);
        System.out.println("Element at index 1: "
                + anArray[1]);
        System.out.println("Element at index 2: "
                + anArray[2]);
        System.out.println("Element at index 3: "
                + anArray[3]);
        System.out.println("Element at index 4: "
                + anArray[4]);
        System.out.println("Element at index 5: "
                + anArray[5]);
        System.out.println("Element at index 6: "
                + anArray[6]);
        System.out.println("Element at index 7: "
                + anArray[7]);
        System.out.println("Element at index 8: "
                + anArray[8]);
        System.out.println("Element at index 9: "
                + anArray[9]);
    }
}
```

The output from this program is:

```
Element at index 0: 100
Element at index 1: 200
Element at index 2: 300
Element at index 3: 400
Element at index 4: 500
Element at index 5: 600
Element at index 6: 700
Element at index 7: 800
Element at index 8: 900
Element at index 9: 1000
```

In a real-world programming situation, you would probably use one of the supported looping constructs to iterate through each element of the array, rather than write each line individually as shown above. However, this example clearly illustrates the array syntax.

Declaring a Variable to Refer to an Array

The above program declares "anArray" with the following line of code:

```
// declares an array of integers
int[] anArray;
```

Like declarations for variables of other types, an array declaration has two components: the array's type and the array's name. An array's type is written astype[], where type is the data type of the contained elements; the square brackets are special symbols indicating that this variable holds an array. The size of the array is not part of its type (which is why the brackets are empty). An array's name can be anything you want, provided that it follows the rules and conventions as previously discussed in the naming section. As with variables of other types, the declaration does not actually create an array — it simply tells the compiler that this variable will hold an array of the specified type.

Similarly, you can declare arrays of other types:

```
byte[] anArrayOfBytes;
short[] anArrayOfShorts;
long[] anArrayOfLongs;
float[] anArrayOfFloats;
double[] anArrayOfDoubles;
boolean[] anArrayOfBooleans;
char[] anArrayOfChars;
String[] anArrayOfStrings;
```

You can also place the square brackets after the array's name:

```
// this form is discouraged
float anArrayOfFloats[];
```

However, convention discourages this form; the brackets identify the array type and should appear with the type designation.

Creating, Initializing, and Accessing an Array

One way to create an array is with the new operator. The next statement in the ArrayDemo program allocates an array with enough memory for ten integer elements and assigns the array to the anArray variable.

```
// create an array of integers
anArray = new int[10];
```

If this statement were missing, the compiler would print an error like the following, and compilation would fail:

ArrayDemo.Java:4: Variable anArray may not have been initialized.

The next few lines assign values to each element of the array:

```
anArray[0] = 100; // initialize first element
anArray[1] = 200; // initialize second element
anArray[2] = 300; // etc.
```

Each array element is accessed by its numerical index:

```
System.out.println("Element 1 at index 0: " + anArray[0]);
System.out.println("Element 2 at index 1: " + anArray[1]);
System.out.println("Element 3 at index 2: " + anArray[2]);
```

Alternatively, you can use the shortcut syntax to create and initialize an array:

```
int[] anArray = {
   100, 200, 300,
   400, 500, 600,
   700, 800, 900, 1000
};
```

Here the length of the array is determined by the number of values provided between{ and}.

You can also declare an array of arrays (also known as a multidimensional array) by using two or more sets of square brackets, such as String[][] names. Each element, therefore, must be accessed by a corresponding number of index values.

In the Java programming language, a multidimensional array is simply an array whose components are themselves arrays. This is unlike arrays in C or Fortran. A consequence of this is that the rows are allowed to vary in length, as shown in the following MultiDimArrayDemo program:

```
class MultiDimArrayDemo {
   public static void main(String[] args) {
      String[][] names = {
         {"Mr. ", "Mrs. ", "Ms. "},
         {"Smith", "Jones"}
      };
      // Mr. Smith
      System.out.println(names[0][0] + names[1][0]);
      // Ms. Jones
      System.out.println(names[0][2] + names[1][1]);
   }
}
```

The output from this program is:

Mr. Smith
Ms. Jones

Finally, you can use the built-in length property to determine the size of any array. The code

System.out.println (anArray.length); -will print the array's size to standard output.

Operators

Now that you've learned how to declare and initialize variables, you probably want to know how to do something with them. Learning the operators of the Java programming language is a good place to start. Operators are special symbols that perform specific operations on one, two, or three operands, and then return a result. The following is a quick reference for operators supported by the Java programming language.

Simple Assignment Operator

= Simple assignment operator

Arithmetic Operators

+ Additive operator (also used for String concatenation)
- Subtraction operator
* Multiplication operator
/ Division operator
% Remainder operator

Unary Operators

+ Unary plus operator; indicates positive value (numbers are positive without this, however)
Unary minus operator; negates an expression
++ Increment operator; increment a value by 1
-- Decrement operator; decrements a value by 1
! Logical complement operator; inverts the value of a boolean

Equality and Relational Operators

== Equal to
!= Not equal to
> Greater than
>= Greater than or equal to
< Less than
<= Less than or equal to

Conditional Operators

&& Conditional-AND
|| Conditional-OR
? Ternary (shorthand for if-then-else statement)

Type Comparison Operator

instanceof Compares an object to a specified type

Control Flow Statements

The statements inside your source files are generally executed from top to bottom, in the order that they appear. Control flow statements, however, break up the flow of execution by employing decision making, looping, and branching, enabling your program to conditionally execute particular blocks of code. This section describes the decision-making statements (if, if-else, switch), the looping statements (for, while, do-while), and the branching statements (break, continue, return) supported by the Java programming language.

The If-elseif-else Statement

```
if(condition A) {// do this       }

elseif( condition B)     { //do this       }

else     { //do this       }
```

The control will go to one of the block of statements depending upon the given condition.

example

```
class IfElseDemo {
   public static void main(String[] args) {

      int testscore = 76;
      char grade;

      if (testscore >= 90) {
         grade = 'A';
      } else if (testscore >= 80) {
         grade = 'B';
      } else if (testscore >= 70) {
         grade = 'C';
      } else if (testscore >= 60) {
         grade = 'D';
      } else {
         grade = 'F';
      }
      System.out.println("Grade = " + grade);
   }
}
```

The output from the program is:

```
Grade = C
```

The Switch Statement

Unlike if-then and if-then-else statements, the switch statement can have a number of possible execution paths. A switch works with the byte, short, char, and int primitive data types. It also works with enumerated types (discussed in Enum Types), the String class, and a few special classes that wrap certain primitive types: Character, Byte, Short, and Integer (discussed in Numbers and Strings).

The following code example, **SwitchDemo**, declares an int named month whose value represents a month. The code displays the name of the month, based on the value of month, using the switch statement.

```
public class SwitchDemo {
   public static void main(String[] args) {

      int month = 8;
      String monthString;
      switch (month) {
         case 1:  monthString = "January";
                  break;
         case 2:  monthString = "February";
                  break;
         case 3:  monthString = "March";
                  break;
         case 4:  monthString = "April";
                  break;
         case 5:  monthString = "May";
                  break;
         case 6:  monthString = "June";
                  break;
         case 7:  monthString = "July";
                  break;
         case 8:  monthString = "August";
                  break;
         case 9:  monthString = "September";
                  break;
         case 10: monthString = "October";
                  break;
         case 11: monthString = "November";
                  break;
         case 12: monthString = "December";
                  break;
         default: monthString = "Invalid month";
                  break;
      }
      System.out.println(monthString);
   }
}
```

In this case, August is printed to standard output.

Looping Statements

The while and do-while Statements

The while statement continually executes a block of statements while a particular condition is true. Its syntax can be expressed as:

```
while (expression) {
    statement(s)
}
```

The while statement evaluates expression, which must return a boolean value. If the expression evaluates to true, the while statement executes the statement(s) in the while block. The while statement continues testing the expression and executing its block until the expression evaluates to false. Using the while statement to print the values from 1 through 10 can be accomplished as in the following **WhileDemo** program:

```
class WhileDemo {
    public static void main(String[] args){
        int count = 1;
        while (count < 11) {
            System.out.println("Count is: " + count);
            count++;
        }
    }
}
```

You can implement an infinite loop using the while statement as follows:

```
while (true){
    // your code goes here
}
```

The Java programming language also provides a do-while statement, which can be expressed as follows:

```
do {
    statement(s)
} while (expression);
```

The difference between do-while and while is that do-while evaluates its expression at the bottom of the loop instead of the top. Therefore, the statements within the do block are always executed at least once, as shown in the following **DoWhileDemo** program:

```
class DoWhileDemo {
   public static void main(String[] args){
      int count = 1;
      do {
         System.out.println("Count is: " + count);
         count++;
      } while (count < 11);
   }
}
```

The for Statement

The **for statement** provides a compact way to iterate over a range of values. Programmers often refer to it as the "for loop" because of the way in which it repeatedly loops until a particular condition is satisfied. The general form of **for statement** can be expressed as follows:

```
for (initialization; termination;
     increment) {
    statement(s)
}
```

When using this version of for statement, keep in mind that:

- The initialization expression initializes the loop; it's executed once, as the loop begins
- When the termination expression evaluates to false, the loop terminates
- The increment expression is invoked after each iteration through the loop; it is perfectly acceptable for this expression to increment or decrement a value

The following program, **ForDemo**, uses the general form of for statement to print the numbers 1 through 10 to standard output:

```
class ForDemo {
   public static void main(String[] args){
      for(int i=1; i<11; i++){
         System.out.println("Count is: " + i);
      }
   }
}
```

The output of this program is:

Count is: 1
Count is: 2
Count is: 3
Count is: 4
Count is: 5
Count is: 6
Count is: 7
Count is: 8
Count is: 9
Count is: 10

Miscellaneous Statements

The return Statement

The last of the branching statements is the return statement. The return statement exits from the current method, and control flow returns to where the method was invoked. The return statement has two forms: one that returns a value, and one that doesn't. To return a value, simply put the value (or an expression that calculates the value) after the return keyword.

```
return ++count;
```

The data type of the returned value must match the type of the method's declared return value. When a method is declared void, return doesn't return a value or doesn't have a return statement at all.

```
return;
```

Comments in Java

Java supports single-line and multi-line comments very similar to C and C++. All characters available inside any comment are ignored by the Java compiler. You can use double forward slashes - // or forward slash and star combination - /* comments */ for comments

```
public class MyCommentsProgram{
  /* This is my comments program.
   This will print 'Hello World' as the output
   This is an example of multi-line comments.
   */
   public static void main(String []args){
     // This is an example of single line comment
     /* This is also an example of single line comment. */
     System.out.println("Hello World");
   }
```

Basic Syntax:

About Java programs, it is very important to keep in mind the following points.

- **Case Sensitivity** - Java is case sensitive, which means identifier **Hello** and **hello** would have different meanings in Java
- **Class Names** - All class names should begin in upper case
  If several words are used to form a name of the class, each inner word's first letter should be in upper case.
  Example class MyFirstJavaClass
- **Method Names** - All method names should begin in lower case
  If several words are used to form the name of the method, then each inner word's first letter should be in upper case.
- Example public void myMethodName()
- **Program File Name** – The name of the program file should exactly match the class name
  When saving the file, you should save it using the class name (Remember Java is case sensitive) and append '.Java' to the end of the name (if the file name and the class name do not match your program will not compile).
  Example : Assume 'MyFirstJavaProgram' is the class name. Then the file should be saved as'MyFirstJavaProgram.Java'
- **public static void main(String args**[]) - Java programs begin processing from the main() method which is a mandatory part of every Java program

Java Identifiers:

All Java components require names. Names used for classes, variables and methods are called identifiers.

In Java, there are several points to remember about identifiers. They are as follows:

- All identifiers should begin with a letter (A to Z or a to z), currency character ($) or an underscore (_)
- After the first character identifiers can have any combination of characters
- A keyword cannot be used as an identifier
- Most importantly identifiers are case sensitive
- Examples of legal identifiers: age, $salary, _value, __1_value
- Examples of illegal identifiers: 123abc, -salary

Java Keywords:

The following list shows the reserved words in Java. These reserved words may not be used as a constant or variable or any other identifier name.

abstract	assert	boolean	break
byte	case	catch	char
class	const	continue	default
do	double	else	enum
extends	final	finally	float
for	goto	if	implements
import	instanceof	int	interface
long	native	new	package
private	protected	public	return
short	static	strictfp	super
switch	synchronized	this	throw
throws	transient	try	void
volatile	while		

8.3. Working with Classes, Objects and Methods

More on Classes and Objects

With the knowledge you now have of the basics of the Java programming language, you can learn to write your own classes. In this lesson, you will find information about defining your own classes, including declaring member variables, methods, and constructors.

You will learn to use your classes to create objects, and how to use the objects you create.

This lesson also covers nesting classes within other classes, and enumerations.

Declaring Classes

You've seen classes defined in the following way:

```
class MyClass {
    // field, constructor, and
    // method declarations
}
```

This is a class declaration. The class body (the area between the braces) contains all the code that provides for the lifecycle of the objects created from the class: Constructors for initializing new objects, declarations for the fields that provide the state of the class and its objects, and methods to implement the behavior of the class and its objects.

The preceding class declaration is a minimal one. It contains only those components of a class declaration that are required. You can provide more information about the class, such as the name of its superclass, whether it implements any interfaces, and so on, at the start of the class declaration. For example,

```
class MyClass extends MySuperClass implements YourInterface {
    // field, constructor, and
    // method declarations
}
```

means that MyClass is a subclass of MySuperClass and that it implements the YourInterface interface.

You can also add modifiers like public or private at the very beginning—so you can see that the opening line of a class declaration can become quite complicated. The modifiers public and private, which determine what other classes can access MyClass, are discussed later in this lesson. The lesson on interfaces and inheritance will explain how and why you would use extends and implements keywords in a class declaration. For the moment you do not need to worry about these extra complications.

In general, class declarations can include these components, in order:

1. Modifiers such as public, private, and a number of others that you will encounter later
2. The class name, with the initial letter capitalized by convention
3. The name of the class's parent (superclass), if any, preceded by the keyword extends. A class can only extend (subclass) one parent.
4. A comma-separated list of interfaces implemented by the class, if any, preceded by the keyword implements. A class can implement more than one interface.
5. The class body, surrounded by braces, {}

Declaring Member Variables

There are several kinds of variables:

- Member variables in a class—these are called fields.
- Variables in a method or block of code—these are called local variables.
- Variables in method declarations—these are called parameters.

The Bicycle class uses the following lines of code to define its fields:

public int cadence;

(Access modifier) (Data type) (Name)

public int gear;
public int speed;
Access Modifiers

The first (left-most) modifier used lets you control what other classes have access to a member field. For the moment, consider only public and private. Other access modifiers will be discussed later.

- public modifier—the field is accessible from all classes
- private modifier—the field is accessible only within its own class

Defining Methods or Functions

Here is an example of a typical method declaration:

```
public double calculateAnswer(double wingSpan, int numberOfEngines,
                              double length, double grossTons) {
    //do the calculation here
}
```

The only required elements of a method declaration are the method's return type, name, a pair of parentheses, (), and a body between braces, {}.

More generally, method declarations have six components, in order:

1. Modifiers—such as public, private, and others you will learn about later
2. The return type—the data type of the value returned by the method, or void if the method does not return a value
3. The method name—the rules for field names apply to method names as well, but the convention is a little different
4. The parameter list in parenthesis—a comma-delimited list of input parameters, preceded by their data types, enclosed by parentheses, (). If there are no parameters, you must use empty parentheses.
5. An exception list—to be discussed later
6. The method body, enclosed between braces—the method's code, including the declaration of local variables, goes here

Naming a Method or Function

Although a method name can be any legal identifier, code conventions restrict method names. By convention, method names should be a verb in lowercase or a multi-word name that begins with a verb in lowercase, followed by adjectives, nouns, etc. In multi-word names, the first letter of each of the second and following words should be capitalized. Here are some examples:

run
runFast
getBackground
getFinalData
compareTo
setX
isEmpty

Typically, a method has a unique name within its class. However, a method might have the same name as other methods due to method overloading.

Overloading Methods

The Java programming language supports overloading methods, and Java can distinguish between methods with different method signatures. This means that methods within a class can have the same name if they have different parameter lists (there are some qualifications to this that will be discussed in the lesson titled "Interfaces and Inheritance").

Suppose that you have a class that can use calligraphy to draw various types of data (strings, integers, and so on) and that contains a method for drawing each data type. It is cumbersome to use a new name for each method—for example, drawString, drawInteger, drawFloat, and so on. In the Java programming language, you can use the same name for all the drawing methods but pass a different argument list to each method. Thus, the data drawing class might declare four methods named draw, each of which has a different parameter list.

```
public class DataArtist {
    ...
    public void draw(String s) {
        ...
    }
    public void draw(int i) {
        ...
    }
    public void draw(double f) {
        ...
    }
    public void draw(int i, double f) {
        ...
    }
}
```

Overloaded methods are differentiated by the number and the type of the arguments passed into the method. In the code sample, draw(String s) and draw(int i) are distinct and unique methods because they require different argument types.

You cannot declare more than one method with the same name and the same number and type of arguments, because the compiler cannot tell them apart.

The compiler does not consider return type when differentiating methods, so you cannot declare two methods with the same signature even if they have a different return type.

Providing Constructors for Your Classes

A class contains constructors that are invoked to create objects from the class blueprint. Constructor declarations look like method declarations—except that they use the name of the class and have no return type. For example, Bicycle has one constructor:

```
public Bicycle(int startCadence, int startSpeed, int startGear) {
   gear = startGear;
   cadence = startCadence;
   speed = startSpeed;
}
```

To create a new Bicycle object called myBike, a constructor is called by the new operator:

Bicycle myBike = new Bicycle(30, 0, 8);

new Bicycle(30, 0, 8) creates space in memory for the object and initializes its fields.

Although Bicycle only has one constructor, it could have others, including a no-argument constructor:

```
public Bicycle() {
   gear = 1;
   cadence = 10;
   speed = 0;
}
```

Bicycle yourBike = new Bicycle(); invokes the no-argument constructor to create a new Bicycle object called yourBike.

Both constructors could have been declared in Bicycle because they have different argument lists. As with methods, the Java platform differentiates constructors on the basis of the number of arguments in the list and their types. You cannot write two constructors that have the same number and type of arguments for the same class, because the platform would not be able to tell them apart. Doing so causes a compile-time error.

You don't have to provide any constructors for your class, but you must be careful when doing this. The compiler automatically provides a no-argument, default constructor for any class without constructors. This default constructor will call the no-argument constructor of the superclass. In this situation, the compiler will complain if the superclass doesn't have a no-argument constructor so you must verify that it does. If your class has no explicit superclass, then it has an implicit superclass of Object, which does have a no-argument constructor.

You can use a superclass constructor yourself. The MountainBike class at the beginning of this lesson did just that. This will be discussed later, in the lesson on interfaces and inheritance.

You can use access modifiers in a constructor's declaration to control which other classes can call the constructor.

Returning a Value from a Method

A method returns to the code that invoked it when it

- completes all the statements in the method,
- reaches a return statement, or
- throws an exception (covered later),

whichever occurs first.

You declare a method's return type in its method declaration. Within the body of the method, you use the return statement to return the value.

Any method declared void doesn't return a value. It does not need to contain a return statement, but it may do so. In such a case, a return statement can be used to branch out of a control flow block and exit the method and is simply used like this:

return;

If you try to return a value from a method that is declared void, you will get a compiler error.

Any method that is not declared void must contain a return statement with a corresponding return value, like this:

return returnValue;

The data type of the return value must match the method's declared return type; you can't return an integer value from a method declared to return a boolean.

The **getArea()** method below returns an integer:

```
// a method for computing the area of the rectangle
   public int getArea() {
      return width * height;
   }
```

This method returns the integer that the expression'width*height' evaluates to.

The getArea method returns a primitive type. A method can also return a reference type. For example, in a program to manipulate Bicycle objects, we might have a method like this:

```
public Bicycle seeWhosFastest(Bicycle myBike, Bicycle yourBike,
                              Environment env) {
   Bicycle fastest;
   // code to calculate which bike is
   // faster, given each bike's gear
   // and cadence and given the
   // environment (terrain and wind)
   return fastest;
}
```

Using the 'this' Keyword

Within an instance method or a constructor, 'this' is a reference to the current object— the object whose method or constructor is being called. You can refer to any member of the current object from within an instance method or a constructor by using 'this'.

Using 'this' with a Field:

The most common reason for using the 'this' keyword is because a field is shadowed by a method or constructor parameter.

For example, the Point class was written like this

```
public class Point {
    public int x = 0;
    public int y = 0;

    //constructor
    public Point(int a, int b) {
        x = a;
        y = b;
    }
}
```

but it could have been written like this:

```
public class Point {
    public int x = 0;
    public int y = 0;

    //constructor
    public Point(int x, int y) {
        this.x = x;
        this.y = y;
    }
}
```

Each argument to the constructor shadows one of the object's fields — inside the constructor x is a local copy of the constructor's first argument. To refer to thePoint field x, the constructor must use this.x.

Using 'this' with a Constructor:

From within a constructor, you can also use the 'this' keyword to call another constructor in the same class. Doing so is called an explicit constructor invocation. Here's another Rectangle class, with a different implementation from the one in the Objects section.

```
public class Rectangle {
    private int x, y;
    private int width, height;

    public Rectangle() {
        this(0, 0, 0, 0);
    }
    public Rectangle(int width, int height) {
        this(0, 0, width, height);
    }
    public Rectangle(int x, int y, int width, int height) {
        this.x = x;
        this.y = y;
        this.width = width;
        this.height = height;
    }
    ...
}
```

This class contains a set of constructors. Each constructor initializes some or all of the rectangle's member variables. The constructors provide a default value for any member variable whose initial value is not provided by an argument. For example, the no-argument constructor calls the four-argument constructor with four 0 values and the two-argument constructor calls the four-argument constructor with two 0 values. As before, the compiler determines which constructor to call, based on the number and the type of arguments.

If present, the invocation of another constructor must be the first line in the constructor.

Class Variables

When a number of objects are created from the same class blueprint, they each have their own distinct copies of instance variables. In the case of the Bicycleclass, the instance variables are cadence, gear, and speed. Each Bicycle object has its own values for these variables, stored in different memory locations.

Sometimes, you want to have variables that are common to all objects. This is accomplished with the static modifier. Fields that have the static modifier in their declaration are called static fields or class variables. They are associated with the class, rather than with any object. Every instance of the class shares a class variable, which is in one fixed location in memory. Any object can change the value of a class variable, but class variables can also be manipulated without creating an instance of the class.

For example, suppose you want to create a number of Bicycle objects and assign each a serial number, beginning with 1 for the first object. This ID number is unique to each object and is therefore an instance variable. At the same time, you need a field to keep track of how many Bicycle objects have been created so that you know what ID to assign to the next one. Such a field is not related to any individual object, but to the class as a whole. For this you need a class variable, numberOfBicycles, as follows:

```
public class Bicycle {

    private int cadence;
    private int gear;
    private int speed;

    // add an instance variable for the object ID
    private int id;

    // add a class variable for the
    // number of Bicycle objects instantiated
    private static int numberOfBicycles = 0;
        ...
}
```

Class variables are referenced by the class name itself, as in

Bicycle.numberOfBicycles

This makes it clear that they are class variables.

Constants

The static modifier, in combination with the final modifier, is also used to define constants. The final modifier indicates that the value of this field cannot change.

For example, the following variable declaration defines a constant named PI, whose value is an approximation of pi (the ratio of the circumference of a circle to its diameter):

static final double PI = 3.141592653589793;

Summary of Creating and Using Classes and Objects

A class declaration names the class and encloses the class body between braces. The class name can be preceded by modifiers. The class body contains fields, methods, and constructors for the class. A class uses fields to contain state information and uses methods to implement behavior. Constructors that initialize a new instance of a class use the name of the class and look like methods without a return type.

You control access to classes and members in the same way: by using an access modifier such as public in their declaration.

You specify a class variable or a class method by using the static keyword in the member's declaration. A member that is not declared as static is implicitly an instance member. Class variables are shared by all instances of a class and can be accessed through the class name as well as an instance reference.

You create an object from a class by using the new operator and a constructor. The new operator returns a reference to the object that was created. You can assign the reference to a variable or use it directly.

Instance variables and methods that are accessible to code outside of the class that they are declared in can be referred to by using a qualified name. The qualified name of an instance variable looks like this:

objectReference.variableName

The qualified name of a method looks like this:

objectReference.methodName(argumentList)
or:
objectReference.methodName()

The garbage collector automatically cleans up unused objects. An object is unused if the program holds no more references to it. You can explicitly drop a reference by setting the variable holding the reference to null.

8.4. Exception Handling

An exception is an event, which occurs during the execution of a program that disrupts the normal flow of the program's instructions.

When an error occurs within a method, the method creates an object and hands it off to the runtime system. The object, called an exception object, contains information about the error, including its type and the state of the program when the error occurred. Creating an exception object and handing it to the runtime system is called throwing an exception. After a method throws an exception, the runtime system attempts to find something to handle it. The code to handle this exception is what we provide.

To catch an exception we first put the code which we suspect to throw an error into a try block like

```
WebElement txtbox_username = driver.findElement(By.id("username"));

try
{
if(txtbox_username.isEnabled())
{
txtbox_username.sendKeys("tutorial");}
  }

 catch(NoSuchElementException e)
{
System.out.println(e.toString());
}
```

Followed by a catch block of code where we tell the system what should be done when the exception occurs. Generally this is where we display the message of the exception object so that we know which exception has occurred and why.

Multiple catch Blocks

A try block can be followed by multiple catch blocks. The syntax for multiple catch blocks looks like the following:

```
try
{
  //Protected code
}catch(ExceptionType1 e1)
{
  //Catch block
}catch(ExceptionType2 e2)
{
  //Catch block
}catch(ExceptionType3 e3)
{
  //Catch block
}
```

The previous statements demonstrate three catch blocks, but you can have any number of them after a single try. If an exception occurs in the protected code, the exception is thrown to the first catch block in the list. If the data type of the exception thrown matches ExceptionType1, it gets caught there. If not, the exception passes down to the second catch statement. This continues until the exception either is caught or falls through all catches, in which case the current method stops execution and the exception is thrown down to the previous method on the call stack.

Example: Here is code segment showing how to use multiple try/catch statements.

```
try
{
  file = new FileInputStream(fileName);
  x = (byte) file.read();
}catch(IOException i)
{
  i.printStackTrace();
  return -1;
}catch(FileNotFoundException f) //Not valid!
{
  f.printStackTrace();
  return -1;
}
```

The throws/throw Keywords:

If a method does not handle a checked exception, the method must declare it using the throws keyword. The throws keyword appears at the end of a method's signature. You can throw an exception, either a newly instantiated one or an exception that you just caught, by using the throw keyword. Let us try to understand the difference in throws and throw keywords.

The following method declares that it throws a RemoteException. Since the exception is not handled from within the code, we are using throw keyword to throw RemoteException.

```
import Java.io.*;
public class className
{
  public void deposit(double amount) throws RemoteException
  {
    // Method implementation
    throw new RemoteException();
  }
  //Remainder of class definition
}
```

A method can declare that it throws more than one exception, in which case the exceptions are declared in a list separated by commas. For example, the following method declares that it throws a RemoteException and an InsufficientFundsException:

```
import Java.io.*;
public class className
{
  public void withdraw(double amount) throws RemoteException,
                  InsufficientFundsException
  {
    // Method implementation
  }
  //Remainder of class definition
}
```

Note- This chapter has referenced Java documentation available on http://docs.oracle.com/javase/tutorial/. Please refer to this link for more detailed tutorials on Java Language

§

9

Creating First Selenium WebDriver Script

Now that we understand Selenium Basics (Selenium IDE and Locators) and basics of Java we are ready to jump into the real Selenium automation tool – Selenium WebDriver.

In this chapter we will see how to create a WebDriver script. Also, we will configure Eclipse environment.

Key objectives

- Exporting Selenium IDE script as a Java Selenium WebDriver script
- Configuration of project structure in Eclipse and use Selenium WebDriver script
- Running of Selenium WebDriver script

9.1. Recording and Exporting Script from IDE

In this section we will record the test case using Selenium IDE and then export the test case using **Java JUnit** option. Follow the steps given below:

1. Open **Selenium IDE** and verify that recording mode is **ON**
2. Assuming that application login page is already open in **Chrome browser** with login page visible, perform the following steps (in IDE recording mode):
    a. Login (Use the username/password with which you have registered earlier)
    b. Search for a Hotel
        i. Select a location, e.g., Sydney
        ii. Select number of rooms, e.g., 2-Two
        iii. Select adults per rooms, e.g., 2-Two
        iv. Click on Search button
    c. Select a Hotel
        i. Select one of the Hotel Radio buttons, e.g., select radio button next to Hotel Creek

d. Book a Hotel
    i. Enter First Name
    ii. Enter Last Name
    iii. Enter Address
    iv. Enter 16-digit Credit Card No.
    v. Enter Credit Card type
    vi. Enter Expiry Month
    vii. Enter Expiry Year
    viii. Enter CVV number
    ix. Click on Book Now
e. After you see the Booking confirmation page, click on Logout link in the top right corner
f. Click on "Click here to Login again" link to go back to Home page

3. Stop recording by clicking on **Stop Recording** button in record toolbar

4. Verify the steps below that are recorded Selenium ID

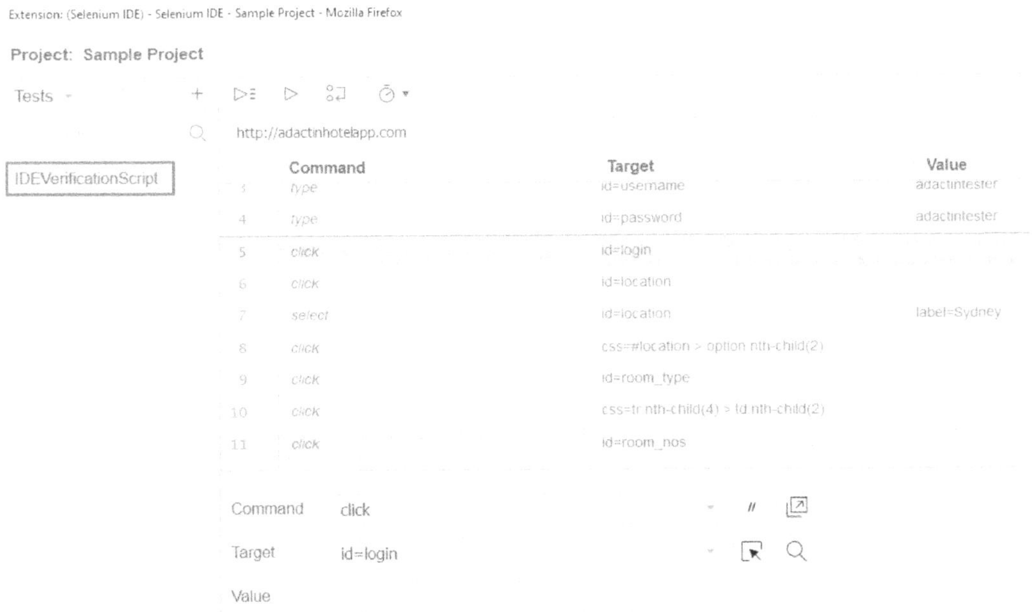

Figure 9-1 - IDE Script

5. Click on Export option after clicking on the '...' symbol to the right of IDE Script

Figure 9-2 - Export IDE Test

Note: We will focus on Selenium WebDriver with Java but as you can see Selenium supports export as C#, Python and Ruby languages as well.

6. Select Java Junit and click on Export. Save it as MyFirstWebDriverTest.java
7. Try to open the script you have saved in an editor like NotePad++ (you can download this freely from internet)

```
import org.openqa.selenium.By;
import org.openqa.selenium.WebDriver;
import org.openqa.selenium.firefox.FirefoxDriver;
import org.openqa.selenium.chrome.ChromeDriver;
import org.openqa.selenium.remote.RemoteWebDriver;
import org.openqa.selenium.remote.DesiredCapabilities;
import org.openqa.selenium.Dimension;
import org.openqa.selenium.WebElement;
import org.openqa.selenium.interactions.Actions;
import org.openqa.selenium.support.ui.ExpectedConditions;
import org.openqa.selenium.support.ui.WebDriverWait;
import org.openqa.selenium.JavascriptExecutor;
import org.openqa.selenium.Alert;
import org.openqa.selenium.Keys;
import java.util.*;
import java.net.MalformedURLException;
import java.net.URL;
public class BookHotelTest {
  private WebDriver driver;
  private Map<String, Object> vars;
  JavascriptExecutor js;
  @Before
  public void setUp() {
    driver = new ChromeDriver();
    js = (JavascriptExecutor) driver;
    vars = new HashMap<String, Object>();
  }
  @After
  public void tearDown() {
    driver.quit();
  }
  @Test
  public void bookHotel() {
    driver.get("https://adactinhotelapp.com/");
    driver.manage().window().setSize(new Dimension(1068, 816));
```

Figure 9-3 - Java code for WebDriver test

Next step will be to review this code and use it in Eclipse.

8. **Let us review the exported Java code. The exported test is Junit test.** JUnit is a unit testing framework for the Java programming language.

9. We will see a class "MyFirstWebDriverTest" shown in the snapshot below

The highlighted lines in the snapshot below will acquire an instance of a new Chrome browser and assign it to the driver (WebDriver) object which we will use to perform all of our browser actions.

```
import org.openqa.selenium.By;
import org.openqa.selenium.WebDriver;
import org.openqa.selenium.firefox.FirefoxDriver;
import org.openqa.selenium.chrome.ChromeDriver;
import org.openqa.selenium.remote.RemoteWebDriver;
import org.openqa.selenium.remote.DesiredCapabilities;
import org.openqa.selenium.Dimension;
import org.openqa.selenium.WebElement;
import org.openqa.selenium.interactions.Actions;
import org.openqa.selenium.support.ui.ExpectedConditions;
import org.openqa.selenium.support.ui.WebDriverWait;
import org.openqa.selenium.JavascriptExecutor;
import org.openqa.selenium.Alert;
import org.openqa.selenium.Keys;
import java.util.*;
import java.net.MalformedURLException;
import java.net.URL;
public class BookHotelTest {
  private WebDriver driver;
  private Map<String, Object> vars;
  JavascriptExecutor js;
  @Before
  public void setUp() {
    driver = new ChromeDriver();
    js = (JavascriptExecutor) driver;
    vars = new HashMap<String, Object>();
  }
  @After
  public void tearDown() {
    driver.quit();
  }
  @Test
  public void bookHotel() {
    driver.get("https://adactinhotelapp.com/");
    driver.manage().window().setSize(new Dimension(1068, 815));
```

Figure 9-4 - Code View

The code written next to **@Test** annotation is performing all our test step actions. The first line fetches the Webpage for us according to the given URL. Then we enter username, password and click Login button, all using our driver instance.

On a Web page, driver finds an element by using its unique identifier which in this case is the id of the element. Every element is expected to have a unique id on the page. Selenium uses various identifiers/locators to recognize a page element.

```
driver.get("http://www.adactinhotelapp.com" );

        driver.findElement(By.id("username")).clear();

        driver.findElement(By.id("username")).sendKeys("adactin123");
        driver.findElement(By.id("password")).clear();
        driver.findElement(By.id("password")).sendKeys("xxxxxxxx");
        driver.findElement(By.id("login")).click();
```

Available JUnit annotations

Annotations are part of syntax in JUnit tests. With annotations, creating and running a JUnit test becomes easier and more readable. They enable Java interpreter to understand when test method starts, what needs to be executed before test method starts and what needs to be executed after test method finishes.

Following table gives an overview of the available annotations in JUnit 4.x.

Annotations

Annotation	Description
@Test public void method()	The @Test annotation identifies a method as a test method.
@Before public void method()	This method is executed before each test. It is used to prepare the test environment (e.g., read input data, initialize the class).
@After public void method()	This method is executed after each test. It is used to clean up the test environment (e.g. delete temporary data, restore defaults). It can also save memory by cleaning up expensive memory structures.
@BeforeClass public static void method()	This method is executed once before the start of all tests. It is used to perform time intensive activities, for example to connect to a database. Methods annotated with this annotation need to be defined as static to work with JUnit.
@AfterClass public static void method()	This method is executed once, after all tests have been finished. It is used to perform clean-up activities, for example to disconnect from a database. Methods annotated with this annotation need to be defined as static to work with JUnit.
@Ignore	Ignores the test method. This is useful when the underlying code has been changed and the test case has not yet been adapted. Or if the execution time of this test is too long to be included.

9.2. Configure Eclipse to Work with Selenium

Note: We will not be using Maven based Java projects as part of this book, instead we will create our own project structure which would be much simpler to implement and understand.

1. Open link https://www.selenium.dev/downloads/
2. Select Download option corresponding to Java under the Selenium Client & WebDriver Language Bindings and unzip after download

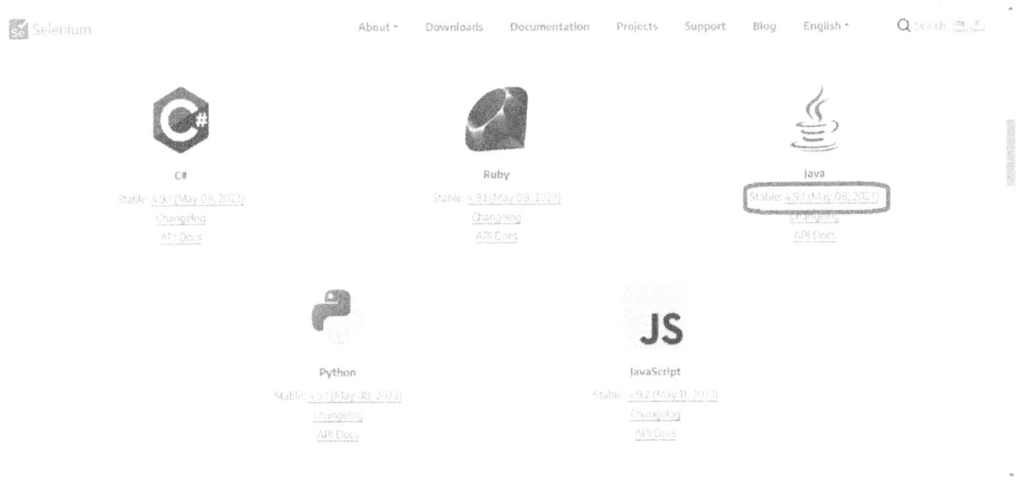

Figure 9-5 - Selenium Client and WebDriver Download

Note: You might find an updated version on this link as Selenium server jar file versions are constantly being updated. You can download the latest version available.

3. Launch **Eclipse** from location C:\eclipse\eclipse.exe or from location where you have extracted eclipse folder in previous section
4. Open the **default workspace** (setup in previous section), C:\Selenium\workspace\
5. Click on **Workbench icon** (if you see default view)

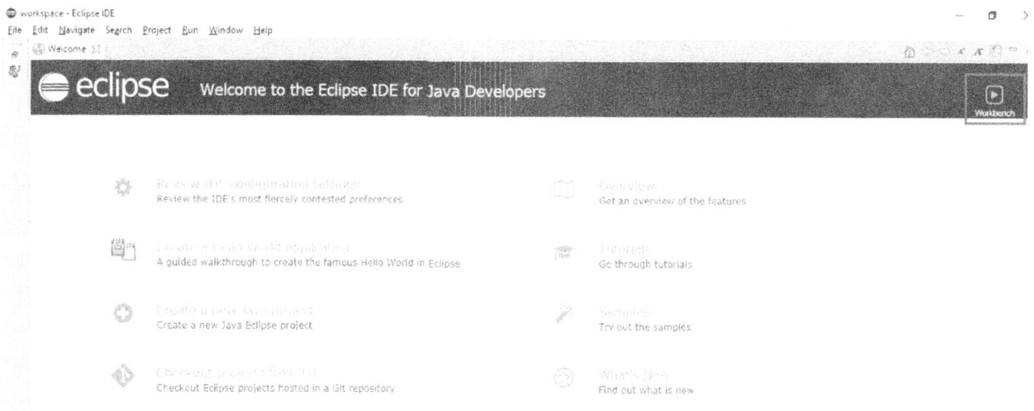

Figure 9-6 - WorkBench

6. You will see **Package Explorer View**

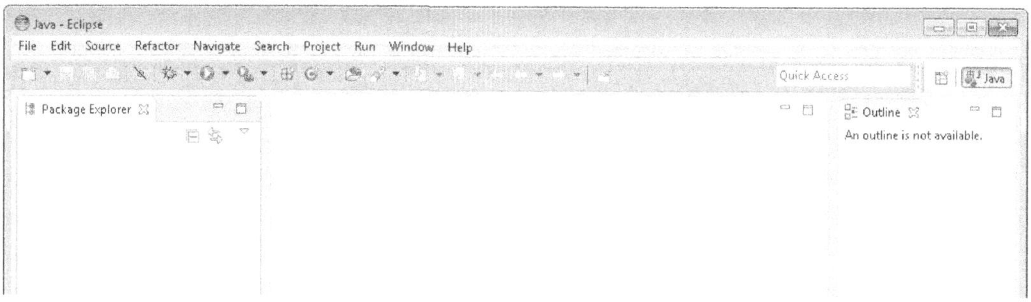

Figure 9-7 – Package Explorer

7. Go to **File > New > Java Project** and enter the name as **HotelApp_TestAutomation**

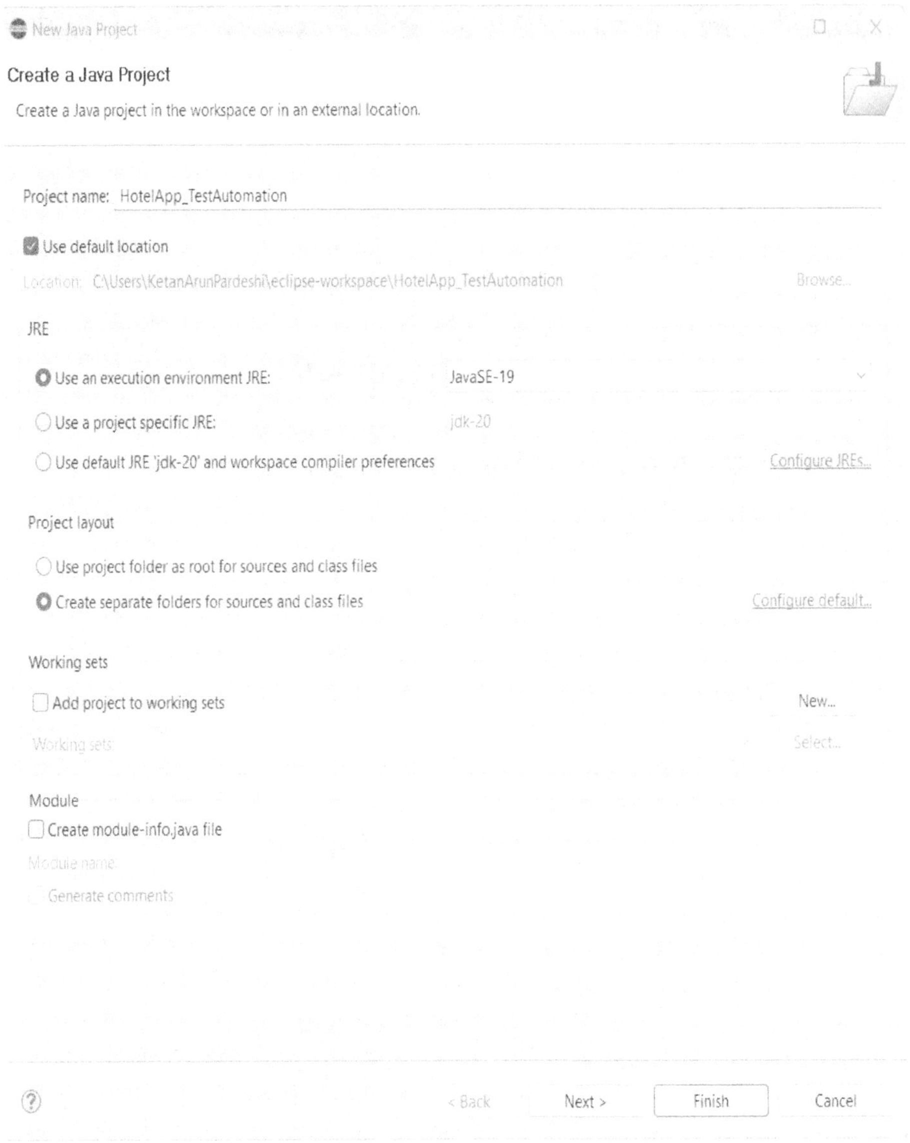

Figure 9-8 – Create Java Project

8. Uncheck Create module-info.java checkbox .

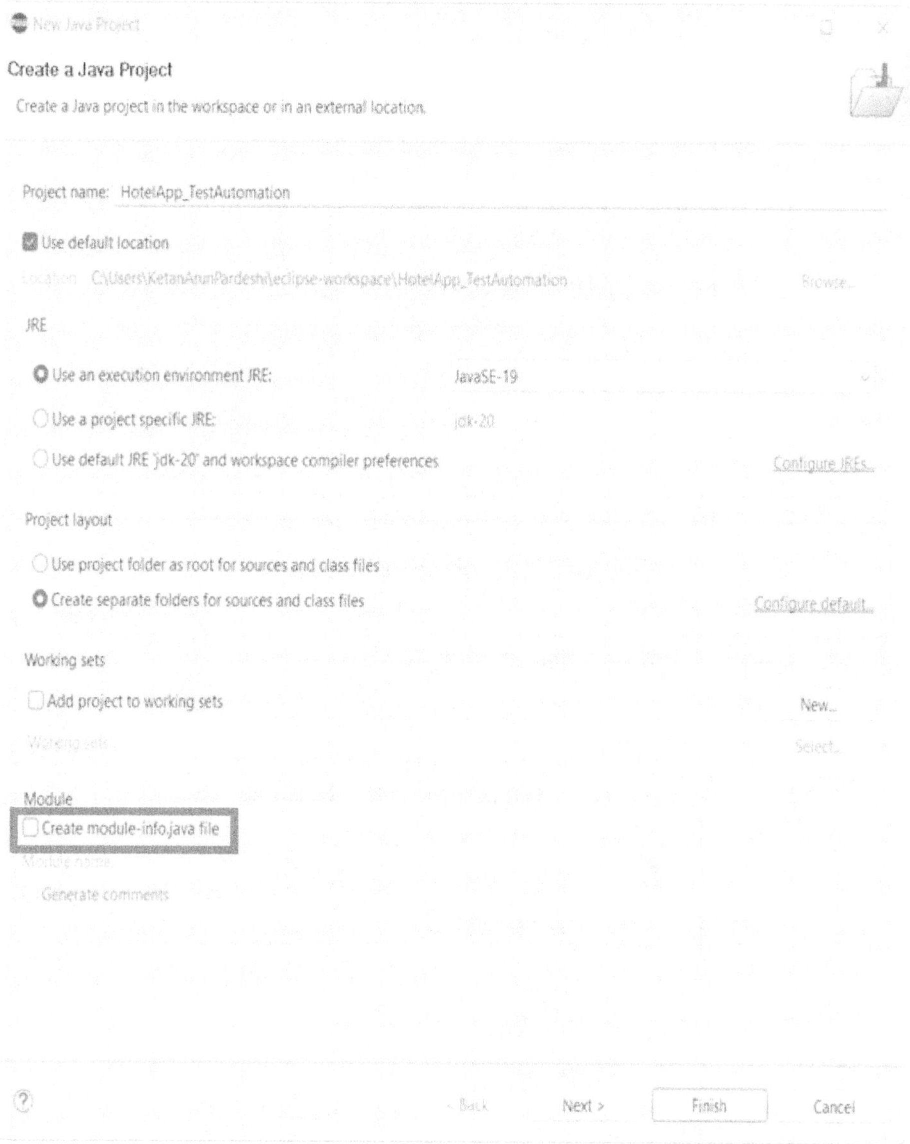

Figure 9-9 - Do not create module-info

9. Click on **Finish**

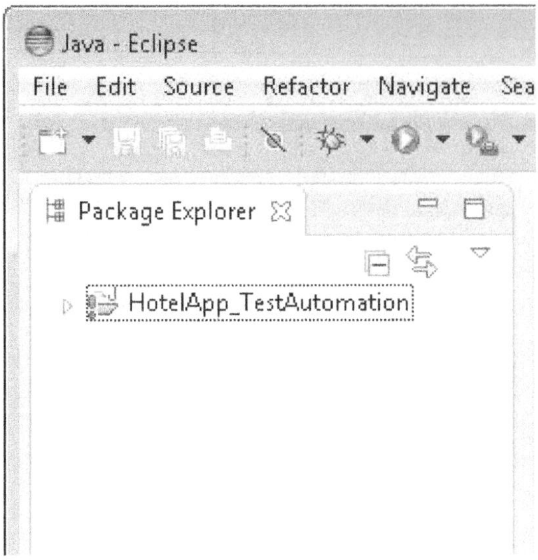

Figure 9-10 – Project Structure

10. Expand the project and select **src folder**

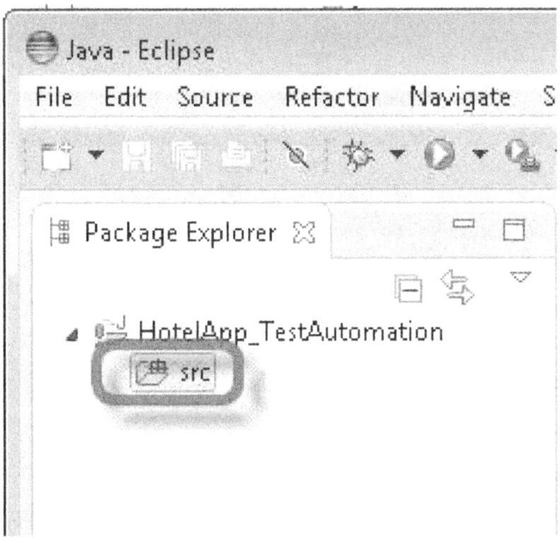

Figure 9-11– Source Folder Selection

11. Now we will create a new package which will contain all tests. Right click on the **src folder**. Select **New → Package**.

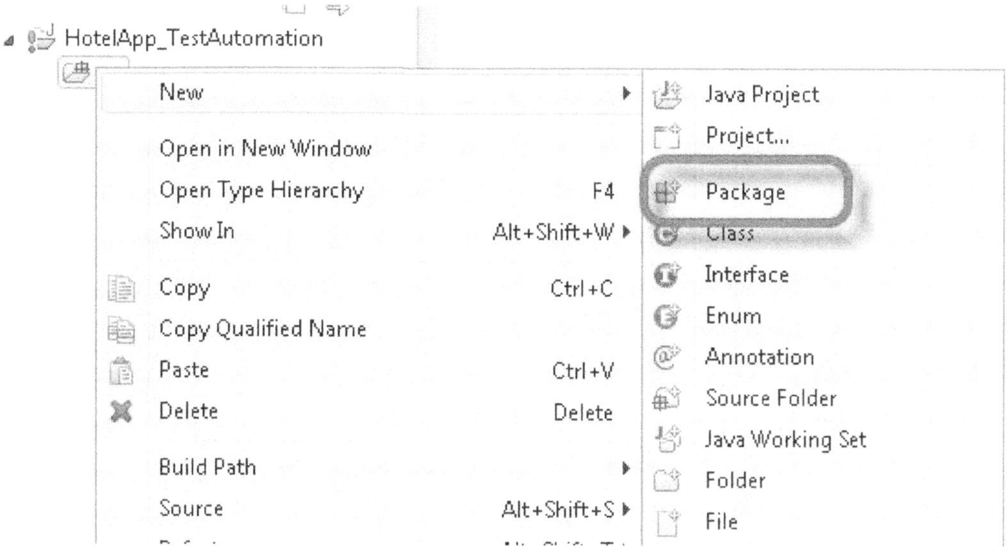

Figure 9-12– New Package Creation

12. Name the Package as **tests** and Click **Finish**

Figure 9-13– Java Test Package creation

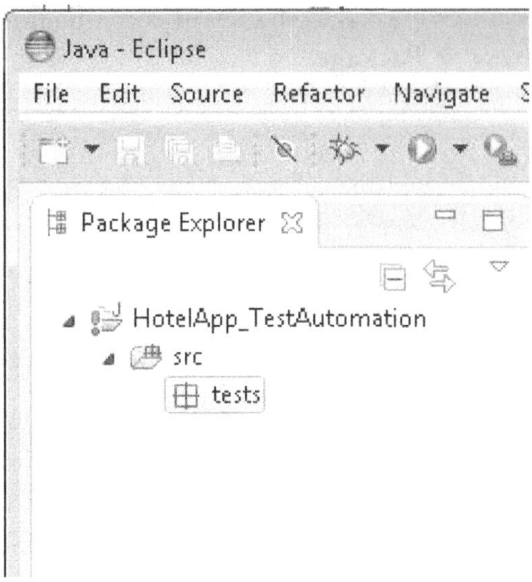

Figure 9-14 Project Structure

13. **Right click** on this tests package and select **New → Class**, in the New Java Class dialog enter Name: **MyFirstWebDriverTest**.

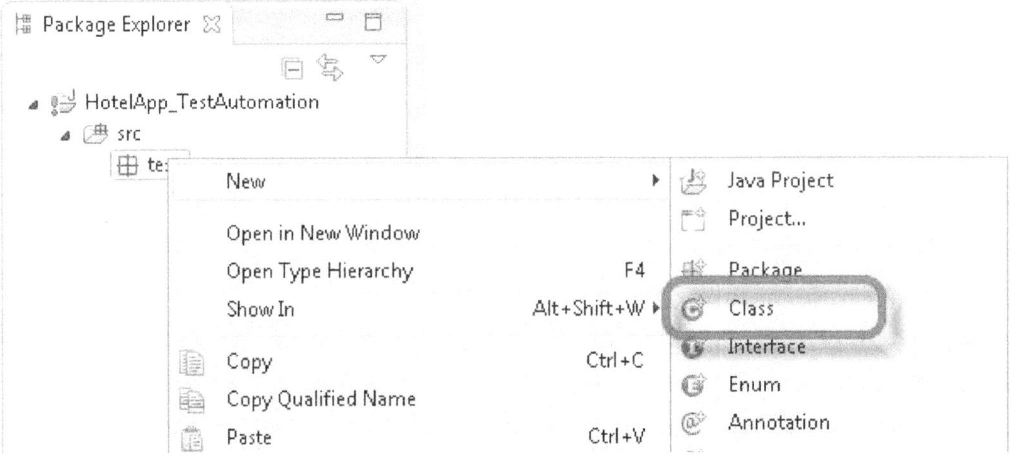

Figure 9-15 - Class Creation option

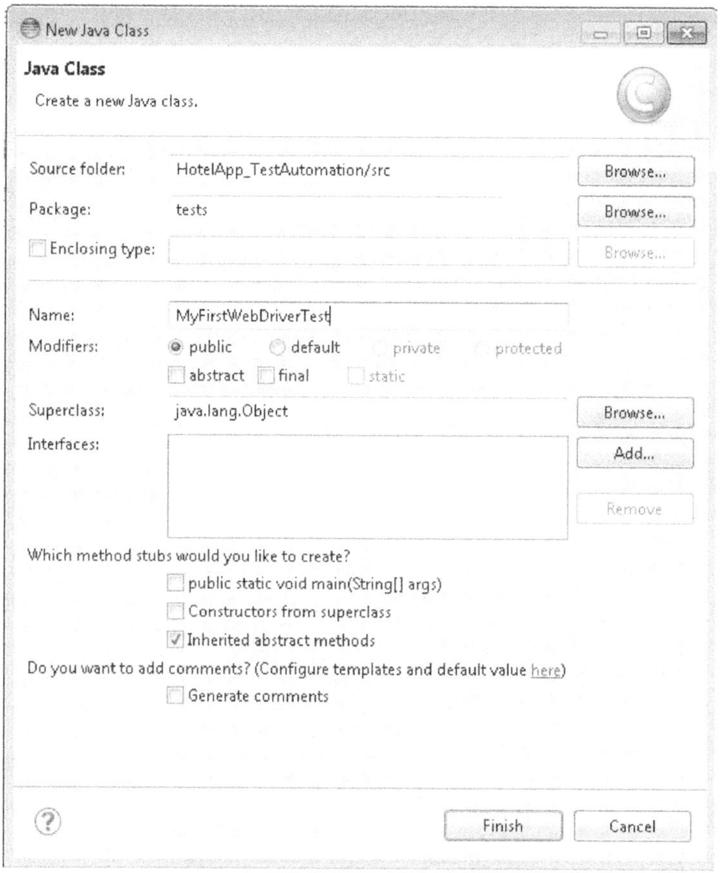

Figure 9-16 – Create Java Test

14. Click **Finish**.

15. We have made some changes in the exported code. Replace the class content with the below code.

```
package tests;

import java.time.Duration;
import org.junit.After;
import org.junit.Before;
import org.junit.Test;
import org.openqa.selenium.By;
import org.openqa.selenium.WebDriver;
import org.openqa.selenium.chrome.ChromeDriver;
import org.openqa.selenium.support.ui.Select;

public class MyFirstWebDriverTest {

        private WebDriver driver;

        @Before
        public void setUp()
```

```
	{
		driver = new ChromeDriver();
		driver.get("https://www.adactinhotelapp.com");
		driver.manage().window().maximize();
		driver.manage().timeouts().implicitlyWait(Duration.ofSeconds(10);
	}

	@Test
	public void testMyFirstWebDriver() throws Exception {

	    driver.findElement(By.xpath(".//*[@id='username']")).clear();

	    driver.findElement(By.xpath(".//*[@id='username']")).sendKeys("adactin123");

	    driver.findElement(By.id("password")).clear();

	    driver.findElement(By.id("password")).sendKeys("xxxxxxxxx");

	    driver.findElement(By.id("login")).click();

	   new Select(driver.findElement(By.id("location"))).selectByVisibleText("Sydney");

	    driver.findElement(By.id("Submit")).click();

	    driver.findElement(By.id("radiobutton_1")).click();

	    driver.findElement(By.id("continue")).click();

	    driver.findElement(By.id("first_name")).clear();

	    driver.findElement(By.id("first_name")).sendKeys("test");

	    driver.findElement(By.id("last_name")).clear();

	    driver.findElement(By.id("last_name")).sendKeys("test");

	    driver.findElement(By.id("address")).clear();

	    driver.findElement(By.id("address")).sendKeys("test");

	    driver.findElement(By.id("cc_num")).clear();

	    driver.findElement(By.id("cc_num")).sendKeys("1212121212121212");

	     new Select(driver.findElement(By.id("cc_type"))).selectByVisibleText("American
Express");

	  newSelect(driver.findElement(By.id("cc_exp_month"))).selectByVisibleText("March");

	   new Select(driver.findElement(By.id("cc_exp_year"))).selectByVisibleText("2015");

	    driver.findElement(By.id("cc_cvv")).clear();

	    driver.findElement(By.id("cc_cvv")).sendKeys("111");

	    driver.findElement(By.id("book_now")).click();

	    driver.findElement(By.linkText("Logout")).click();

	    driver.findElement(By.linkText("Click here to login again")).click();
	  }

	@After
	public void cleanUp()
	{
		driver.quit();
	}

}
```

Figure 9-17 – Project Structure

16. Select **HotelApp_TestAutomation** folder, right click and select **New → Folder**

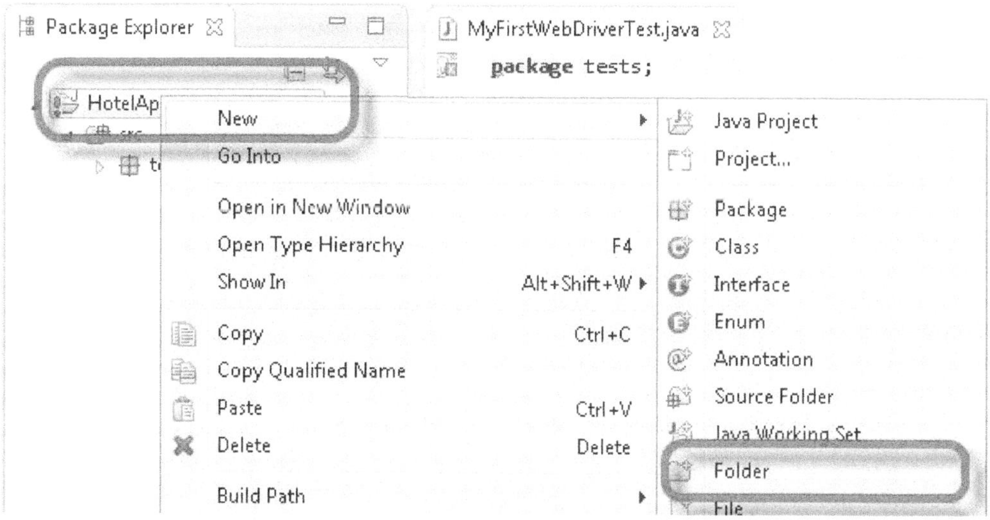

Figure 9-18 – Folder Creation

17. Name the folder Lib and Click **Finish**

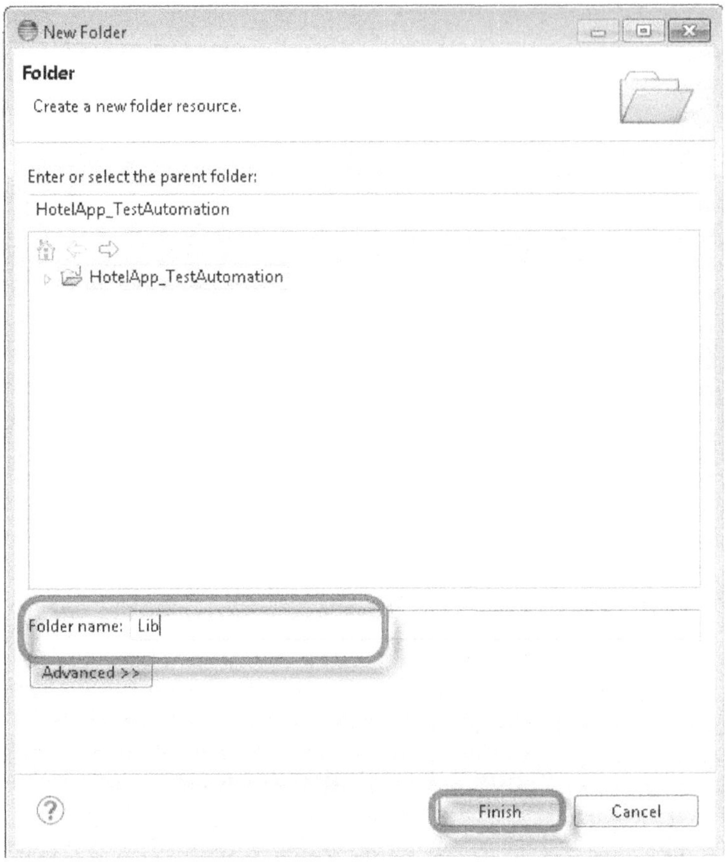

Figure 9-19 – Folder Name

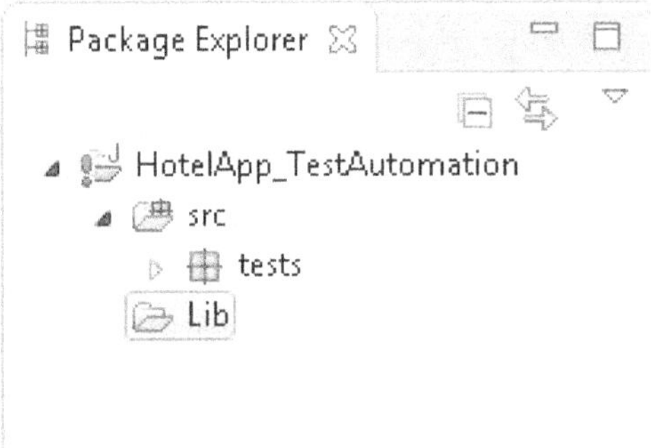

Figure 9-20 – Project Structure

18. Go to the location where your workspace Folder is stored. In our case it is C:\Selenium\workspace\HotelApp_TestAutomation. You will see Lib folder available. Double click to get in **Lib folder**

19. Copy all the jar files present inside selenium-server-4.***.** folder into the Lib folder which we had copied earlier in location C:\Selenium\.

20. Come back to Eclipse and click on your **project** and **press F5**. You should be able to see the Lib folder and Jar file added to your project structure.

Figure 9-21 – Lib folder and file

21. Now click on the **Project → Properties** and in opened window click on **'Java Build Path'**. Then click the **'Libraries'** tab

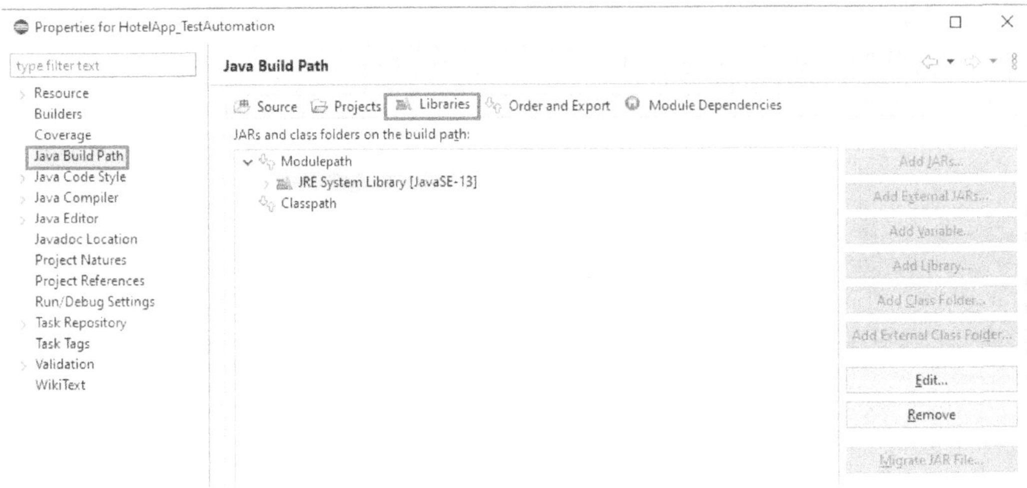

Figure 9-22 - Configure Build Path

22. Select Classpath, click on Add External Jars button and Add all jar files from C:\Selenium\workspace\HotelApp_TestAutomation\Lib

Figure 9-23 – Add External Jars

23. Click Apply and Close

*MyFirstWebDriverTest.java ×

```java
package tests;

import java.time.Duration;
import org.junit.After;
import org.junit.Before;
import org.junit.Test;
import org.openqa.selenium.By;
import org.openqa.selenium.WebDriver;
import org.openqa.selenium.chrome.ChromeDriver;
import org.openqa.selenium.support.ui.Select;

public class MyFirstWebDriverTest {

    private WebDriver driver;

    @Before
    public void setUp()
    {
        driver = new ChromeDriver();
        driver.get("https://www.adactinhotelapp.com");
        driver.manage().window().maximize();
        driver.manage().timeouts().implicitlyWait(Duration.ofSeconds(10));
    }

    @Test
     public void testMyFirstWebDriver() throws Exception {

        driver.findElement(By.xpath(".//*[@id='username']")).clear();

        driver.findElement(By.xpath(".//*[@id='username']")).sendKeys("adactin123");

        driver.findElement(By.id("password")).clear();

        driver.findElement(By.id("password")).sendKeys("xxxxxxxx");

        driver.findElement(By.id("login")).click();

        new Select(driver.findElement(By.id("location"))).selectByVisibleText("Sydney");

        driver.findElement(By.id("Submit")).click();

        driver.findElement(By.id("radiobutton_1")).click();

```

Writable Smart Insert 82 : 23 : 2384

Figure 9-24 – Project Structure

- Make sure the package name is corrected to the right name of the package
- Also make sure the Class name is the same as the name of our newly created class in eclipse

24. See below updated code after fixes

```
MyFirstWebDriverTest.java
package tests;
import org.junit.Test;
import org.junit.Before;
import org.junit.After;
import static org.junit.Assert.*;
import static org.hamcrest.CoreMatchers.is;
import static org.hamcrest.core.IsNot.not;
import org.openqa.selenium.By;
import org.openqa.selenium.WebDriver;
import org.openqa.selenium.firefox.FirefoxDriver;
import org.openqa.selenium.chrome.ChromeDriver;
import org.openqa.selenium.remote.RemoteWebDriver;
import org.openqa.selenium.remote.DesiredCapabilities;
import org.openqa.selenium.Dimension;
import org.openqa.selenium.WebElement;
import org.openqa.selenium.interactions.Actions;
import org.openqa.selenium.support.ui.ExpectedConditions;
import org.openqa.selenium.support.ui.WebDriverWait;
import org.openqa.selenium.JavascriptExecutor;
import org.openqa.selenium.Alert;
import org.openqa.selenium.Keys;
import java.util.*;
import java.net.MalformedURLException;
import java.net.URL;
public class MyFirstWebDriverTest {
  private WebDriver driver;
  private Map<String, Object> vars;
  JavascriptExecutor js;
```

Figure 9-26– Fixed Code

Note: **Troubleshooting Errors**

- If you get any errors regarding classes not found then you have not imported the Selenium jar correctly
- Also make sure your JRE is not throwing an error. You can go to **Project** --> **Properties→ Java Build Path** and check the JRE is not throwing an error. If yes, you can double click on JRE to select correct Java execution environment.

9.3. Running the Test

Now that we have configured our first WebDriver test let us run it and verify if it correctly executes. You have various options to run this Java code in Eclipse. You can run it as:

- A Java program using main method of a class.
- using JUnit Test Framework
- using TestNG Framework

As part of this exercise we will run this script as JUnit script.

Project Clean-up

Before you run the script it is always good to clean up any existing .class files and re-build the project. To clean project files perform below steps.

Note: This is not a mandatory step but is always a recommended step as at times you might see abnormal behavior or errors due to conflicts with old class files.

1. Select **Project → Clean**...

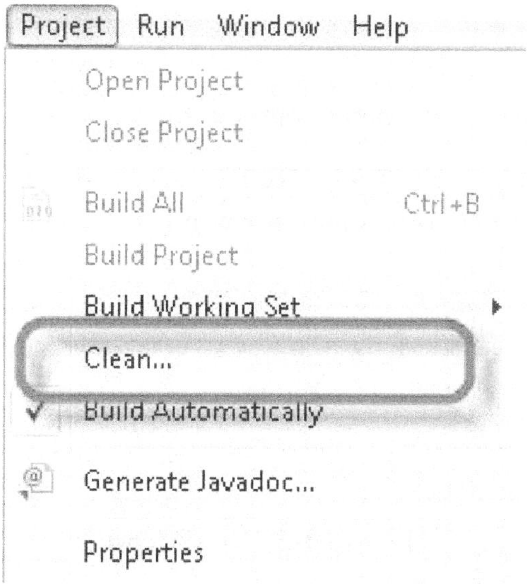

Figure 9-27- Clean Project Class files

2. Select "**Clean all projects**" radio button on the Clean dialog box and click OK

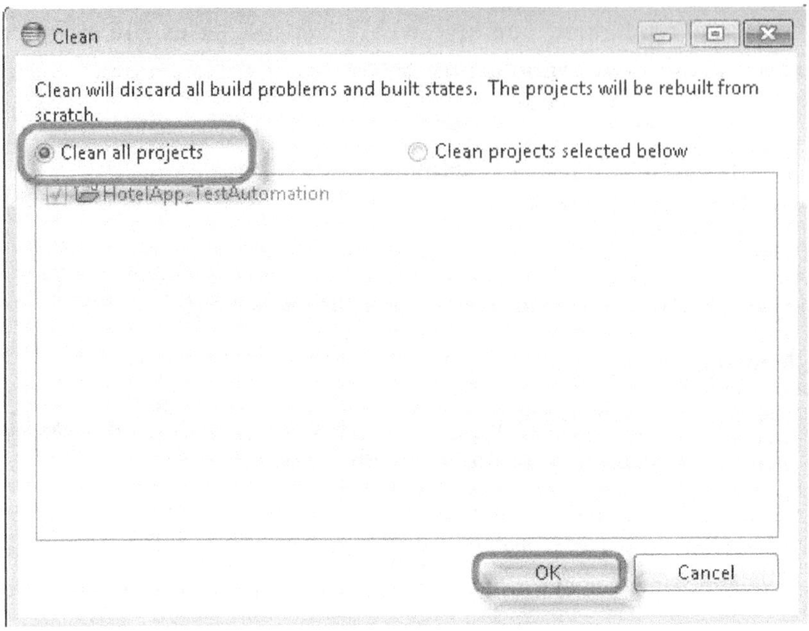

Figure 9-28 – Clean Project Dialog

Script Execution

There are two ways to execute code in Eclipse IDE.

1. On Eclipse's menu bar, click **Run** > **Run** or press **Ctrl+F11** to run the entire code (If you are running as a normal Java program).

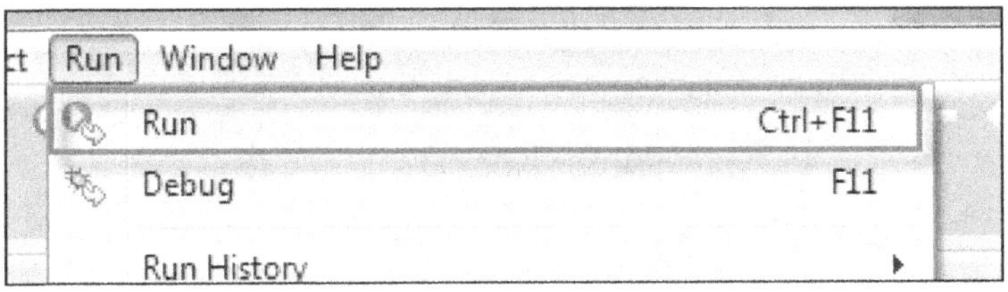

Figure 9-29 Run Menu

2. Or to execute your test, you right click script name in the project explorer and **Run As → JUnit Test.**

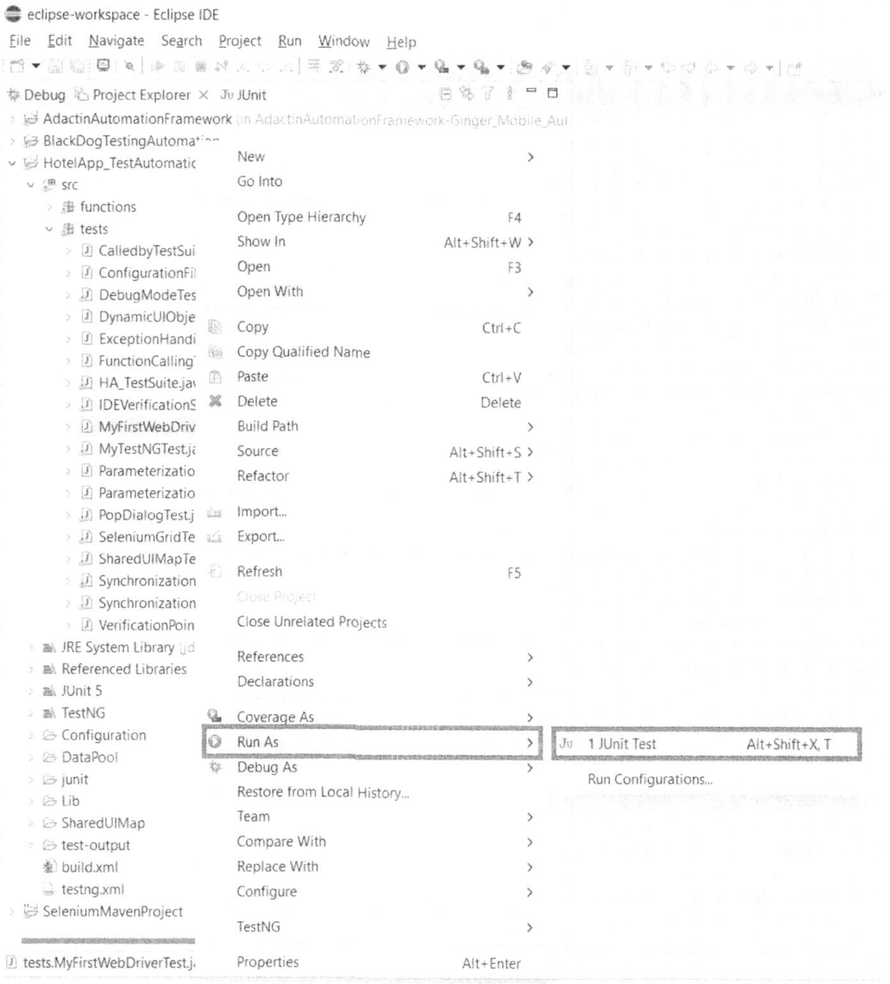

Figure 9-30 Run As Junit

EXERCISE

1. Open your second Selenium IDE script "**IDEVerificationScript**"
   - Export the script as Junit WebDriver test
   - Create a new test in Eclipse and use the exported script
   - Review the exported code, classes and methods (specially review code exported verification method)
   - Run the test and view the results.

§

10

Selenium Methods

While working with Selenium we need to use Java as well as selenium WebDriver functions. WebDriver provides various methods to perform different actions on Web elements. In this chapter we'll have an insight through these.

Let us look at the script we previously exported from IDE and configured in Eclipse. This script performs the following actions:

- Login.
- Search a hotel.
- Book a hotel room.
- Finally log out.

Consider the following code:

```
@Test
  public void testMyFirstWebDriver() throws Exception {

    driver.get("https://www.adactinhotelapp.com");

    driver.findElement(By.xpath(".//*[@id='username']")).clear();

    driver.findElement(By.xpath(".//*[@id='username']")).sendKeys("adactin123");

    driver.findElement(By.id("password")).clear();

    driver.findElement(By.id("password")).sendKeys("xxxxxxxxx");

    driver.findElement(By.id("login")).click();

    new Select(driver.findElement(By.id("location"))).selectByVisibleText("Sydney");

    driver.findElement(By.id("Submit")).click();

    driver.findElement(By.id("radiobutton_1")).click();

    driver.findElement(By.id("continue")).click();

    driver.findElement(By.id("first_name")).clear();

    driver.findElement(By.id("first_name")).sendKeys("test");

    driver.findElement(By.id("last_name")).clear();

    driver.findElement(By.id("last_name")).sendKeys("test");
```

```
    driver.findElement(By.id("address")).clear();

    driver.findElement(By.id("address")).sendKeys("test");

    driver.findElement(By.id("cc_num")).clear();

    driver.findElement(By.id("cc_num")).sendKeys("1212121212121212");

    new Select(driver.findElement(By.id("cc_type"))).selectByVisibleText("American Express");

    new Select(driver.findElement(By.id("cc_exp_month"))).selectByVisibleText("March");

    new Select(driver.findElement(By.id("cc_exp_year"))).selectByVisibleText("2015");

    driver.findElement(By.id("cc_cvv")).clear();

    driver.findElement(By.id("cc_cvv")).sendKeys("111");

    driver.findElement(By.id("book_now")).click();

    driver.findElement(By.linkText("Logout")).click();

    driver.findElement(By.linkText("Click here to login again")).click();
}
```

Figure 10-1 – Selenium WebDriver Script Code

If we look carefully at these code lines we see different methods are being used for performing different actions. For instance:

sendKeys( )	to send keyboard input to target field
selectByVisibleText( )	to select a drop-down list item
Click( )	to click a button or link

10.1. Common Selenium WebDriver Methods

Common element methods for UI Interaction

The following lists show details of some of the methods that Selenium provides to perform actions on different Web elements (listbox, edit box, checkbox, radio button) depending upon their type.

Method	**Purpose**
clear()	Clears all of the contents if the element is a text entity.
click()	Simulates a mouse click on the element.
getAttribute(String name)	Returns the value associated with the provided attribute name (if present) or null (if not present).
getTagName()	Returns the tag name for this element.

getText()	Returns the visible text contained within this element (including sub elements) if not hidden via CSS.
isEnabled()	Returns true for input elements that are currently enabled; otherwise false.
isSelected()	Returns true if the element (radio buttons, options within a select, and checkboxes) is currently selected; otherwise false.
sendKeys(CharSequence... keysToSend)	Simulates typing into an element.
submit()	Submits the same block if the element is a form (or contained within a form). Blocks until new page is loaded.

Note: **getAttribute** is one of the most common methods used specifically to get property values and used to verify data in the fields.

Search Element methods

In addition to the above methods designed for interacting with the element in hand, WebDriver also provides two methods allowing you to search for elements within the current page's scope:

Method	**Purpose**
findElement(By by)	Finds the first element located by the provided method (based on different location type).
findElements(By by)	Finds all elements located by the provided method.

Select Web Element's Methods

WebDriver provides a support class named Select to greatly simplify interaction with select elements and their association options. It is mostly used with elements of type list boxes.

Method	**Purpose**
selectByIndex(int index)/ deselectByIndex(int index)	Selects/deselects the option at the given index.
selectByValue(String value)/ deselectByValue(String value)	Selects/deselects the option(s) that has a value matching the argument.
selectByVisibleText(String text)/ deselectByVisibleTest(String text)	Selects/deselects the option(s) that displays text matching the argument.

deselectAll()	Deselects all options.
getAllSelectedOptions()	Returns a List<WebElement> of all selected options.
getFirstSelectedOption()	Returns a WebElement representing the first selected option.
getOptions()	Returns a List<WebElement> of all options.
isMultiple()	Returns true if this is a multi-select list; false otherwise.

WebDriver Methods

WebDriver methods are useful when you are working with the browser object itself and you would want to perform certain operations like close browser, get title of browser page, or if working with application which has multiple frames or Web pop-up windows.

Method	**Purpose**
close()	Close the current window, quitting the browser if it's the last window currently open.
get(JString url)	Load a new Web page in the current browser window.
getCurrentUrl()	Gets a string representing the current URL that the browser is looking at.
getPageSource()	Get the source of the last loaded page.
getTitle()	The title of the current page.
getWindowHandle()	Return an opaque handle to this window that uniquely identifies it within this driver instance.
getWindowHandles()	Return a set of window handles which can be used to iterate over all open windows of this WebDriver instance by passing them to switchTo().WebDriver.Options. window()
manage()	Gets the Option interface
navigate()	An abstraction allowing the driver to access the browser's history and to navigate to a given URL.
quit()	Quits this driver, closing every associated window.

We will understand the methods better as we use more of these methods in coming chapters.

§

11

Multiple Choice Questions Set-1

1. What are the three main columns in the Keyword View?
   A. Operation, Target, Value
   B. Command, Target, Data
   C. Command, Target, Value
   D. Command, Item, Value

2. Which of the following tools can be used to find XPath of a UI Element?
   A. Selenium IDE
   B. Chropath add-on
   C. Object Spy
   D. Eclipse

3. Which of these is not a supported language for Selenium WebDriver?
   A. Java
   B. C#
   C. Ruby
   D. C++

4. Which of these components is not part of Selenium Suite?
   A. Selenium IDE
   B. Selenium Grid
   C. Selenium Tester
   D. Selenium WebDriver

5. Which of the following is used to locate a Web element in Selenium?
   A. CSS
   B. XPath
   C. Id
   D. All of the above

6. Select the browser which is supported by Selenium IDE.
   A. Chrome
   B. Firefox
   C. Edge
   D. All Of Above

7. What are Selenium commands in Selenium IDE known as?
   A. SIDE
   B. Selenese
   C. Perl commands
   D. Ruby commands

8. Within Eclipse IDE after setting up Java Project where do we associate project to Selenium jar files?
   A. Java Compiler
   B. Resources
   C. Java Build Path
   D. Project References

9. Which method in WebDriver is used to get property value of any Web element?
   A. get
   B. getattribute
   C. getproperty
   D. getitem

10. Which method is used by Selenium WebDriver to set a value in the edit fields in a Web application?
    A. Set
    B. SendKeys
    C. sendValues
    D. SetValues

11. Which method is used by Selenium WebDriver to find a web element based on locators?
    A. FindElement
    B. SearchElement
    C. GetElement

12. Which method is used to select the option at the given index?
    A. SelectByIndex()
    B. SelectIndex()
    C. SelectByVisibleText()
    D. SelectByValue()

Answers

Q1. Answer: C
Explanation – Selenium IDE's 3 main columns are Command, Target and Value

Q2. Answer: B
Explanation – Chropath add-on is used to get Xpath value of any Web element

Q3. Answer: D
Explanation – C++ is not a supported language in Selenium

Q4. Answer: C
Explanation – Selenium Tester is not part of Selenium Suite

Q5. Answer: D
Explanation – All of the options can be used to locate a Web element in Selenium

Q6. Answer: B
Explanation – Selenium IDE is supported as an add-on with Firefox, Chrome and Edge browser

Q7. Answer: D
Explanation – Selenium commands are known as Selenese

Q8. Answer: C
Explanation – Within Eclipse IDE after setting up Java Project, selenium jar files are associated to project in Java Build Path. Refer to section - configure Eclipse to work with Selenium.

Q9. Answer: B
Explanation – The getattribute method is used to get the property value of any Web element in Selenium WebDriver

Q10. Answer: B
Explanation – The sendKeys method is used to set values in edit fields in Web forms using Selenium

Q11. Answer: A
Explanation – The FindElement method is used to find a Web element based on locator

Q12. Answer: A
Explanation – The SelectByIndex method is used to select an option at a given Index

§

12

Verification Point in Selenium

Introduction

Till now, we have just executed a set of actions a functional tester will perform. But we have not compared any actual values on the application against expected values to validate application functionality.

To accomplish this we will insert verification points in Selenium. In this chapter we will understand the need for verification and how to insert a verification point in Selenium WebDriver. Also we will understand the different kinds of validations or checks we can perform in Selenium.

Key objectives:

- Need for verification points
- Insert a verification point in Selenium script
- Type of Verification points

12.1. Need for a Verification Point

A verification point is a specialized step that compares two values and reports the result. A verification point compares the actual results from the test run, with the expected results in the test case.

A basic test cannot be considered a valid functional test without some form of validation.

You use a checkpoint to:

- Verify the state of an object
- Confirm that an application performs as expected

A verification point checks whether an application responds appropriately when a user performs tasks correctly while testing the application. A verification point ensures that a user is barred from performing certain tasks and confirms that invalid or incomplete data are flagged with appropriate messages.

- Examples of validation are: Specifying limits, conditions, or boundaries example

We were once testing an investment banking application where we could create varied instruments such as bonds, ADRs, etc. When creating an instrument, we would receive a "Save was successful" message and a unique instrument number. As part of our test case, we had to verify that "Save was successful" message appears. We also had to verify that we could search using the same instrument number and verify that all details were saved correctly. As part of our automation scripts, we had to verify both these conditions, so we used a verification point to verify this.

12.2. Inserting a Verification Point

A Verification point can be inserted in the WebDriver script. In this section we will lay out a test scenario in which we have expected result and we will try to automate that test scenario.

Test Scenario

Let us take a simple test case for automation from our Hotel Application.

Test Objective: To verify that when a location is selected in Search Hotel page, same location is displayed in Select Hotel page

Test Steps:

1. Login to the application using User credentials
2. Select Location as "Sydney" in Location field in Search Hotel Page
3. Keep all the default selections
4. Click on Search Button
5. Verify that in the next Select Hotel Page the correct Location is displayed

Expected Result

1. Correct location "Sydney" should appear in location column of select hotel search results

How to insert a verification point

Let us see how to insert a verification point.

Pre-conditions

1. **Right click** on your existing **MyFirstWebDriverTest.java** script, select **Copy**

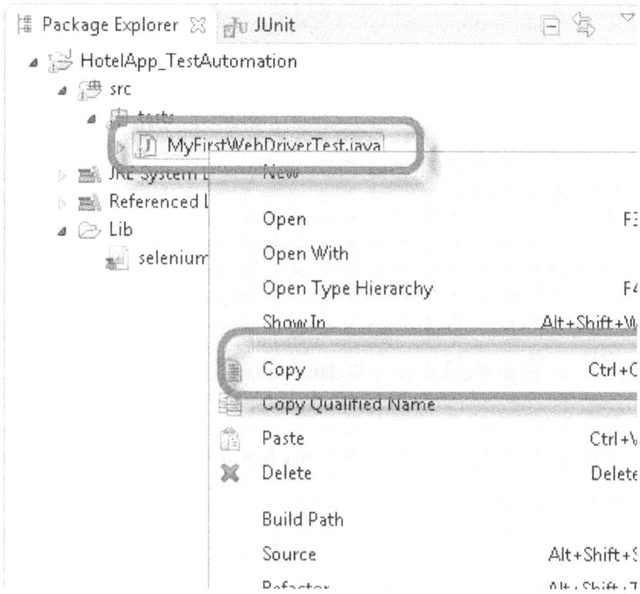

Figure 12-1 – Copy Script

2. Select the **tests** package folder, **right click** and select **Paste**

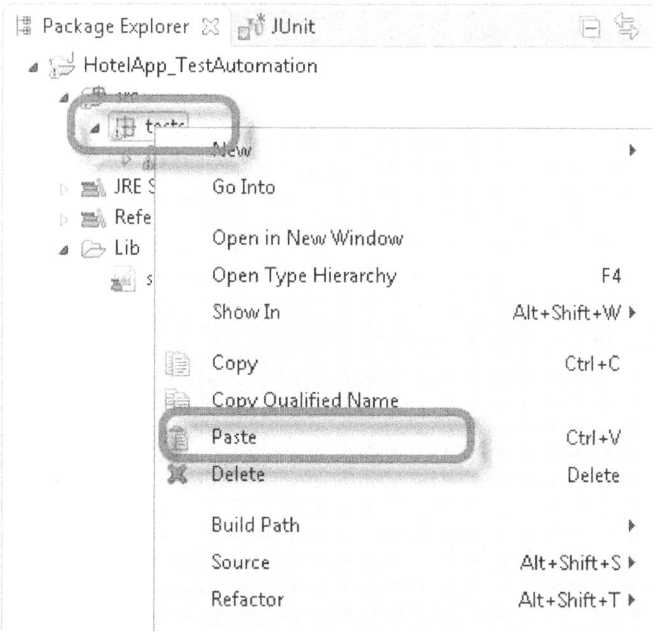

Figure 12-2 – Paste Script

3. In the Name Conflict dialog box , enter name of the script as **VerificationPointTest** and click **OK**

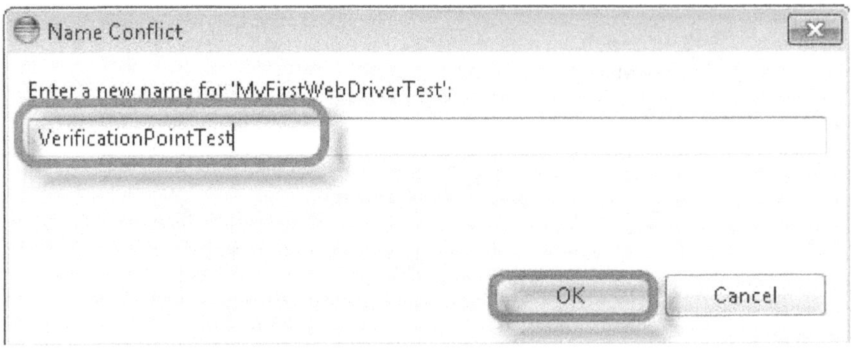

Figure 12-3 – Name Conflict dialog

4. Double click on the newly created "**VerificationPointTest**.**java**" **script** so that you can see the script

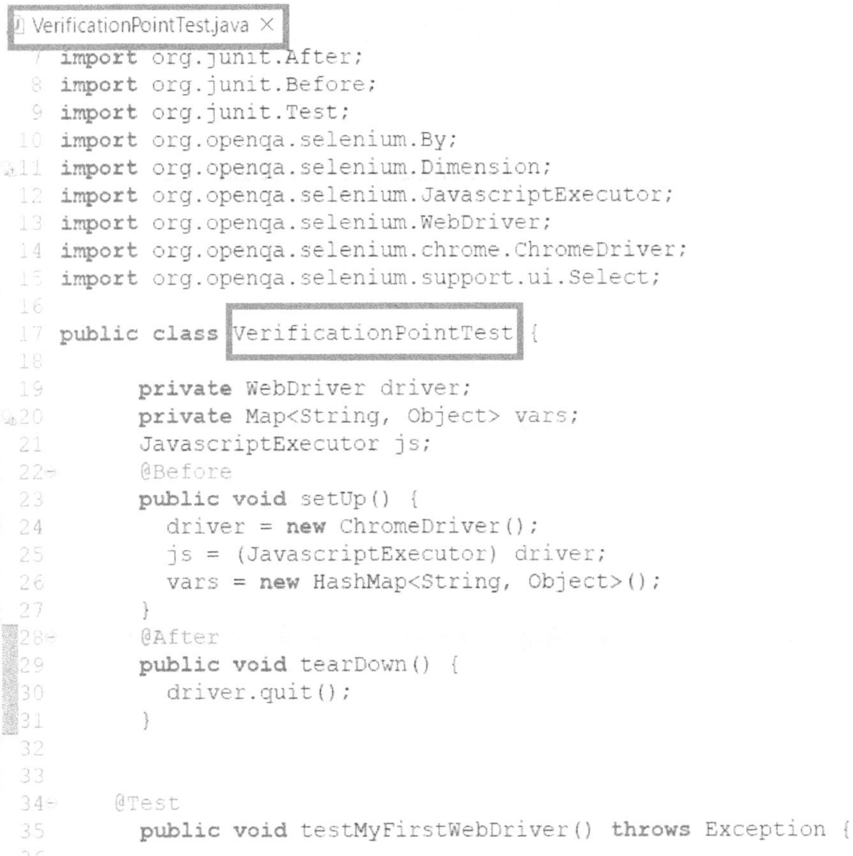

Figure 12-4 – Verification Point Test

Now we will insert a verification point.

Steps to insert a verification point

Objective – We need to verify that in Select Hotel Page we see location Sydney which we selected in previous search step

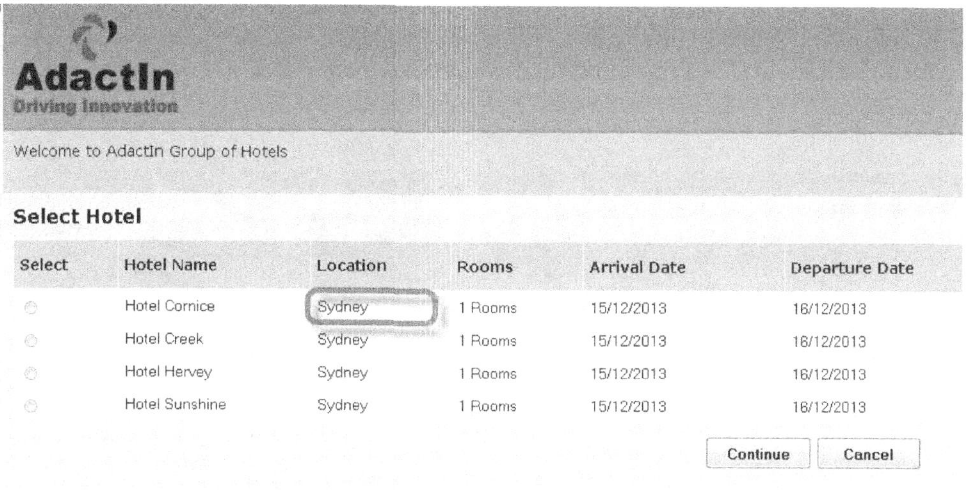

Figure 12-5 – Expected location value

1. In the script go to the step after which user submits search request for location Sydney

```
@Test
  public void testMyFirstWebDriver() throws Exception {

    driver.findElement(By.xpath(".//*[@id='username']")).clear();

    driver.findElement(By.xpath(".//*[@id='username']")).sendKeys("Adactin");

    driver.findElement(By.id("password")).clear();

    driver.findElement(By.id("password")).sendKeys("Adactin123");

    driver.findElement(By.id("login")).click();

    new Select(driver.findElement(By.id("location"))).selectByVisibleText("Sydney");

    driver.findElement(By.id("Submit")).click();

    driver.findElement(By.id("radiobutton_1")).click();

    driver.findElement(By.id("continue")).click();
```

Figure 12-6 – Step of verification point insertion

2. Now let us find out the XPath or id value for the Location field in the Select Hotel Page. To do this, open your Chrome browser, open application url, login and navigate to the Select Hotel Page.

Note: We can ignore step 2 if application is already open and we are on the Select Hotel page

3. In Chrome browser, right click and click on inspect. Inspect Location field and open the Chropath tab to view the properties

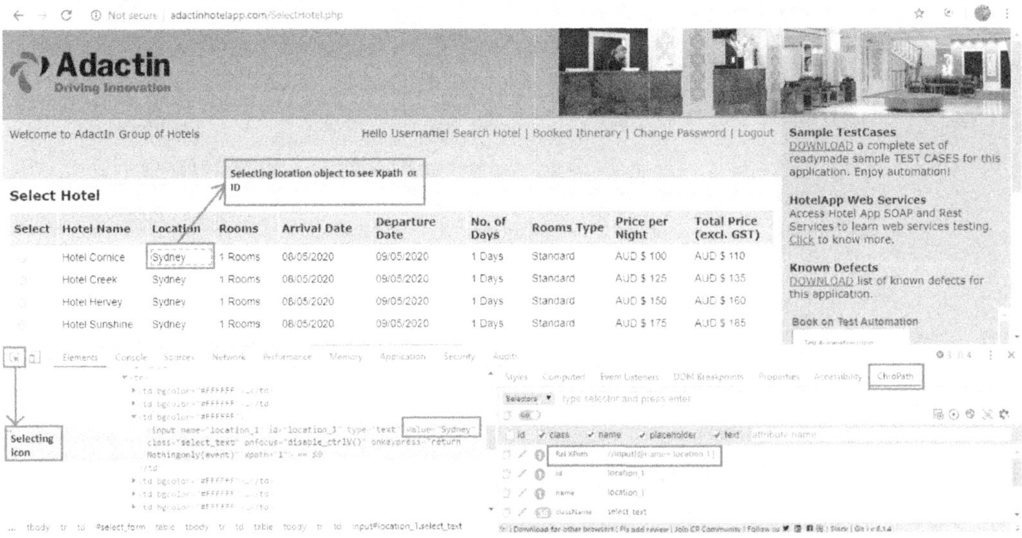

Figure 12-7 – Xpath_Id finder using ChroPath

We notice in the above snapshot that Xpath for location field is

Xpath = .//*[@id='location_1']

And

Id = location_1

Now we know the XPath or id of the object to locate the object.

In the left hand side, inside the Page source, it can be seen that the 'Sydney' value is stored in the attribute: value

Note: There might be other properties as well which might store "Sydney" (like defaultValue) but always try to pick up a property which sounds more stable and reasonable to store property value.

4. Now to insert a verification point in the script before clicking on logout, we will insert the statement below:

```
String slocation = driver.findElement(By.xpath("//input[contains(@id,
'location_')]")).getAttribute("value");
if(slocation.equalsIgnoreCase("Sydney"))
{
       System.out.println("Results are correct");
}
else
{
      System.out.println("Results are incorrect");
}
```

Figure 12-8 – Using getAttribute method

In the above statement

driver.findElement – Help to find the element based on locator

By.xpath(" //input[contains(@id, 'location_')]")) – helps to locate element based on xpath. This is the same xpath we got using ChroPath in the previous step

getAttribute – As discussed in the previous chapter, getAttribute helps us to get the desired property value of the Web element

value - is the property name which stores the actual value for location

```
driver.findElement(By.cssSelector("#adult_room > option:nth-child(3)")).click();
driver.findElement(By.id("Submit")).click();

String slocation = driver.findElement(By.xpath("//input[contains(@id, 'location_')]")).getAttribute("value");

driver.findElement(By.id("radiobutton_1")).click();
driver.findElement(By.id("continue")).click();
driver.findElement(By.id("first_name")).click();
driver.findElement(By.id("first_name")).sendKeys("Ketan");
driver.findElement(By.id("last_name")).sendKeys("Pardeshi");
driver.findElement(By.id("address")).sendKeys("D508, Pebbles");
driver.findElement(By.id("cc_num")).click();
driver.findElement(By.id("cc_num")).sendKeys("3156312321321313");
```

Figure 12-9 – Get Attribute Value

5. Now we would need to compare it with our expected value and report Pass or Fail. Insert the following statements after the getattribute statement. You can type the statements.

```
@Test
  public void testMyFirstWebDriver() throws Exception {

    driver.findElement(By.xpath(".//*[@id='username']")).clear();

    driver.findElement(By.xpath(".//*[@id='username']")).sendKeys("Adactin");

    driver.findElement(By.id("password")).clear();

    driver.findElement(By.id("password")).sendKeys("Adactin123");

    driver.findElement(By.id("login")).click();

    new Select(driver.findElement(By.id("location"))).selectByVisibleText("Sydney");

    driver.findElement(By.id("Submit")).click();

    String slocation = driver.findElement(By.xpath("//input[contains(@id, 'location_')]")).getAttribute("value");

    if(slocation.equalsIgnoreCase("Sydney"))
    {
        System.out.println("Results are correct");
    }
    else
    {
          System.out.println("Results are incorrect");
    }
    driver.findElement(By.id("radiobutton_1")).click();

    driver.findElement(By.id("continue")).click();

    driver.findElement(By.id("first_name")).clear();
    driver.findElement(By.id("first_name")).sendKeys("test");
    driver.findElement(By.id("last_name")).clear();
    driver.findElement(By.id("last_name")).sendKeys("test");
    driver.findElement(By.id("address")).clear();
    driver.findElement(By.id("address")).sendKeys("test");
```

Figure 12-10 – Compare Results

If/else - This statement is used as a conditional structure

equalsIgnoreCase – This method is used for string comparison in Java after ignoring the case of the string

System.out.println – Is used to print output to the console

6. That's it. Now we can perform **Project → Clean**... and run the script by selecting the script, right click and select **Run As Junit Test**. Verify the result in the Eclipse Console

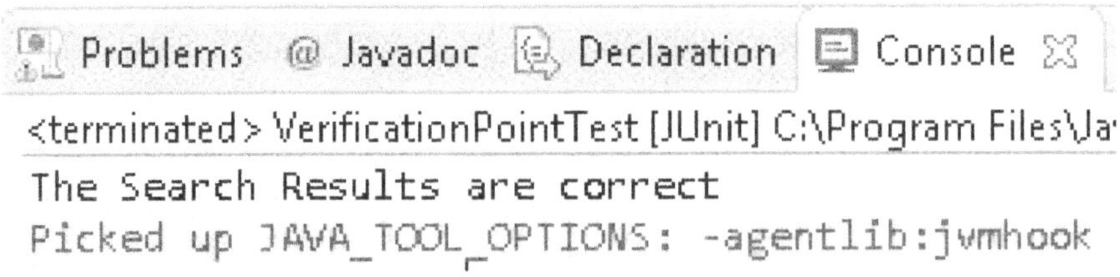

Figure 12-11 – Test Results

12.3. Understand how to Implement a Few Common Validations

The common types of validations are:

- Is the page title as expected?
- Does text exist on the Page/Webelement and is as expected?

Let us look at how to achieve each of these using the WebDriver API.

Get Page Title (using GetTitle Method)

You can get the current page title simply by calling method getTitle() on the WebDriver instance. Here is a simple test that would verify the title

```
String spageTitle = driver.getTitle();
if (spageTitle.equalsIgnoreCase("AdactIn.com - Select Hotel"))
        System.out.println("Page Title is correct. Actual page title is " + spageTitle);
else
        System.out.println("Page Title is incorrect. Actual page title is " + spageTitle);
```

Figure 12-12 – Get Page Title code sample

Validations against an element on the page (using GetText method)

You can use the following approach for validating text within an element. Here we find the element using one of the locator strategies and then call getText() or getAttribute() on the element object returned.

For instance if we do not select Location field and click the Search button you get the following error message in the application

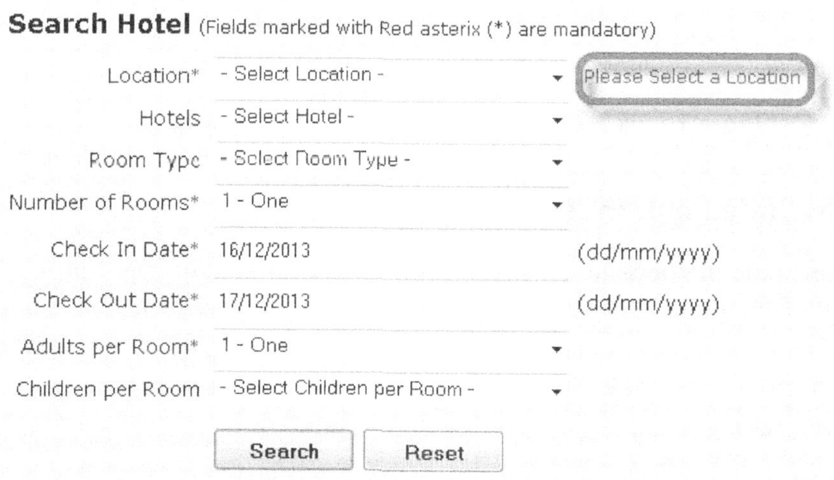

Figure 12-13– Location Field blank Error

We can use gettext method on the message field "Please Select a Location".

Using ChroPath verify the XPath for the error message.

Figure 12-14 – XPath for Location field error

Xpath = .//*[@id='location_span']

```
String sLocationFieldError = driver.findElement(By.xpath("//span[contains((text(),'Please Select a Location')]")).getText();

  if (sLocationFieldError.equalsIgnoreCase("Please Select a Location"))
        System.out.println("Mandatory Error check for Location field passed. Actual Location Field Error is " + sLocationFieldError);
  else
        System.out.println("Mandatory Error check for Location field passed. Actual Location Field Error is " + sLocationFieldError);
```

Figure 12-15 – Verification Point for text on page

12.4. Assert Statements in Junit

JUnit provides static methods in the Assert class to test for certain conditions. These assertion methods typically start with assert and allow you to specify the error message, the expected and the actual result. An assertion method compares the actual value returned by a test to the expected value, and throws an AssertionException if the comparison test fails.

When we use the Assert statement we do not have to use an 'If-Else' logical statement as the Assert statement will verify the result for us and return the correct value.

But important to note is that in case of failure, the Assert statement will Abort and exit the script. There are ways of implementing the Assert statement to avoid stopping and exiting the script. We will see a sample implementation in the script below.

The following table gives an overview of these methods. Parameters in [] brackets are optional.

Test methods

Statement	**Description**
fail(String)	Let the method fail. Might be used to check that a certain part of the code is not reached. Or to have a failing test before the test code is implemented. The String parameter is optional.
assertTrue([message], boolean condition)	Checks that the boolean condition is true.
assertFalse([message], boolean condition)	Checks that the boolean condition is false.
assertEquals([String message], expected, actual)	Tests that two values are the same. Note: for arrays the reference is checked, not the content of the arrays.
assertEquals([String message], expected, actual, tolerance)	Test that float or double values match. The tolerance is the number of decimals which must be the same.
assertNull([message], object)	Checks that the object is null.
assertNotNull([message], object)	Checks that the object is not null.
assertSame([String], expected, actual)	Checks that both variables refer to the same object.
assertNotSame([String], expected, actual)	Checks that both variables refer to different objects.

```
@Test
  public void testMyFirstWebDriver() throws Exception {


    driver.findElement(By.xpath(".//*[@id='username']")).clear();

    driver.findElement(By.xpath(".//*[@id='username']")).sendKeys("Adactin");

    driver.findElement(By.id("password")).clear();

    driver.findElement(By.id("password")).sendKeys("Adactin123");

    driver.findElement(By.id("login")).click();
    try
    {
        Assert.assertTrue(driver.findElement(By.linkText("Logout")).isDisplayed());
    }catch(Exception ex)
    {
        System.out.println(ex.toString());
    }
```

Figure 12-16 – Junit Assert Statements

Let us see an example where we had earlier used the Assert statement.

1. Go to your Selenium IDE and make sure script "IDEVerificationScript" is Open

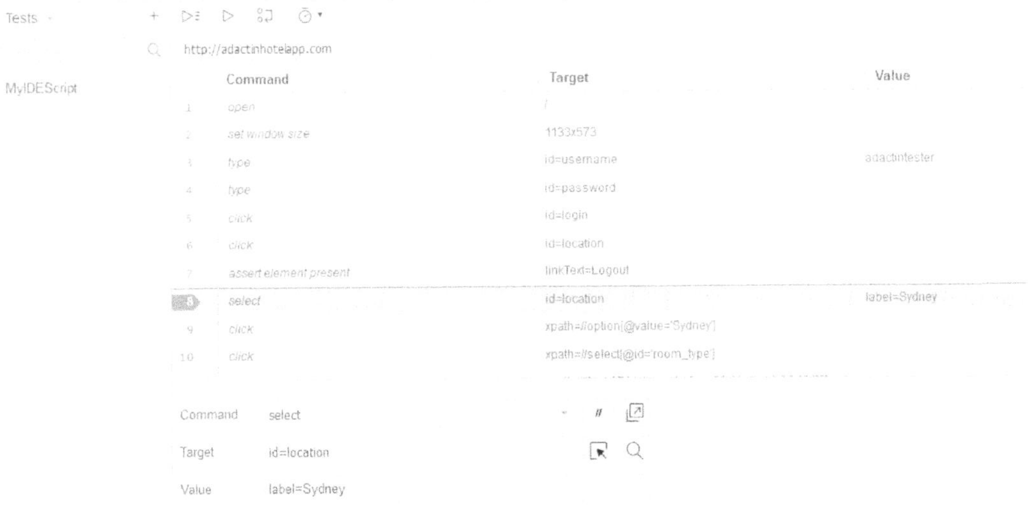

Figure 12-17– Verification Point IDE Script

2. Go to test case name. Click on '...'. Click on Export>Export as Java/JUnit4/WebDriver and export the script as a WebDriver Junit test

3. Open the exported WebDriver test in NotePad++

```
@After
public void tearDown() {
  driver.quit();
}
@Test
public void searchHotel() {
  driver.get("https://adactinhotelapp.com/");
  driver.manage().window().setSize(new Dimension(1550, 830));
  driver.findElement(By.id("username")).click();
  driver.findElement(By.id("login")).click();
  driver.findElement(By.id("location")).click();
  {
    List<WebElement> elements = driver.findElements(By.linkText("Logout"));
    assert(elements.size() > 0);
  }
  {
    WebElement dropdown = driver.findElement(By.id("location"));
    dropdown.findElement(By.xpath("//option[. = 'Sydney']")).click();
  }
  driver.findElement(By.cssSelector("#location > option:nth-child(2)")).click();
  driver.findElement(By.id("room_nos")).click();
  driver.findElement(By.id("room_nos")).click();
  driver.findElement(By.id("room_nos")).click();
  {
    WebElement dropdown = driver.findElement(By.id("room_nos"));
    dropdown.findElement(By.xpath("//option[. = '2 - Two']")).click();
  }
  driver.findElement(By.cssSelector("#room_nos > option:nth-child(3)")).click();
  driver.findElement(By.id("adult_room")).click();
  {
    WebElement dropdown = driver.findElement(By.id("adult_room"));
    dropdown.findElement(By.xpath("//option[. = '2 - Two']")).click();
  }
```

Figure 12-18 - Exported Verification Point Script

If you notice a new statement, assert has been added which validates that the logout link is present.

Now note that in the above implementation if the logout link does not appear the script will abort. But there would be scenarios where we would not want to abort the script on verification failure. To avoid aborting the script we can include the assert command in the Try-Catch block. See sample code below.

```
try {
    assertTrue(isElementPresent(By.linkText("Logout")));
} catch (Error e) {
  verificationErrors.append(e.toString());
}
```

Figure 12-19- Assert statement in Try/Catch block

In the sample code above, the exception thrown by the failed assert statement will be captured in the catch block and the script will move to the next step and will not abort.

EXERCISE

1. According to the functional requirements of our sample application, the user should see a message saying "Hello username!" on a successful login as shown in the highlight text in the figure below.

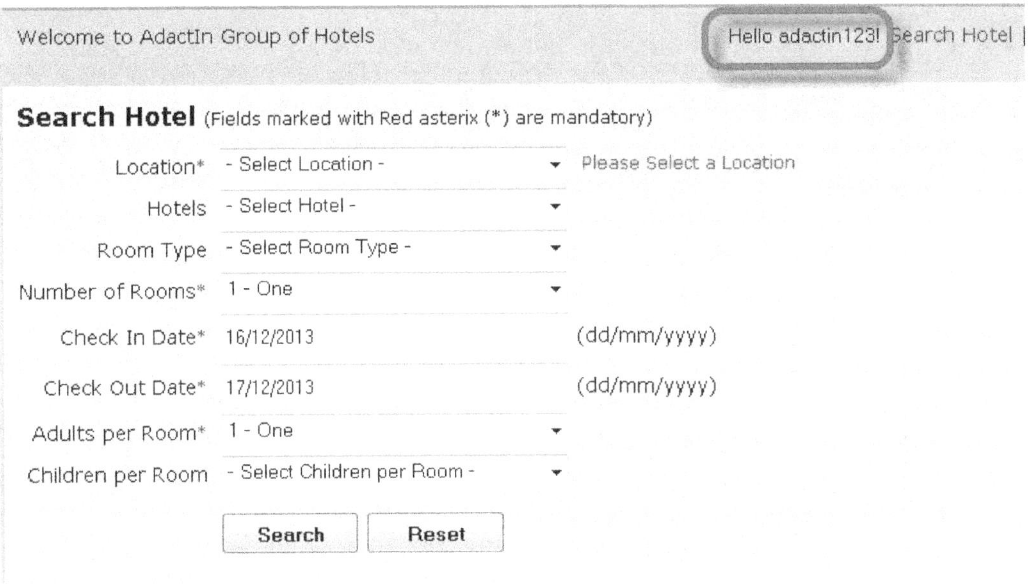

Figure 12-20– Welcome user message

Create a new script with a verification point which verifies a successful login with the correct message "Hello username"

Hint

1. Using ChroPath we find the Xpath for the highlighted area of the page where the expected message is displayed.
2. Code snippet should look as below

```
String sWelcometext = driver.findElement(By.xpath(".//*[@id='username_show']")).getAttribute("value");
if(sWelcometext.equalsIgnoreCase ("Hello adactin123"))
{
            System.out.println("Login Test Pass");
            }
else
{
            System.out.println("Login Test Pass");
            }
```

Now we can run this test again for the same user i.e. 'adactin123'. For a successful run we get a console message: Login Test Pass.

If we want to run the same test for different users we have to parameterize our test

§

13

Shared UI Map

Introduction

We now know how Selenium works as a tool, and recognizes Web elements based on properties and its values. In a WebDriver script, the Web element property values are located within the script. For instance, when we login, the username Web element's Xpath property gets added locally within the script statement.

```
driver.findElement(By.xpath(".//*[@id='username']")).clear();
```

In the above statement, the property value for Xpath (Xpath=.//*[@id='username']) is located within the script.

What if due to business requirements, the developer changes the property value of these Web elements? For instance, change 'username' to 'customername'. This would result in all of our scripts that use this Web element to fail. In order to fix this issue, we will need to go to each and every script, and change its properties information. This can be a nightmare to maintain!

Worst can be if the object is being used multiple times within the same script. We would need to make the same change multiple times in every script.

A better solution would be if we keep all the required Web elements and their properties in an external location/file and all the scripts could just use Web element properties from this shared location. This will certainly avoid redundancy. Also, if any Web element changes, we would only need to change the Web element once and the script will be working again.

example

At one of our clients, where we had to implement automation, we were given 100 existing automation scripts. We were told that the scripts used to work 3 months earlier, but now, they fail on new builds. We were asked to fix the scripts. Guess what we found? All the scripts were using local Web element properties. When we identified the Web elements that were causing the script to fail, we discovered that the same

Web element was used in all 100 scripts. So the Web element had to be modified at least 100 times as it was being used in all the scripts. But there were at least 100 Web elements, which had changed. Adding to this, they informed us that UI (user interface) changes were still happening and that the Web elements will change again. Our recommendation to them was to hold on and to re-do the scripts using a Shared UI Map. The advice stemmed from the fact that the same effort invested now would be required again when we get a new application build, with updated Web elements. Yes, it did mean that most of the previous efforts already made had been wasted. But our re-scripting approach, using a Shared Web Element Map, ensured that script maintenance was future proofed.

In this chapter we will discuss how to create and add objects to a shared UI Map, and how to use objects from a shared UI Map.

Key objectives:

- What is a Shared UI Map
- Create a Shared UI Map
- Add Web element and properties in a Shared UI Map
- Using a Shared UI Map in the WebDriver script

13.1. What is a Shared UI Map?

What makes a Shared UI Map helpful?

Its primary purpose is making test script management much easier. When a locator needs to be edited, there is a central location from where you easily find that object, rather than having to search through the entire script code. Also, it allows changing the Identifier in a single place, rather than having to make the change in multiple places within a test script, or for that matter, in multiple test scripts.

In a nutshell, a Shared UI Map has two significant advantages:

- Using a centralized location for UI Web elements instead of having them scattered throughout the script. This makes script maintenance more efficient.
- Cryptic HTML Identifiers and names can be given more human-readable names, improving the readability of test scripts.

Consider the following difficult to understand example of our Login Test.

```
public void Login() throws Exception {
driver.get("http://www.adactinhotelapp.com");
driver.findElement(By.xpath(".//*[@id='username']")).clear();
driver.findElement(By.xpath(".//*[@id='username']")).sendKeys("adactin123");
driver.findElement(By.xpath(".//*[@id='password']")).clear();
driver.findElement(By.xpath(".//*[@id='password']")).sendKeys("xxxxxx");
driver.findElement(By.xpath(".//*[@id='login']")).click();
}
```

Figure 13-1 - Login script with local Web element properties

This script would be hard to follow for anyone not familiar with the AUT's (Application Under Test) page source especially when the locator references are of more complex XPath or CSS types. Even regular users of the application might have difficulty understanding the script. A better script could be:

```
public void Login() throws Exception {
driver.get("http://www.adactinhotelapp.com");
driver.findElement(By.xpath(readMap("login.username")).clear();
driver.findElement(By.xpath(readMap("login.username")).sendKeys("adactin123");
driver.findElement(By.xpath(readMap("login.password")).clear();
driver.findElement(By.xpath(readMap("login.password")).sendKeys("xxxxxx");
driver.findElement(By.xpath(readMap("login.submit")).click();
}
```

Figure 13-2 - Login script with Shared Web element properties

readMap( ) could be a method we use to read the element locator from our Shared UI Map. Now, using some comments and whitespace along with the UI Map identifiers makes a very readable script.

```
public void Login() throws Exception {
        // Open app url.
            driver.get(baseUrl);

        // Provide admin username.
        driver.findElement(By.xpath(readMap("login.username")).clear();
    driver.findElement(By.xpath(readMap("login.username")).sendKeys("adactin123");

        // Provide admin Password.
        driver.findElement(By.xpath(readMap("login.password")).clear();
    driver.findElement(By.xpath(readMap("login.password")).sendKeys("xxxxxx");

        // Click on Login button.
        driver.findElement(By.xpath(readMap("login.submit")).click();   }
```

Figure 13-3 - Login script with Shared Web element properties with comments

There are various ways a Shared UI Map can be implemented. One, we could create a class which only has public String variables each storing a locator. Alternatively, a text/properties file storing key value pairs could be used. In Java, a properties file containing key/value pairs is probably a convenient and efficient method as this uses built-in Java functions.

13.2. Add a Shared UI Map to Selenium Project

Let's add a UI Map to our project.

1. Select the Project Folder **HotelApp_TestAutomation, Right Click** and select **New → Folder**
2. Name the Folder **SharedUIMap**

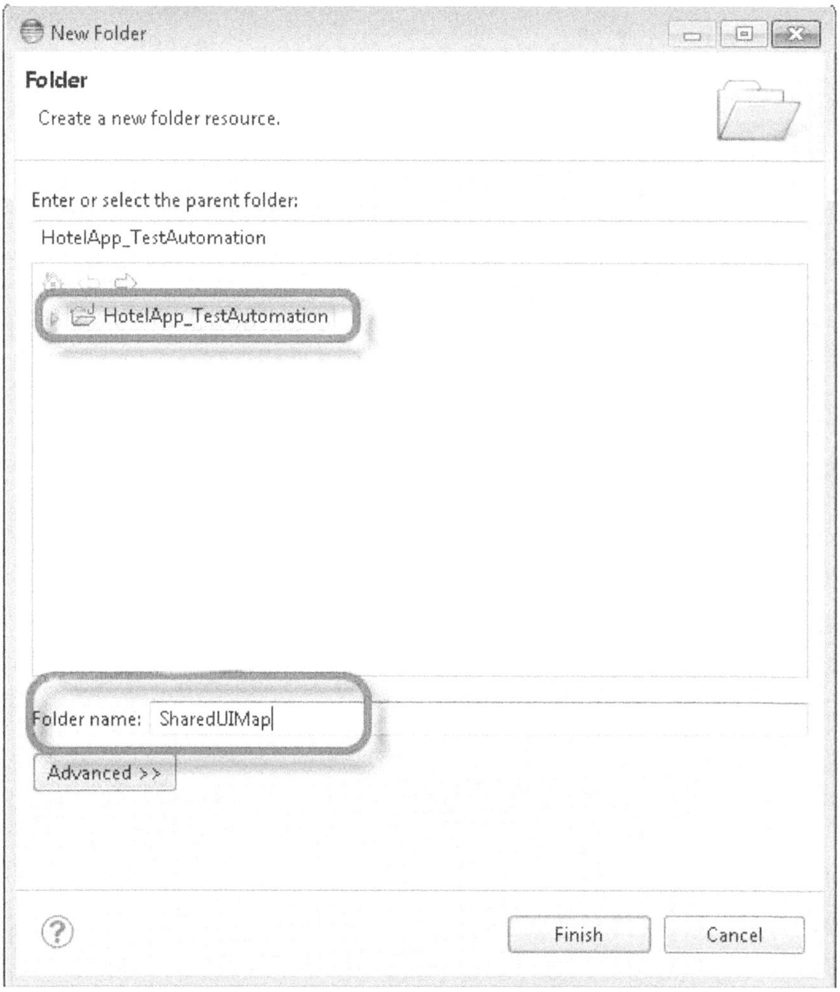

Figure 13-4 – Shared UI Map Folder

3. **Right Click** on newly created **SharedUI Map folder** and select **New → File**

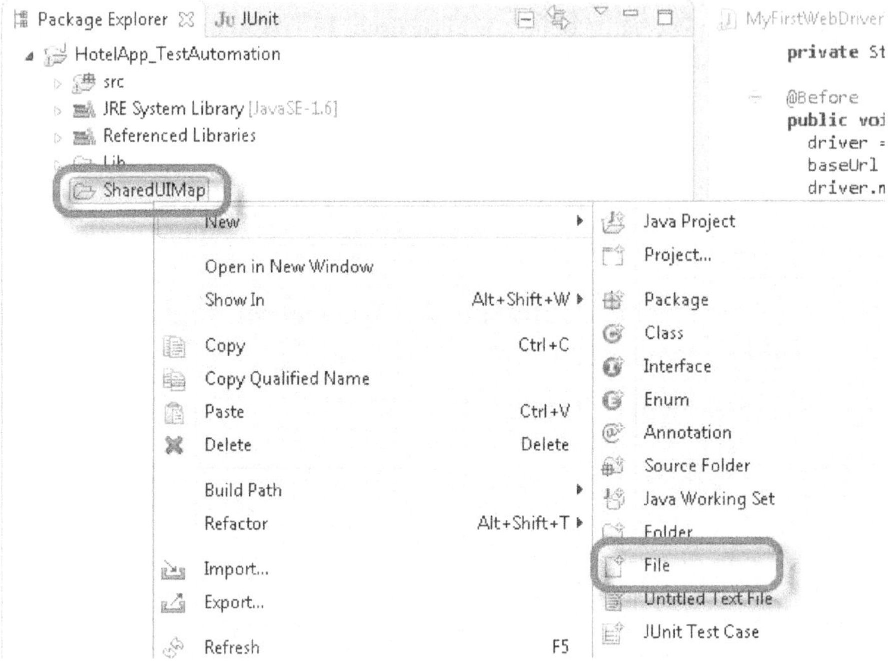

Figure 13-5 – Create New File

4. Give the name **SharedUIMap.properties** and click **Finish**

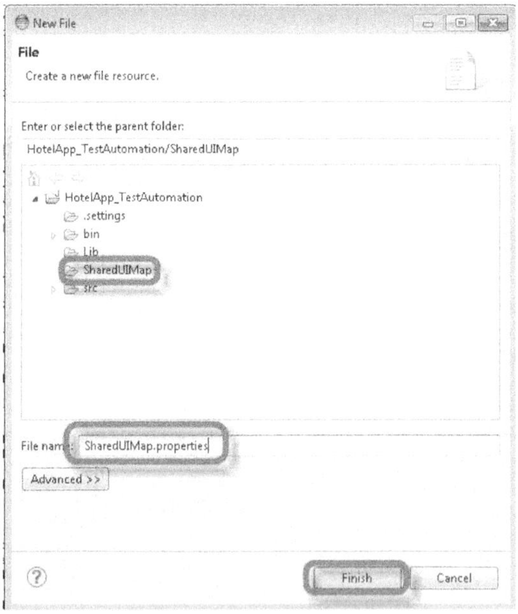

Figure 13-6 – SharedUIMap.properties file

5. Copy all of your Web element locator references to this file. Each element is stored as a key value pair, where value is the locator and key is a name for it, which you can give according to your choice. You can refer to your MyFirstWebDriverTest.java or alternatively use ChroPath for locator values.

Note: We suggest strictly following some naming conventions while naming Web Elements. For e.g.

- ElementType_page_name_element_name

So we have our property file **SharedUIMap.properties** looking like below. Here we assign 'aliases' reader-friendly identifiers for UI elements.

```
################ Login Page Elements #########

# Referenced by XPath Locator
Txt_Login_Username = .//*[@id='username']

# Referenced by Id Locator
Txt_Login_Password = password

# Referenced by Id Locator
Btn_Login_Login =login


################ Search Hotel Page Elements #########

# Referenced by Id Locator
Lst_SearchHotel_Location = location

# Referenced by Id Locator
Btn_SearchHotel_Search = Submit

################ Select Hotel Page Elements #########

# Referenced by Id Locator
Rad_SelectHotel_RadioButton_1 = radiobutton_1

# Referenced by Id Locator
Btn_SelectHotel_Continue = continue
```

```
############### Booking Hotel Page Elements #########

# Referenced by Id Locator
Txt_BookingHotel_FirstName = first_name

# Referenced by Id Locator
Txt_BookingHotel_LastName = last_name

# Referenced by Id Locator
Txt_BookingHotel_Address = address

# Referenced by Id Locator
Txt_BookingHotel_CCNumber = cc_num

# Referenced by Id Locator
Lst_BookingHotel_CCType = cc_type

# Referenced by Id Locator
Lst_BookingHotel_CCExpMonth = cc_exp_month


# Referenced by Id Locator
Lst_BookingHotel_CCExpYear = cc_exp_year


# Referenced by Id Locator
Txt_BookingHotel_CCCvvNumber = cc_cvv


# Referenced by Id Locator
Btn_BookingHotel_BookNow = book_now

# Referenced by Id Locator
Lnk_BookingHotel_Logout = Logout

############### Logout Elements #########

# Referenced by Id Locator
Lnk_Logout_ClickHeretoLoginAgain = Click here to login again
```

Figure 13-7 – Web Elements Name Value pairs

The locators will still refer to HTML objects, but we have introduced a layer of abstraction between the test script and the UI elements. Values are read from the properties file and used in the Test Class to implement the Shared UI Map.

6. See the **SharedUIMap.properties** file in the below snapshot from Eclipse

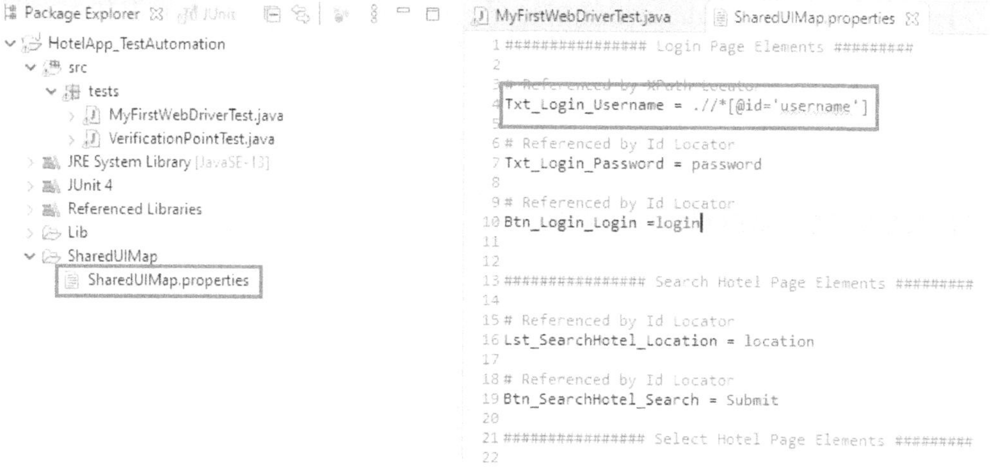

Figure 13-8 – Eclipse view of SharedUIMap.properties

13.3. Using a Shared UI Map file in Script

Now that we have created our external Shared UI map file, we would want to use it in our script. Let us follow these steps to create a script using a Shared UI map.

Pre-conditions

1. **Right click** on your existing **MyFirstWebDriverTest.java** script, select **Copy**

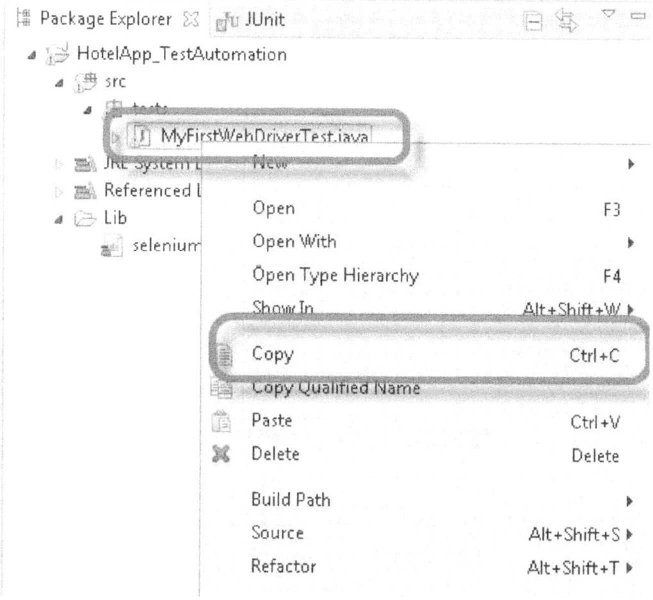

Figure 13-9 – Copy Script

2. Select the **tests package** folder, **right click** and select **Paste**
3. In the Name Conflict dialog box, enter the name of the script as **SharedUIMapTest.Java** and click **OK**

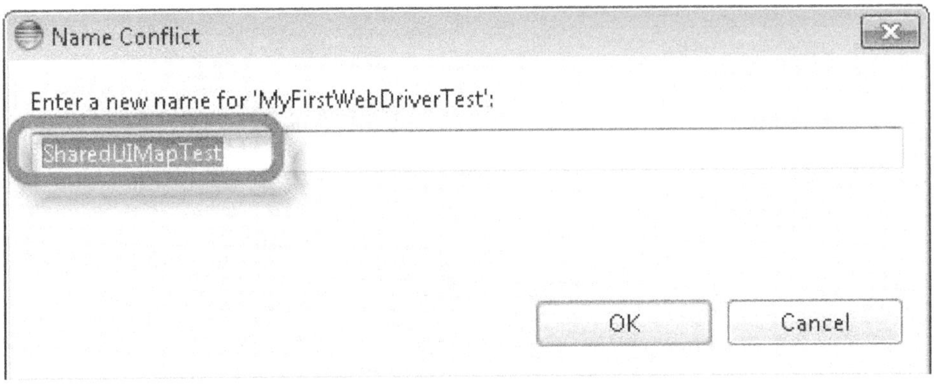

Figure 13-10 – Name Conflict dialog

4. Double click on your new "**SharedUIMapTest.Java**" script to see the script

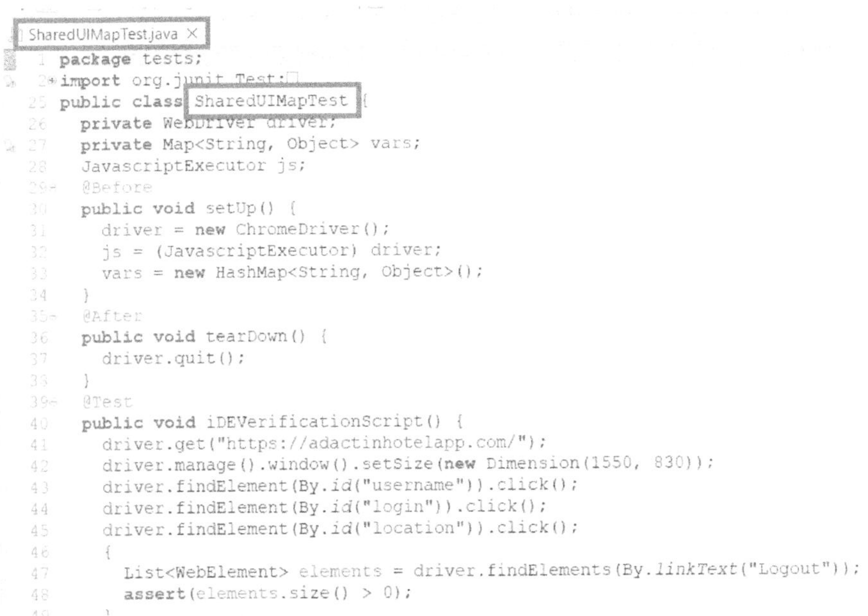

Figure 13-11 – SharedUITest

5. Let us design our script to read locators from our Shared UI Map file. Add the following code to the Login Test Class paying careful attention to the **@ Before** section

```
import java.util.Properties;

import java.io.FileInputStream

public Properties prop;

@Before
public void setUp() throws Exception {
 prop = new Properties();
prop.load(new FileInputStream("./SharedUIMap/SharedUIMap.properties"));

driver = new ChromeDriver();
   baseUrl = "http://www.adactinhotelapp.com/";
   driver.manage().timeouts().implicitlyWait( Duration.ofSeconds(30));
```

Figure 13-12 – Properties code to be added

In the above code

prop = new Properties();

The prop statement helps to create a new object of type Properties which is a inherited from class Java.util. It is used to store key/value pairs.

prop.load(new FileInputStream(".*/SharedUIMap/SharedUIMap.properties*"));

Above statement loads key/value pairs into a Properties object from a stream.

In above code we are importing two java libaries Java.Util.Properties and Java.io.FileInputStream. Eclipse will automatically prompt us to add those files when we use **Properties** and **FileInputStream** objects.

Note: If Eclipse is throwing error icon, it means that there might be syntax/compilation error in the code.

```
package tests;

import java.io.FileInputStream;
import java.io.FileNotFoundException;
import java.io.IOException;
import java.time.Duration;
import org.junit.After;
import org.junit.Before;
import org.junit.Test;
import org.openqa.selenium.By;
import org.openqa.selenium.WebDriver;
import org.openqa.selenium.chrome.ChromeDriver;
import org.openqa.selenium.support.ui.Select;

public class SharedUIMapTest {

    private WebDriver driver;
    public Properties prop;
    @Before
    public void setUp() throws Exception, IOException
    {
        prop = new Properties();
        prop.load(new FileInputStream("./SharedUIMap/SharedUIMap.properties"));
        driver = new ChromeDriver();
        driver.get("https://www.adactinhotelapp.com");
        driver.manage().window().maximize();
        driver.manage().timeouts().implicitlyWait(Duration.ofSeconds(20));
    }
```

Figure 13-13 – Error thrown in Eclipse

To resolve this place your cursor on the error icon and click on appropriate action.'

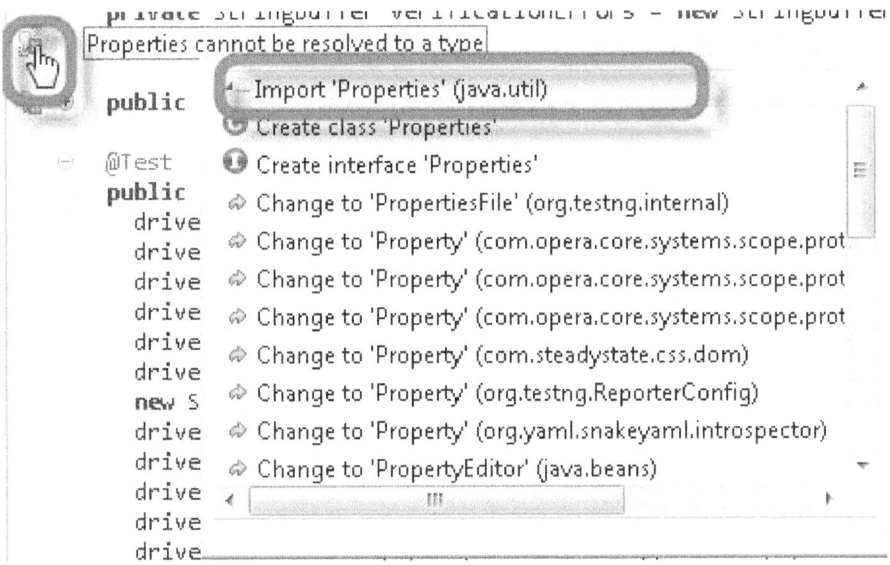

Figure 13-14 – Resolve Eclipse Error

Eclipse is a smart IDE and helps you resolve issues with the code.

6. Now let us replace local UI properties with property names mentioned in the Shared UI Map file.

We will use the method **getProperty**

prop.getProperty(String val) – This method will read data value for the specified variable from the properties file.

Argument Val is replaced by the Web element's name in the Shared UI map file. Copy the following code under the @Test annotation into SharedUIMapTest.java file

```
@Test
public void testMyFirstWebDriver() throws Exception {
  driver.get(baseUrl);
  driver.findElement(By.xpath(prop.getProperty("Txt_Login_Username"))).clear();
  driver.findElement(By.xpath(prop.getProperty("Txt_Login_Username"))).sendKeys("adactin123");
  driver.findElement(By.id(prop.getProperty("Txt_Login_Password"))).clear();
  driver.findElement(By.id(prop.getProperty("Txt_Login_Password"))).sendKeys("adactin123");
  driver.findElement(By.id(prop.getProperty("Btn_Login_Login"))).click();
  new Select(driver.findElement(By.id(prop.getProperty("Lst_SearchHotel_Location")))).selectByVisibleText("Sydney");
  driver.findElement(By.id(prop.getProperty("Btn_SearchHotel_Search"))).click();
  driver.findElement(By.id(prop.getProperty("Rad_SelectHotel_RadioButton_1"))).click();
  driver.findElement(By.id(prop.getProperty("Btn_SelectHotel_Continue"))).click();
  driver.findElement(By.id(prop.getProperty("Txt_BookingHotel_FirstName"))).clear();
  driver.findElement(By.id(prop.getProperty("Txt_BookingHotel_FirstName"))).sendKeys("test");
  driver.findElement(By.id(prop.getProperty("Txt_BookingHotel_LastName"))).clear();
  driver.findElement(By.id(prop.getProperty("Txt_BookingHotel_LastName"))).sendKeys("test");
  driver.findElement(By.id(prop.getProperty("Txt_BookingHotel_Address"))).clear();
  driver.findElement(By.id(prop.getProperty("Txt_BookingHotel_Address"))).sendKeys("test");
  driver.findElement(By.id(prop.getProperty("Txt_BookingHotel_CCNumber"))).clear();
  driver.findElement(By.id(prop.getProperty("Txt_BookingHotel_CCNumber"))).sendKeys("1212121212121212");
  new Select(driver.findElement(By.id(prop.getProperty("Lst_BookingHotel_CCType")))).selectByVisibleText("American Express");
  new Select(driver.findElement(By.id(prop.getProperty("Lst_BookingHotel_CCExpMonth")))).selectByVisibleText("March");
  new Select(driver.findElement(By.id(prop.getProperty("Lst_BookingHotel_CCExpYear")))).selectByVisibleText("2015");
  driver.findElement(By.id(prop.getProperty("Txt_BookingHotel_CCCvvNumber"))).clear();
  driver.findElement(By.id(prop.getProperty("Txt_BookingHotel_CCCvvNumber"))).sendKeys("111");
  driver.findElement(By.id(prop.getProperty("Btn_BookingHotel_BookNow"))).click();
    driver.findElement(By.linkText(prop.getProperty("Lnk_BookingHotel_Logout"))).click();
  driver.findElement(By.linkText(prop.getProperty("Lnk_Logout_ClickHeretoLoginAgain"))).click();

}
```

Figure 13-15 – Test script with reference to Shared UI map

7. Perform Project clean-up using **Project → Clean Up**… and clean all selected projects.

8. Select the test and Run the test using **Run → Run As JUnit Test**. Confirm that the test can run fine.

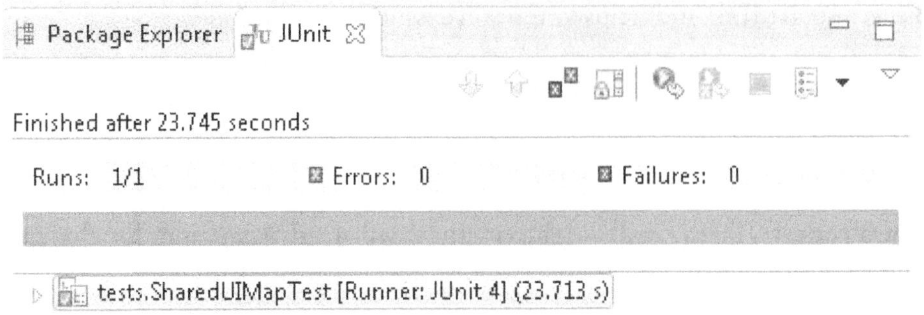

Figure 13-16 – Test Results

EXERCISE

1. Replace VerificationPointTest.Java script with UI elements from the Shared UI Map file.

§

14

Using Functions

Functions help divide your test into logical units, such as areas of key functionalities of the application. Functions help make our scripts modular and reusable, which will save us maintenance effort and also help us improve productivity. These functions can then be re-used in different scripts.

For example, all of our scripts will have to login to the application. Now, instead of recording login steps repeatedly in every script, we can keep an external login function and re-use that function in all of our scripts.

example Let us see another practical example here:

At one of our client engagements, we were automating an investment banking application. As a first step of every test case, we had to create investment instruments after which we had to validate, and add details in later steps (we had more than 100 test cases for each instrument type). Creating an instrument was a tedious step with up to 50 field values to be entered. Based on the test scenario, input data would change. Now recording the steps of investment instrument creation in each and every script would have been a nightmare and time consuming. It would have also been a maintenance issue, if in later development stages the application workflow is changed or new fields were added.

So we created functions to create instruments and for each of the test cases that were automated, we just invoked the same function in every script. This helped us reduce the overall time to automate. This also assisted in maintenance down the line, when the investment instrument creation workflow changed.

Key objectives:

- Create Functions
- Calling Functions in WebDriver script

14.1. Creating Functions in WebDriver

Key steps in creating Functions in WebDriver using Eclipse IDE include:

1. Create a separate Package and Class for Functions
2. Create Function definition and import any required Java libraries in the class

3. Add steps to functions based on function's objective
4. Replace any data within functions with arguments from that function
5. Within the script, import Functions Package and Extend class to use function within your scripts

Pre-conditions

1. Select **HotelApp_TestAutomation/src** folder, **right click** and select **New → Package**

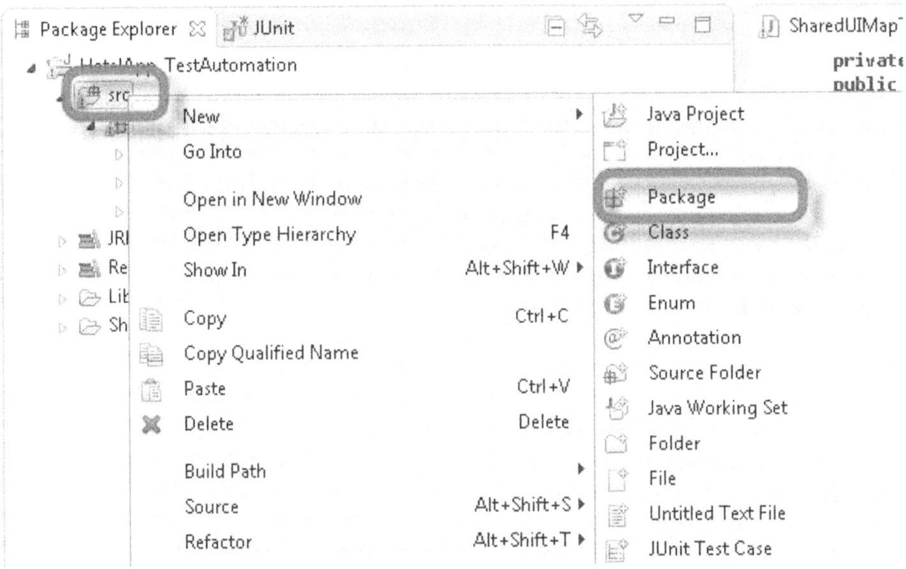

Figure 14-1 – New Package Creation

2. In **Java Package** dialog enter the Name **functions** and click **Finish**

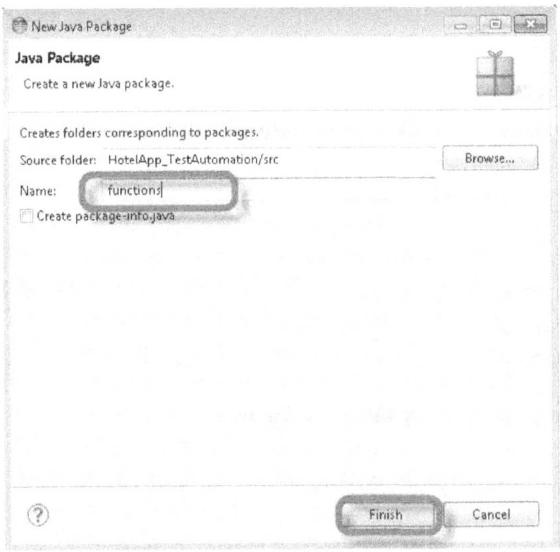

Figure 14-2 – Package Name

A new package is created with the name functions.

3. **Right click** on package **functions** and select **New → Class**

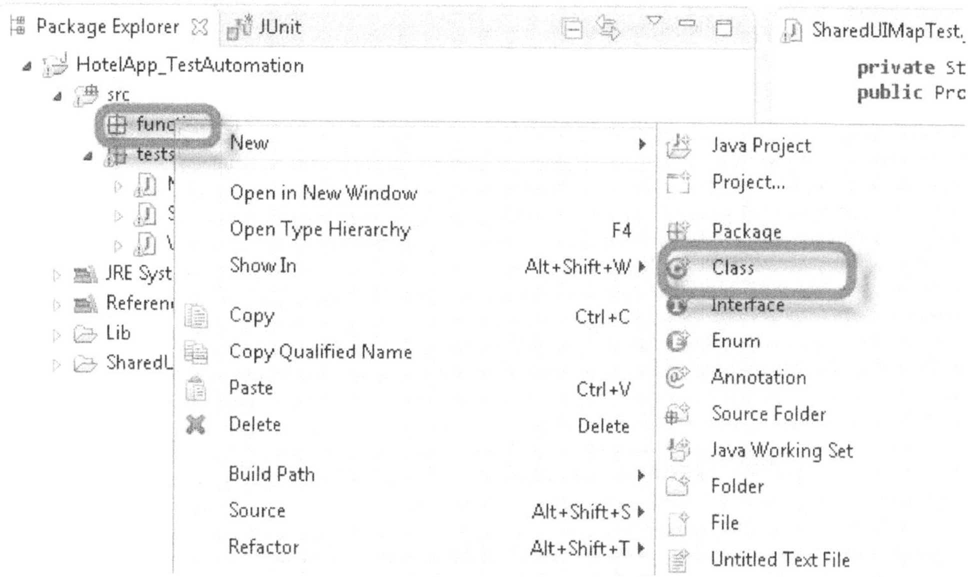

Figure 14-3 – New Class Creation

4. Give it the name **HotelApp_BusinessFunctions** and Click **Finish**

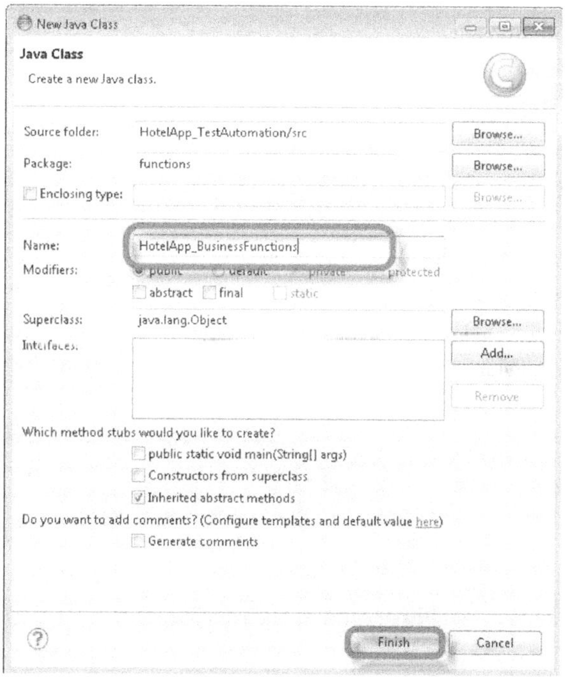

Figure 14-4 – New Class name

5. Double click on the newly created **function file**. You will see the following script

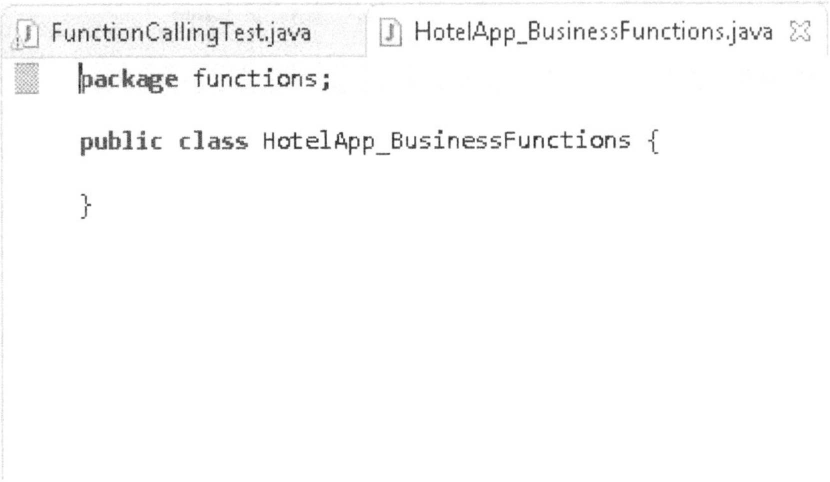

Figure 14-5 – Default functions script

6. **Right click** on your existing **SharedUIMap.java** script, select **Copy**

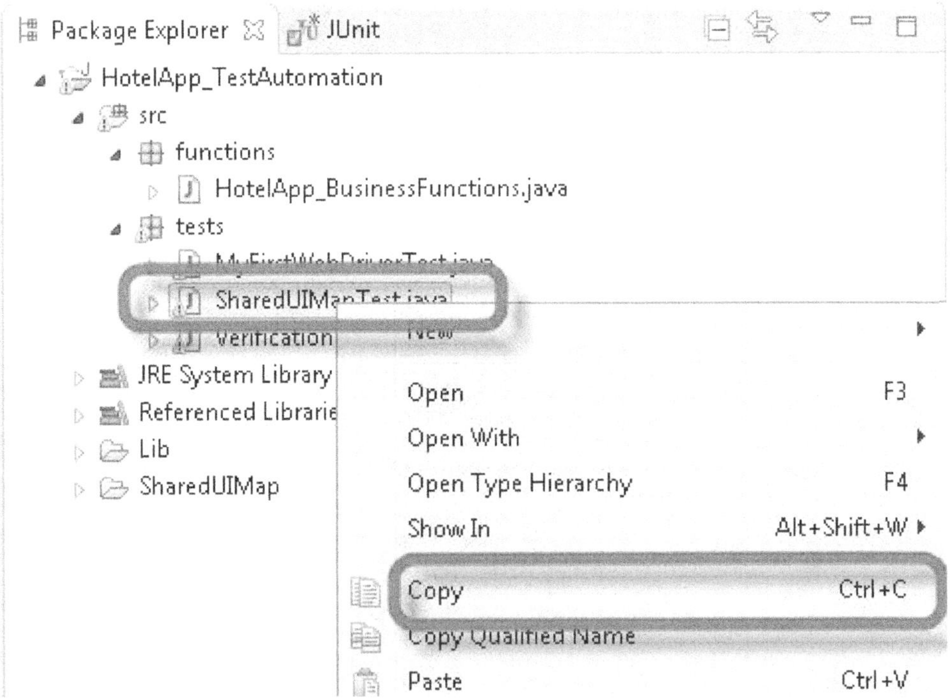

Figure 14-6– Copy Script

7. Select the **tests** package folder, **right click** and select **Paste**

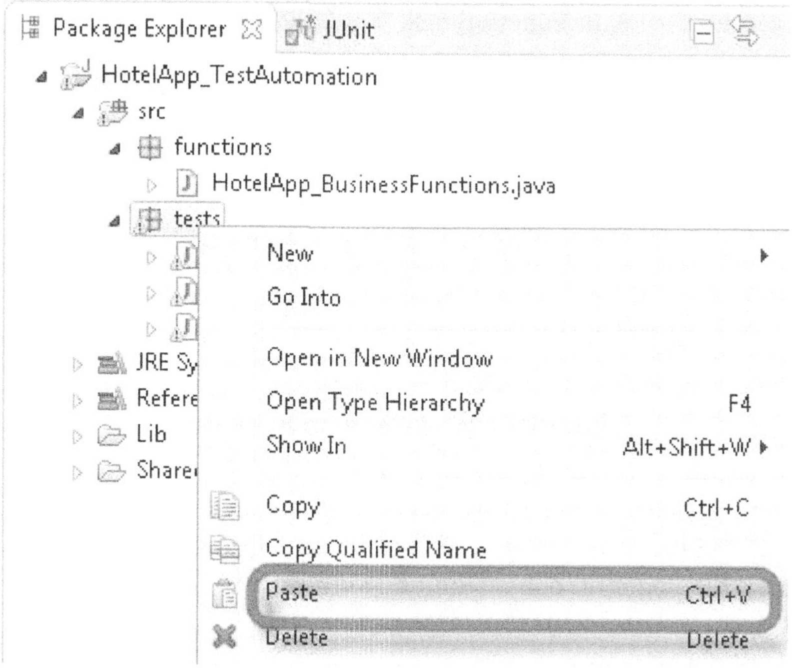

Figure 14-7 – Paste Script

8. In the Name Conflict dialog box , enter "**FunctionCallingTest**" and click **OK**

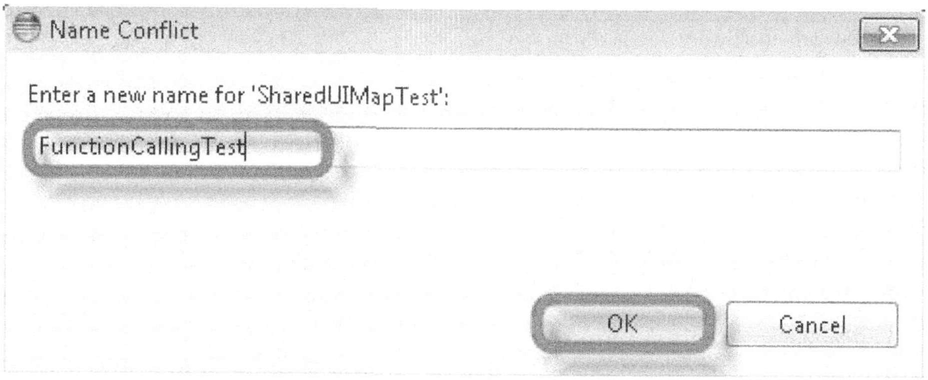

Figure 14-8 – Name Conflict dialog

9. Double click on "FunctionCallingTest.java" script so that you can see the script

Create Function Definitions

Now let us create the Login Function first

As discussed in the Basic Java chapter, syntax for creating a method or function is below

Here is an example of a typical method declaration:

```
modifier returnValueType methodName(list of parameters) {
  // Method body;
}
```

Figure 14-9 – Java Method Syntax

10. Based on the above syntax, let us create the following Function for Login. We will copy steps to login to the application from our CallingFunctionsTest.java script and create the Login function definition with its arguments.

```
public void  HA_BF_Login ( WebDriver driver, String sUserName, String sPassword ) {

  driver.findElement(By.xpath(prop.getProperty("Txt_Login_Username"))).clear();
  driver.findElement(By.xpath(prop.getProperty("Txt_Login_Username"))).sendKeys("adactin123");
  driver.findElement(By.id(prop.getProperty("Txt_Login_Password"))).clear();
  driver.findElement(By.id(prop.getProperty("Txt_Login_Password"))).sendKeys("xxxxxx");
  driver.findElement(By.id(prop.getProperty("Btn_Login_Login"))).click();
}
```

Figure 14-10 – Login Method

Note: The Login method has to be copied within the Class. Method is always part of a class in Java.

Note: We are trying to follow a few naming conventions on how to name methods.

- HA – HotelApp
- BF – Business Function
- Login – Name of the function

We are going to cover more on naming conventions in further chapters.

11. Once you copy the steps you will notice quite a few errors popping up in the Eclipse interface

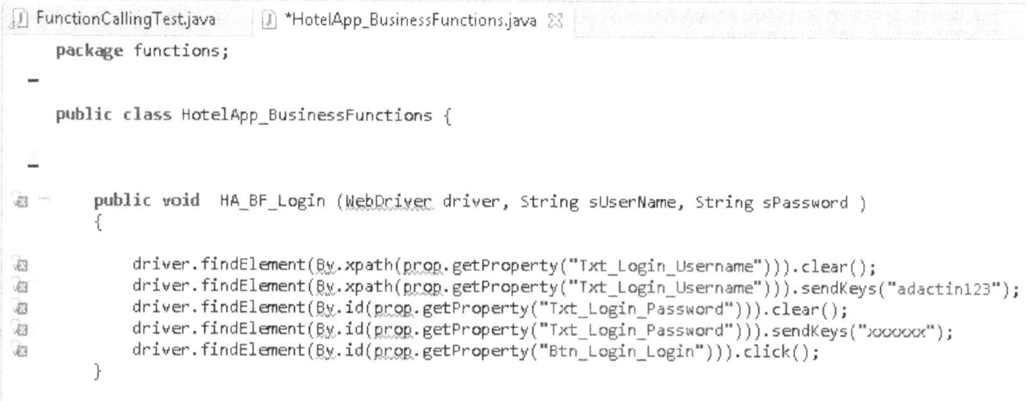

Figure 14-11 – Errors in Eclipse Interface

If you look at the errors (place your cursor above the red 'x' before each line) you will notice quite a few logical reasons for the errors.

- Variable WebDriver is unresolved
- Variable By is unresolved
- Variable prop is unresolved
- Variable baseURL is unresolved

12. To resolve the errors we need to

- Import the WebDriver library
- Import the Properties Library
- Import org.openqa.selenium.By
- Declare driver and prop variables as public (we can copy their declarations from the FunctionCallingTest.java file)

Your script should appear as follows after resolving the Eclipse errors

Figure 14-12 – Resolve Eclipse Errors

13. We can now replace the values of arguments within the function

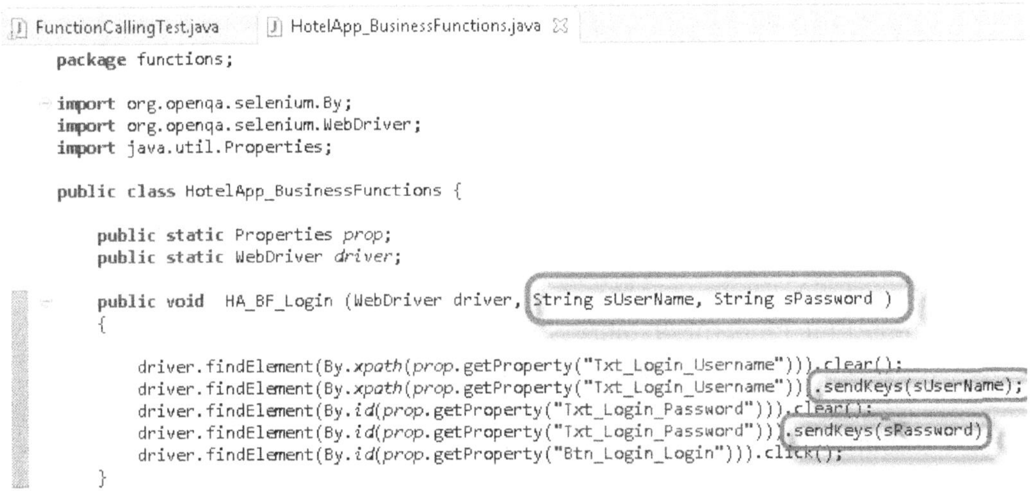

Figure 14-13– Replace arguments

Note: You would have noticed that we are not returning any value as part of this function; hence the return type is void. If we were to return any variable value, we could have used appropriate return type (int, String etc) and used the return statement to return the value.

14.2. Calling a Function in WebDriver Script

1. Double click on copied script "**FunctionCallingTest.java**" to see the script

```
package tests;

import java.io.FileInputStream;

public class FunctionCallingTest {

    private WebDriver driver;
    public Properties prop;
    @Before
    public void setUp() throws Exception, IOException
    {
        prop = new Properties();
        prop.load(new FileInputStream("./SharedUIMap/SharedUIMap.properties"));
        driver = new ChromeDriver();
        driver.get("https://www.adactinhotelapp.com");
        driver.manage().window().maximize();
        driver.manage().timeouts().implicitlyWait(Duration.ofSeconds(20));
    }

    @Test
    public void testMyFirstWebDriver() throws Exception {
        driver.findElement(By.xpath(prop.getProperty("Txt_Login_Username"))).clear();
```

Figure 14-14– FunctionCallingTest script

2. In order to use the functions just created, we will need to import the functions class into this script. Use the statement below to import the class HotelApp_BusinessFunctions

```
import functions.HotelApp_BusinessFunctions;
```

Figure 14-15– Import Function Class

3. The next step is to **extend** the existing test case to inherit functions from our function library file **HotelApp_BusinessFunctions**.

MyFirstWebDriverTest.java VerificationPointTest.java SharedUIMap.properties SharedUIMapTest.java HotelApp_BusinessFunctions.java *FunctionCallingTest.java

```
package tests;

import java.io.FileInputStream;
import java.io.IOException;
import java.time.Duration;
import java.util.Properties;
import org.junit.After;
import org.junit.Before;
import org.junit.Test;
import org.openqa.selenium.By;
import org.openqa.selenium.WebDriver;
import org.openqa.selenium.chrome.ChromeDriver;
import org.openqa.selenium.support.ui.Select;

import functions.HotelApp_BusinessFunctions;

public class FunctionCallingTest extends HotelApp_BusinessFunctions{

    private WebDriver driver;
    public Properties prop;

    @Before
    public void setUp() throws Exception, IOException
    {
        prop = new Properties();
        prop.load(new FileInputStream("./SharedUIMap/SharedUIMap.properties"));
        driver = new ChromeDriver();
        driver.get("https://www.adactinhotelapp.com");
        driver.manage().window().maximize();
        driver.manage().timeouts().implicitlyWait(Duration.ofSeconds(20));
    }
```

Figure 14-16– Extend Test class to inherit functions

4. Comment/Remove the declarations for variables prop, driver and base URL which have already been declared in Functions file

Figure 14-17– Comment already defined variables

5. Make a call **to HA_BF_Login** function and comment the existing steps from the script

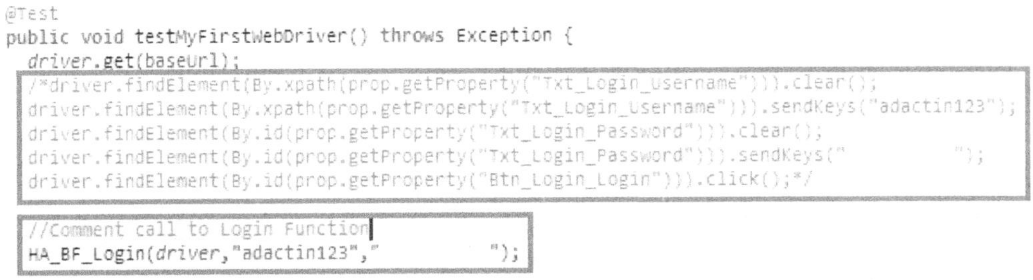

Figure 14-18– Call Function

6. Perform Project clean-up using **Project → Clean Up**... and clean all selected projects.

7. Select the test and Run the test using **Run → Run As JUnit Test**. Confirm the test can run fine.

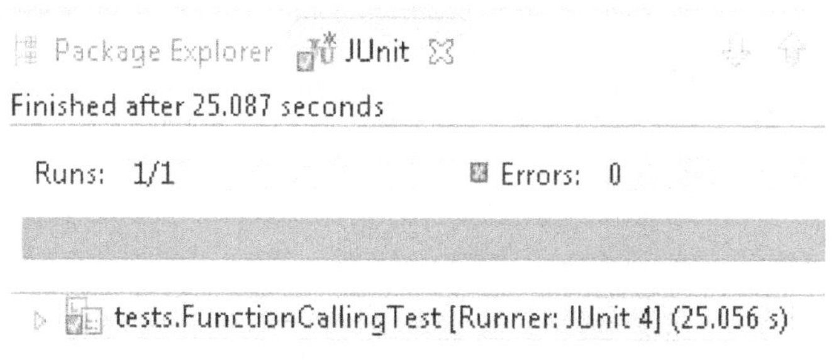

Figure 12-19– Run Results

EXERCISE

1. Create a function for each of the below modules and save them in the HotelApp_BusinessFunctions file

- Search Hotel
- Select Hotel
- Hotel Booking
- Logout

Replace existing script with function calls to these functions

§

15

Using a Configuration File

During automation we work with various parameters which are dynamic in nature. For instance, application path or the URL may vary based on the environment we are running the scripts (stage or test environment) on. This would mean we would need to change parameter values based on our execution needs.

Secondly, if we have a lot of scripts, we would need to change these parameter values in each and every script.

example

In the previous script that we created, we had the application URL (www.adactinhotelapp.com) defined within our script. Now assume we have automated 100+ scripts and our project manager wants us to execute these test cases in the staging environment or Build2 (which has a different URL, www.adactinhotelapp.com/HotelAppBuild2). In order to run our scripts in different environments we would need to go into each of the scripts and modify the URL value.

And guess what, if next time our project manager wants us to again re-run the scripts in a different test environment, we would need to update the scripts again. That can be a nightmare!

A better solution would be to keep these dynamic values in a configuration file which is centrally located and shared between all scripts. So we make a change at a single place and it will help all of the scripts get updated parameter values.

Also another fundamental automation rule is we would want to avoid making code/script changes for these parameter related changes as these can cause regression issues in the script. So making changes in a configuration file will be a safer way to update parameter values.

Key objectives:

- Create a configuration file
- Using a configuration file parameter within the WebDriver script

15.1. Create a Configuration File

Let's add a UI Map to our project.

1. Select the Project Folder **HotelApp_TestAutomation**, Right Click and select **New → Folder**
2. Name the Folder **Configuration**

Figure 15-1 – Configuration Folder

3. Right Click on the new **Configuration** folder and select **New → File**

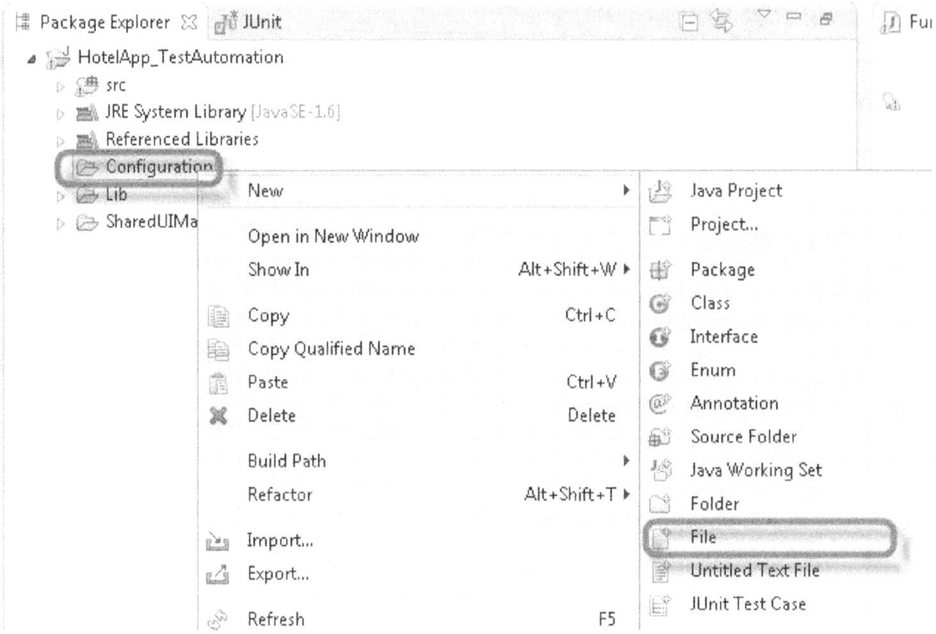

Figure 15-2 – Create New File

4. Give the name **HA_Configuration.properties** and click **Finish**

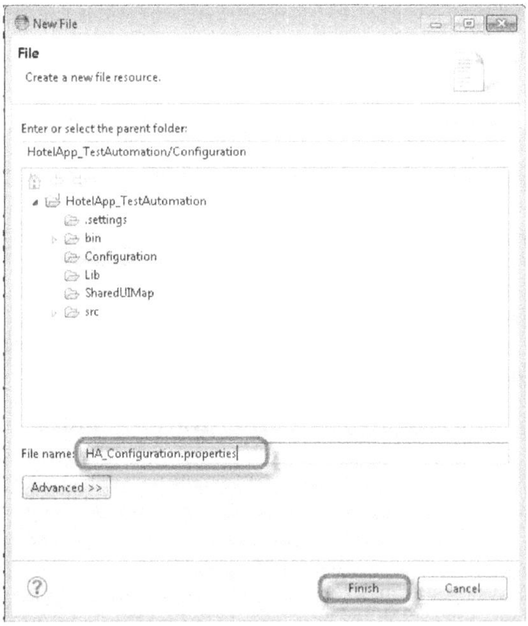

Figure 15-3 – HA_Configuration.properties file

5. Double click and open the configuration file and add all the dynamic parameters that can affect our script.

So our property file **HA_Configuration.properties** should look like this.

```
#//////////Application URL//////////

  AppUrl=http://www.adactinhotelapp.com/

#//////////Shared UI Map File Path //////////

SharedUIMap=./SharedUIMap/SharedUIMap.properties
```

Figure 15-4 – Web Elements Name Value pairs

Note: We will add more parameters in coming chapters.

6. Save the file using File → Save or Ctrl + S

15.2. Using Configuration File Parameters in a Script

Now that we have created an external configuration file, we want to use it in our script. Let us follow these steps to create a script which uses a Configuration file.

1. **Right click** on your existing FunctionCallingTest.java script, select **Copy**.
2. Select the **tests** package folder, **right click** and select **Paste**.
3. In the Name Conflict dialog box, enter name of the script as **ConfigurationFileTest** and click **OK**.
4. Double click on the new "**ConfigurationFileTest.java**" script so that you can see the script.
5. Let us design our script to read parameters from our Configuration file. We will follow these steps to modify the script
   a. Add two variables to store the Application URL and the SharedUIMap path.

```
public static String sAppURL;

  public static String sSharedUIMapPath;
```

b. Use the Prop.load method to read the Configuration file

```
prop.load(new FileInputStream(“./Configuration/HA_Configuration.
properties”));
```

c. Use the **Prop.getProperty** method to retrieve the value from the configuration file

```
sAppURL = prop.getProperty(“AppUrL”);

        sSharedUIMapPath = prop.getProperty(“SharedUIMap”);
```

d. Replace the SharedUI map path and the Application URL with the value retrieved from the Configuration file

```
prop.load(new FileInputStream(sSharedUIMapPath));
driver.get(sAppURL);
```

Your script should look as in the figure below:

```
public class SharedUIMapTest {

    private WebDriver driver;
    public Properties prop;

    @Before
    public void setUp() throws Exception, IOException
    {
        prop = new Properties();

        prop.load(new FileInputStream("./SharedUIMap/SharedUIMap.properties"));

        driver = new ChromeDriver();

        driver.get(prop.getProperty("sAppURL"));
        driver.manage().window().maximize();
        driver.manage().timeouts().implicitlyWait(Duration.ofSeconds(20));
    }
```

Figure 15-5 – Script Changes

6. Perform Project clean-up using **Project → Clean Up**... and clean all selected projects.

7. Save the test. Select the test and Run the test using **Run → Run As JUnit Test**. Confirm that the test can run fine.

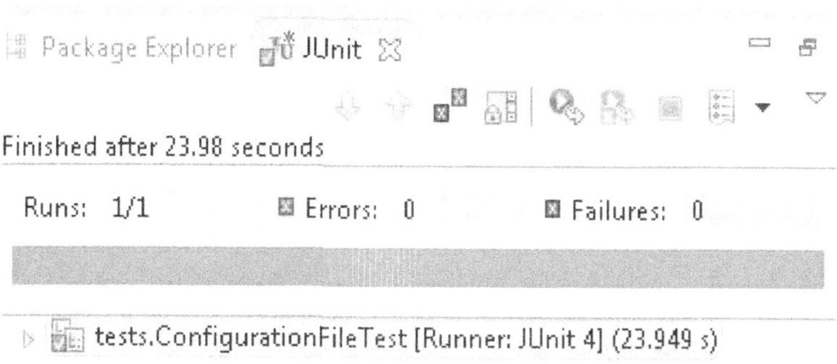

Figure 15-6 – Test Results

EXERCISE

1. Create a new function which will load a configuration file and a **SharedUIMap** file in a function library file
2. Create a copy of an existing script and call this function in your script.

Hint

Add below lines in a new function

```
prop.load(new FileInputStream(“./Configuration/HA_Configuration.properties”));
sSharedUIMapPath = prop.getProperty(“SharedUIMap”);
prop.load(new FileInputStream(sSharedUIMapPath));
```

§

16

Data Driven Testing - Parameterization

Introduction

In many instances, when we are performing regression testing, we need to repeat the same test case with different sets of data. This can be a monotonous and time-consuming task, depending on how many different data sets are required for the test case.

example Let us take an example:

I worked for one of our retail domain clients as part of the testing and automation team which had more than 2000 stores in the country, They had developed a point of sales system. Once this was manually tested, they gave us a list of more than 10000 usernames and password pairs, and asked us to set them up in the system. As a testing team, our task was to verify if all usernames and passwords are set up correctly. We were given a target of .01% failure threshold. If we had to verify all this manually, assuming we would verify 1 username/password combination every 1 minute (as there were a couple of validations we had to do once logged in), it would have taken us 5000 minutes or approximately 20-24 days of man effort. Imagine how laborious and time-consuming that task would have been.

Solution: Wouldn't it be great if you had an automated script to could pick up the first username and password entered from an excel sheet, log the user in and perform all the validations without any manual intervention?

An even better solution would be to create a script to iterate across all the 10000 usernames and passwords. It took us less than 4 hours to develop the script, ran it overnight, got the failed records, again re-tested the records once fixed and delivered it to the customer with 0% issues. The concept of running the same script with multiple dataset values is called **Parameterization**.

Any test case which needs to be executed multiple times with different data values is an ideal candidate for automation.

Another objective of Data Driven tests is to keep all of the test data in one central location.

example Let us taken an example

Suppose you have 200+ scripts and all the scripts start with login which needs a username and password. Where should you be keeping this test data? If this data is residing within the script, and for some reason your username or password changes you would need to go in each script and make these changes.

Is there a better solution?

Wouldn't it be nice if you could have all of this data in one central file so that all of the scripts which need username/password refer to just one file? If any of the usernames/passwords change, then we just make the change(s) at one location and all of the scripts will be good to go.

Key objectives:

- Data driving the script to read specific data from central location.
- How to parameterize a Selenium WebDriver script to execute the same steps for multiple sets of data.

16.1. Data Drive a Script with a Single Value from an Excel Sheet

Selenium does not provide any out-of-the-box solution for data driven testing but leaves it up to the user to implement this on his own. To implement data-driven tests in Selenium we can take multiple approaches which include:

- **Reading test data from an Excel file using JUnit and JXL**
- Creating a data-driven test using JUnit
- Creating a data-driven test using TestNG
- Reading test data from a CSV file using JUnit
- Reading test data from a database using JUnit and JDBC
- Creating a data-driven test in NUnit, Ruby or Python

As a part of this chapter, we will discuss how to read test data from an Excel file using JUnit and JXL for parameterization.

To read data from Excel, we need APIs which support opening files, reading data, and writing data into Excel. We should know various classes and methods which support above mentioned operations. **Jxl.jar** is an open source Java API which supports reading and writing to Excel spreadsheets.

To use Excel sheets as our data source we need to do the following:

1. Download and add Jxl.jar to our project
2. Create a Data file in MS Excel

3. Create a java function to fetch the data from the file
4. Parameterize our script to use the data and java function

We store test data in the form of tables within an Excel sheet. One table can have any number of columns. A table always begins with a column header row. This column header, along with the row number, is used for accessing the data.

Problem Description: Let us try to create a script which will read data from the first row of the data sheet and search hotel based on the retrieved value.

Let us see how to implement this.

Step 1 - Download and add Jxl.jar to our project

1. Download jxl.jar from the internet. You can either search on www.google.com or use the link (https://repo1.maven.org/maven2/net/sourceforge/jexcelapi/jxl/2.6.12/jxl-2.6.12.jar). It will directly download the jxl jar file

Note: jxl.jar is just one of many library files that interact with Excel files. We could alternatively use any another Java library file to read/write data to Excel

2. Copy the **jxl.jar** file in your Lib folder in your workspace. In our project it is at location C:\Selenium\workspace\HotelApp_TestAutomation\Lib

Figure- 16-1 – JXL.jar file

3. Add this Jar file in your Build path. Go to Eclipse. Select your Project **HotelApp_TestAutomation**, right click and select **Build Path→ Configure build Path**

4. Click on **Libraries** tab and select classpath

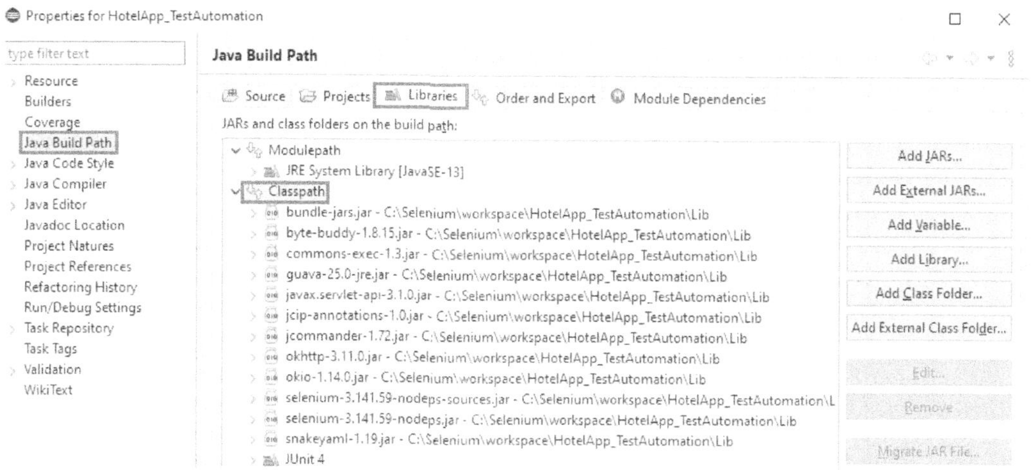

Figure- 16-2 – Libraries Tab

5. Click on Add External JARs… and browse to the location of jxl.jar file and add it to Libraries

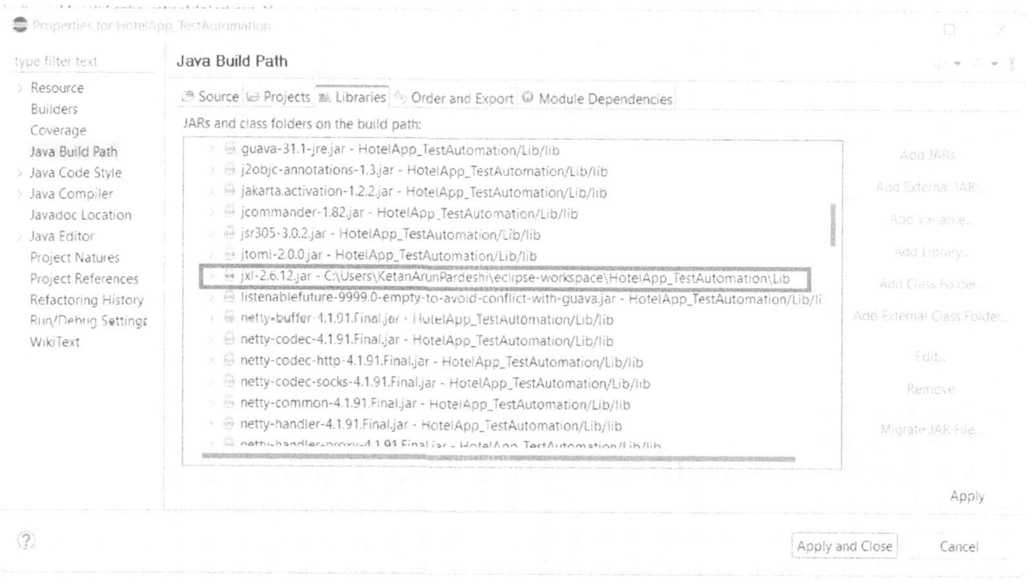

Figure- 16-3 – jxl.jar file added

6. Click **OK** to close the **Java Build Path** dialog

Step 2 - Create a Data file in MS Excel

7. Select Java Project **HotelApp_TestAutomation** and right click and select **New → Folder**

8. Add a new folder called **DataPool** to our project

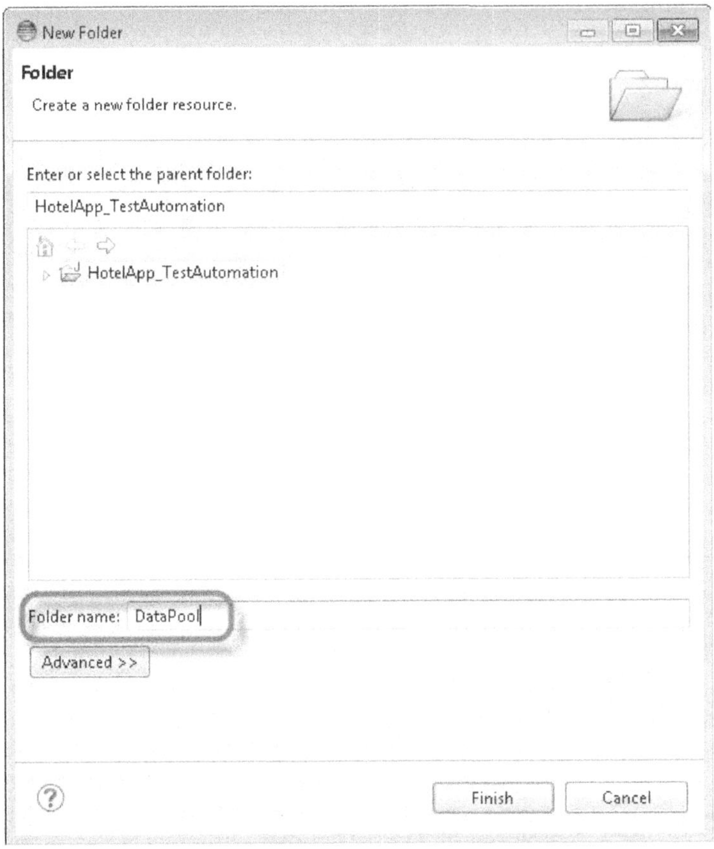

Figure- 16-4– DataPool Folder

This is the folder in which we keep all of our data files for the project.

9. Create an Excel file named **HA_HotelSearch.xls** and store data in the following format

ID	Location
1	Sydney
2	Brisbane
3	Melbourne
4	Adelaide
5	ENDOFROW

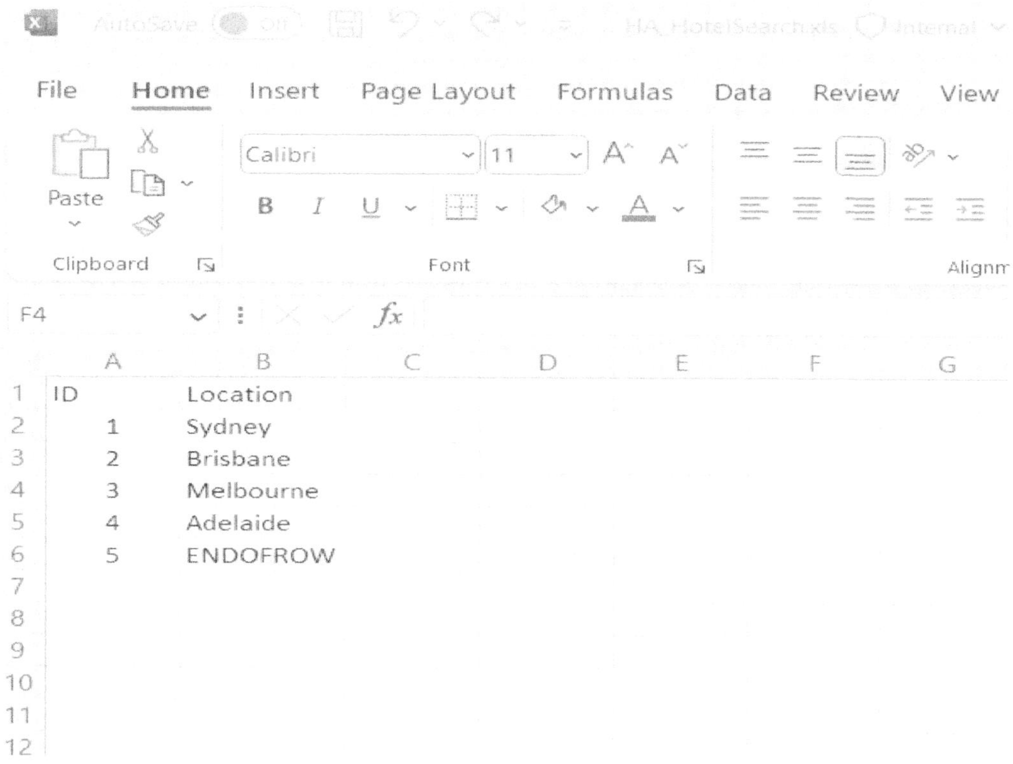

Figure- 16-5– Excel Sheet content

Note that we have entered "ENDOFROW" in the last row to indicate there is no more data left to be read in the datasheet.

Save this file in the **DataPool** folder of our project and close the sheet (C:\Selenium\workspace\HotelApp_TestAutomation\DataPool)

Note: The relative path to our file is ".\DataPool\ HA_HotelSearch.xls"

10 Within your eclipse project, select your **Java Project** and press F5 (Refresh). Expand the DataPool folder and notice that you can see your newly created HA_HotelSearch.xls file

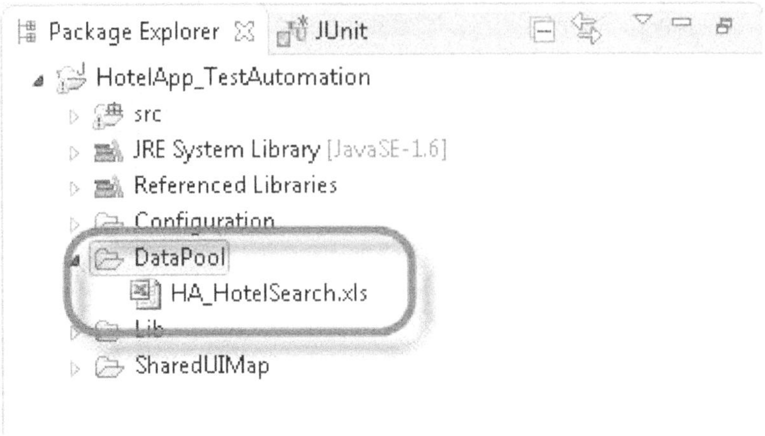

Figure- 16-6– Datasheet in eclipse view

Step 3 - Create a Java function to fetch the data from the file

Now that we have created our external datasheet file, we would want to use it in our script. Let us follow these steps to create a script which uses our Datapool.

11. Double click the newly created "HotelApp_BusinessFunctions.java" to see the functions file

For the sake of simplicity we are going to add this new function in our HotelApp_BusinessFunctions file itself.

Note: The better approach is to separate business functions from generic functions. So, we can even create a separate class say GenericFunctions for keeping all our generic functions which we are going to use in all our test scripts. Business functions are those which are specific to our application under test, like login, booking, logout, etc. Generic functions are those which can be generically used across projects like reading data from Excel, reading data from a database, etc.

12. Add the following function to our functions file code

Below is the function that fetches the table from the Excel sheet. It uses Java Excel API to fetch data from the excel sheet. It takes three parameters-

1. strFilePath - path of the XL file/workbook containing the data, the path is relative to the Java project
2. column - name of the column header from which you want to read data
3. row - row number of the row from which you wish to fetch data from

The function returns a String containing data from the required cell.

```
public static String HA_GF_readXL (int row, String column, String strFilePath)
{
                Cell c= null;
                int reqCol=0;
                WorkbookSettings ws = null;
                Workbook workbook = null;
                Sheet sheet = null;
                FileInputStream fs = null;
try{
        fs = new FileInputStream(new File(strFilePath));
        ws = new WorkbookSettings();
        ws.setLocale(new Locale("en", "EN"));

        // opening the work book and sheet for reading data
        workbook = Workbook.getWorkbook(fs, ws);
        sheet = workbook.getSheet(0);

        // Sanitise given data
        String col = column.trim();

        //loop for going through the given row
        for(int j=0; j<sheet.getColumns(); j++)
        {
                Cell cell = sheet.getCell(j,0);
                if((cell.getContents().trim()).equalsIgnoreCase(col))
                {
                        reqCol= cell.getColumn();
                        //System.out.println("column No:"+reqCol);
                        c = sheet.getCell(reqCol, row);
                        fs.close();
                        return c.getContents();
                }
        }
}
catch(BiffException be)
{
            System.out.println(be.getMessage().toString());
        System.out.println("The given file should have .xls extension.");
}
catch(Exception e)
{
        e.printStackTrace();
            System.out.println (e.getMessage().toString());
}
   System.out.println ("NO MATCH FOUND IN GIVEN FILE: PROBLEM IS
COMING FROM DATA FILE");
return null;
}
```

Figure- 16-7– ReadExcel Function

Note: Above code and other script snippets can be copied from

https://www.adactin.com/products/books-publications/test-automation-using-selenium-webdriver-with-java/ section

13. You will get a few Eclipse errors which are related to importing of jxl and other libraries. Follow the errors and add all the required libraries.

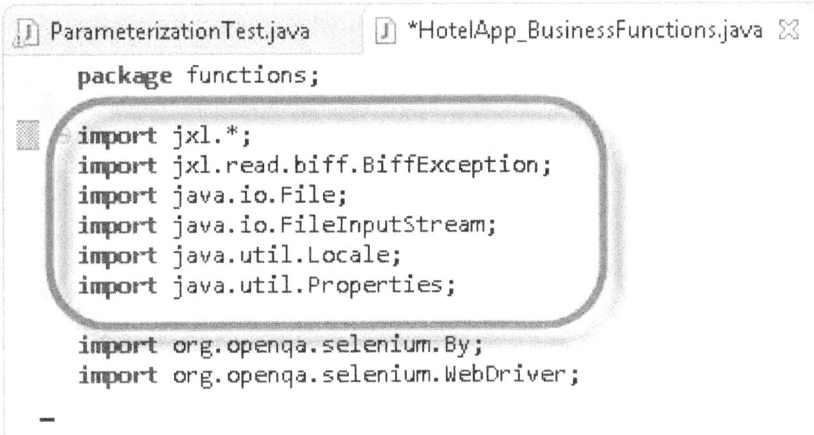

ParameterizationTest.java | *HotelApp_BusinessFunctions.java

```
package functions;

import jxl.*;
import jxl.read.biff.BiffException;
import java.io.File;
import java.io.FileInputStream;
import java.util.Locale;
import java.util.Properties;

import org.openqa.selenium.By;
import org.openqa.selenium.WebDriver;
```

Figure- 16-8– Libraries to be imported

14. Save your Function Library file once you are able to fix all of the errors.

Step 4 - Parameterize our script to use the data and Java function

15. **Right click** on your existing **ConfigurationFileTest.java** script, select **Copy**
16. Select the **tests** package folder, right click and select **Paste**
17. In the Name Conflict dialog box, enter **ParameterizationTest** as the name of the script and click **OK**
18. Double click on **ParameterizationTest**.java file and add the code given below in **@ Test** to read from the Excel sheet. We have to perform two steps:
    a. Read data from the Excel sheet using the **ReadXL** function
    b. Replace constant value with data read from Excel

Use the code below to read data from Excel

```
String strFile = "./DataPool/HA_HotelSearch.xls";

String strLocation =  HA_GF_readXL(1,"Location",strFile);
```

Figure- 16-9– Code snippet to read from Excel sheet

In the above code we are reading from column "Location" and row number 1.

Once location is stored in variable strLocation, replace the value "Sydney" with variable strLocation.

```
new
Select(driver.findElement(By.id(prop.getProperty("Lst_SearchHotel_Location")))).
selectByVisibleText(strLocation );
```

Find below final code snippet

```
@Test
public void testMyFirstWebDriver() throws Exception {

  driver.get(sAppURL);

  //Comment - Call to Login Function
  HA_BF_Login(driver, "adactin123", "adactin123");

  String strFile = "./DataPool/HA_HotelSearch.xls";
  String strLocation =  HA_GF_readXL(1,"Location",strFile);

  new Select(driver.findElement(By.id(prop.getProperty("Lst_SearchHotel_Location")))).selectByVisibleText(strLocation)
  driver.findElement(By.id(prop.getProperty("Btn_SearchHotel_Search"))).click();
  driver.findElement(By.id(prop.getProperty("Rad_SelectHotel_RadioButton_1"))).click();
```

Figure- 16-10– Parameterized script code

19. Save the test and perform Project clean-up using **Project → Clean Up**... and clean all selected projects
20. Select and Run the test using **Run → Run As JUnit Test**. Confirm that the test can run fine with the correct location value

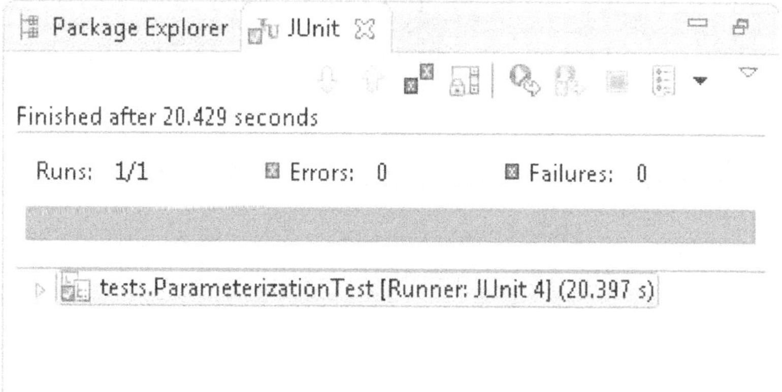

Figure 16-11– Test Results

EXERCISE

1. Run the above script for row value 2 and verify it reads the second location correctly.

16.2. Parameterize the Script with Multiple Values from an Excel Sheet

In the previous example we read data from an Excel data sheet. The advantage we got was to keep all data files at a central location and share them among all of the scripts. If data values need to change, we can change them at one central location instead of repeatedly changing them in all of the scripts.

Another aspect of data-driven testing is parameterization when we can run same script for multiple sets of data values.

Problem Description: Let us try to create a script which will create multiple bookings for different locations (Sydney, Brisbane, Melbourne, and Adelaide) in the Search Hotel page.

Solution – Based on our previous script we now know how to read data from an Excel datasheet. If we know how many rows of data are available in the Excel sheet, we can use any looping statement (like for, while) to loop the script for the number of rows available in the datasheet.

So as a solution we can create a function which will return the maximum number of valid rows in an Excel sheet and use its return value to loop through the script.

Let us see the steps given below to parameterize the script to run for multiple values -

1. Double click on "**HotelApp_BusinessFunctions.java**" to see the code in the functions file
2. Add the following function to our functions file. It fetches Rowcount from the excel sheet using two parameters:
    - strFilePath - the path of the XL file/workbook containing the data, the path is relative to our java project
    - column - name of the column header from which you want to read data

The function returns an Integer containing the row count within the datasheet.

```
public static int HA_GF_XLRowCount (String strFilePath, String sColumn)
{
  int k;
 for (k = 1; k < 999; k++)
 {
 String sParamVal =       HA_GF_readXL(k,sColumn, strFilePath);
 if (sParamVal.equals("ENDOFROW"))
 {
   break;
 }
 }

return k;
}
```

Figure- 16-12– XLRowCount Function

Note: We have made an assumption that the maximum number of rows in any data sheet would not exceed 999. We can update it based on our project needs. Also we are using the string "ENDOFROW" to indicate we have reached the last row in the sheet.

3. Verify that there are no syntax errors. Try to resolve syntax errors if any exist. Save the Function Library file.
4. **Right click** on your existing **ParametrizationTest.java** script, select **Copy**
5. Select the **tests package folder**, right click and select **Paste**
6. In the Name Conflict dialog box, enter the name of the script as **ParametrizationLoopTest.java** and click **OK**
7. Double click the **ParametrizationLoopTest.java** file and add the code below in @ **Test** to read from Excel sheet.

Figure- 16-13– Parametrization script code

Key things to modify include:

- Use the HA_GF_XLRowCount function to retrieve the row count
- Use a For loop to loop though the maximum number of rows
- In function HA_GF_readXL, replace the static row number with a dynamic row number coming from the for loop
- Do not forget to close the for loop with a closing parenthesis

8. Save the test and perform Project clean-up using **Project → Clean Up**… and clean all selected projects.

9. Select and run the test using **Run → Run As** JUnit Test. Confirm that the test can run fine with the correct location value

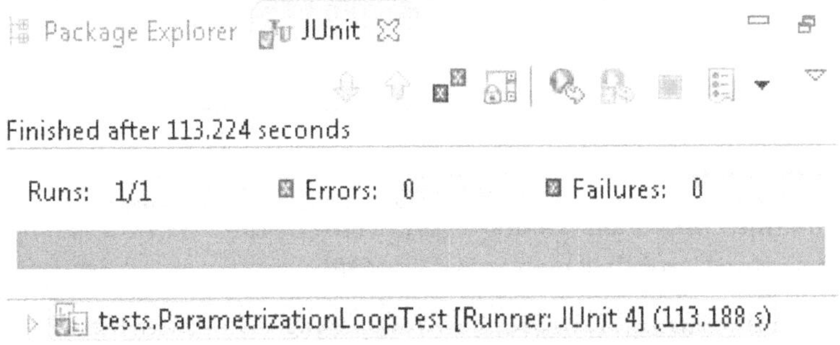

Figure- 16-14– Test Results

EXERCISE

1. Add the DataPool folder path in the Configuration file. Modify the existing script to read the DataPool folder path from the Configuration file and access the Search Hotel datasheet.

2. Create another Datasheet named HA_Booking.xls and parameterize the data in the booking form (first name, last name, address) to be read from HA_Booking.xls. Update the script to run for 4 iterations for each location and different booking detail values.

§

17

Synchronizing WebDriver scripts

Introduction

In any Web automation project, the automation success depends upon the robustness of your scripts; whether that's adaptation of your code to project or software changes, or synchronization of the script with the site's performance.

Many a time, your application performance will vary which will require you to manipulate your WebDriver script's execution speed.

In one of the applications that we tested, it took more than 60 seconds for an application form to save and confirm that save was successful. How does WebDriver support these situations?

Synchronization is a critical issue for any test automation script. You may think that synchronization of test script actions is a built-in ability of today's functional testing tools. Reality shows that many unexpected test script failures are related to synchronization issues generating false negative results. These false negatives make it hard to detect real application problems as each test script failure may be related to a test script synchronization issue. Synchronization errors are timing issues therefore they are non-deterministic, heavily dependent on the HW/SW, the network and their utilization. The biggest challenge in automating a Web application is the loading of a Web page which is always at the mercy of certain conditions, such as:

- Load on the server
- Network speed
- Performance of AUT
- Ajax call to load an element

Key objectives:

- What is Synchronization?
- Approaches to script synchronization
- Synchronizing a script

17.1. What is Synchronization?

What is script synchronization?

Test scripts need to be synchronized in a way that the script driving the application **waits** until the AUT is ready to accept the next user input.

The following are some situations where synchronization is required:

- The creation of a window (more general control) must be completed before it can receive messages/commands
- A page is completely loaded before you can click on a link on the page
- A button must be activated before you can click on it
- A data grid has loaded a row before you can verify the row
- A data grid has loaded completely before you can verify the row count of the grid
- A tree is expanded before you can select one of its children
- Selecting a tree node, the details pane of the node needs to be completely loaded before you can verify text on it.

Ajax specific synchronization problems

Ajax is shorthand for Asynchronous JavaScript and XML. With Ajax, Web applications can retrieve data from the server asynchronously in the background without interfering with the display and behavior of the existing page. The use of Ajax techniques has led to an increase in interactive or dynamic interfaces on Web pages. Data is usually retrieved using the XMLHttpRequest object. The asynchronous nature of Ajax makes it hard to detect when an Ajax request is fulfilled and when the application is ready to proceed with the next UI interaction (as the UI is not blocked during executing the request).

Even worse, the current browsers do not indicate the execution and the end of the execution of Ajax based requests. Processing normal HTML pages is indicated visually by an animated icon in the browser and programmatically by an event that is fired when processing is completed. This means that there is no easy way to decide when the application is ready to process the next UI action. Usually this is not a problem for humans because we have multiple cognitive techniques to detect if an application is ready to proceed. Humans are also not that fast when working with an application compared to "computer programs" like test automation tools that are driving the application. So many of the synchronization problems do not appear when a human is accessing the application. But asynchronous behavior as seen in Ajax applications is a real nightmare for a testing tool.

17.2. Approaches used for Script Synchronization

So the major task in automating Web applications is to wait for the HTML element to appear in the page before your automation test code starts performing an action on them. You need to make sure that the Web element is present before the code begins working on it. This can be achieved by waiting for the element to appear in the page. For synchronization we can implement three different execution control mechanisms:

1. Implicit wait – used to set the default wait time throughout the program
2. Explicit wait – used to set the wait time for a particular instance only
3. The global controlling of wait times

Implicit Wait

An implicit wait is to tell WebDriver to poll the DOM for a certain amount of time when trying to find an element or elements if they are not immediately available. The default setting is 0. Once set, the implicit wait is set for the life of the WebDriver object instance.

Driver.manage().timeouts().implicitlywait(Duration.ofSeconds(10))

This means that you are setting 10 seconds to be your default wait time whenever the driver comes across a situation where it can't find the required element or condition.

Explicit Static Wait

The simplest and easiest way to handle synchronization is to use the **Explicit static wait** statement in scripts. If you know that the application can take up to 60 seconds to respond, you can enter a Thread.sleep statement after the step where you want WebDriver to wait which will '**pause**' the script for 60 seconds. See the statement below.

Thread.sleep(timeToSleep);

The disadvantage of using static wait statements is that script will **wait for the entire 60 seconds**, even though the application might respond quicker.

Automation testers should be careful while adding wait statements, as this increases the overall test execution time of the automation suite.

Note: You can define time to wait as a global variable. The benefit of this is that we can control the wait time centrally according to our AUT performance

public int expectedWaitTime = 10;

And then we use this global variable as an argument for all our explicit wait methods.

Thread.sleep(expectedWaitTime);

Explicit Dynamic Wait

We noticed in the above Static Wait statement that the script will wait for the entire time mentioned in the Wait statement. A better option would be for the WebDriver script to smartly wait or forge ahead based on an expected visual cue which will save overall script execution time

Explicit Dynamic wait can be achieved in multiple ways:

- Using a WebDriverWait Class
- Using a custom function to wait for an element's existence

Using a WebDriverWait Class

A **WebDriverWait** class is used in combination with an **ExpectedCondition** class, which provides an elegant way to implement Explicit Wait.

The Selenium WebDriver provides WebDriverWait and ExpectedConditions classes for implementing an Explicit wait.

The ExpectedCondition class provides a set of predefined conditions to wait before proceeding further in the code. The following table shows some common conditions that we frequently come across when automating Web browsers supported by the ExpectedCondition class:

Selenium method	Detail
elementToBeClickable(By locator)	An element is visible and enabled
elementToBeSelected(WebElement element)	An element is selected
presenceOfElementLocated(By locator)	An expectation for checking that an element is present on the DOM of a page.
textToBePresentInElement(By locator, java.lang.String text)	An expectation for checking if the given text is present in the specified element.
textToBePresentInElementValue(By locator, java.lang.String text)	An expectation for checking if the given text is present in the specified elements value attribute.
titleContains(java.lang.String title)	An expectation for checking that the title contains a case-sensitive substring
visibilityOf(WebElement element)	An expectation for checking that an element, known to be present on the DOM of a page, is visible.
visibilityOfElementLocated(By locator)	An expectation for checking that an element is present on the DOM of a page and visible.

Figure-17-1 – Expected Class Conditions

For the following example, we shall wait up to 10 seconds for an element whose id is "Logout" to become visible before proceeding to the next command.

1. Declare a WebDriverWait variable. In this example, we will use "myWaitVar" as the name of the variable.

   WebDriver driver = new ChromeDriver();
   WebDriverWait myWaitVar = new WebDriverWait(driver, Duration.ofSeconds(10));

2. Import required packages (suggested by Eclipse). In this case we need to **import** org.openqa.selenium.support.ui.WebDriverWait;

3. Use myWaitVar with expected condition:

Use myWaitVar with ExpectedConditions on portions where you need the explicit wait to occur. In this case, we will use explicit wait to wait on for logout button appears after you book a hotel and get an order number for the order (AdactIn Hotel App Booking Confirmation Page) before we click onto it.

myWaitVar.until(ExpectedConditions.visibilityOfElementLocated(By.id("Logout"));
driver.findElement(By.id("Logout")).click();

Using a Custom function

Function to explicitly wait for WebElement Presence

We can even create a custom function to wait for the presence of a Webelement using WebDriver FindElements method.

```
//Function to dynamically wait for element presence

public void HA_GF_WaitForElementPresent (WebDriver driver , By by, int iTimeOut) throws
InterruptedException
        {
                int iTotal = 0;
                int iSleepTime = 5000;
                while(iTotal < iTimeOut)
                {
                        List<WebElement> oWebElements = driver.findElements(by);
                        if(oWebElements.size()>0)
                                        return;
                                else
                                {

                                        Thread.sleep(iSleepTime);
                                        iTotal = iTotal + iSleepTime;
                        System.out.println(String.format("Waited for %d milliseconds.[%s]", iTotal, by));


                                }
                }
        }
```

Figure-17-2– Sample Code for Explicit Wait for Element Present

Note: Above code and other script snippets can be copied from https://www.adactin.com/products/books-publications/test-automation-using-selenium-webdriver-with-java/

section.

The above code uses the following logic:

- Input arguments: driver object, locator and timeout
- Function findElements returns a list of objects as defined by the locator argument
- If the Webelement is found, then the function returns back to the main script
- If the Webelement is not found, then the function waits for iSleepTime (defined in the function) and again searches for the element
- iSleeptime repeats until either the Webelement is found or the Wait time exceeds iTimeout

Note: You will need to import two more libraries using **import java.util.List**; and **import org.openqa.selenium.WebElement**. Eclipse will automatically suggest this after you paste the code into the Eclipse IDE.. If Eclipse throws any other error try to resolve it using basic java knowledge.

The above custom function can also be enhanced to wait until a specific object property value is achieved using the **getAttribute** method.

Function to explicit wait for WebElement property value matches expected value

```
//Function to dynamically wait for Webelement to achieve its property value

public void HA_GF_WaitForPropertyValue (WebDriver driver , By by, String ExpPropertyVal, String
PropertName, int iTimeOut) throws InterruptedException
    {
        int iTotal = 0;
        int iSleepTime = 5000;
        while(iTotal < iTimeOut)
        {
            List<WebElement> oWebElements = driver.findElements(by);
            if(oWebElements.size()>0)
                for (WebElement weOption : oWebElements)
                {
                    if(weOption.getAttribute(PropertName).
equalsIgnoreCase(ExpPropertyVal))

                    {
                        return;
                    }
                    else

                    {
                        Thread.sleep(iSleepTime);
                        iTotal = iTotal + iSleepTime;
                        System.out.println(String.
format("Waited for %d milliseconds.[%s]", iTotal, by));
                    }

                }
            else
            {

                Thread.sleep(iSleepTime);
                iTotal = iTotal + iSleepTime;
                System.out.println(String.format("Waited for %d
milliseconds.[%s]", iTotal, by));

            }
        }
    }
```

Figure-17 3 - Sample Code for Explicit Wait for Property Value Existence

Note: Above code and other script snippets can be copied from https://www.adactin.com/products/books-publications/test-automation-using-selenium-webdriver-with-java/

section

17.3. Using Script Synchronization in a Script

Let us take a practical scenario to illustrate script synchronization.

Problem Description – As part of our Hotel booking workflow, when a user clicks on Book Now an order number is generated. We need to write this order number in the result.

The challenge here is that when a user clicks on the "Book Now' button it takes approximately 6-8 seconds before the order number is generated. So to resolve this we need script synchronization.

Solution – We can follow multiple solutions to this problem.

- Use Static Wait, which is the easiest and simplest of all, before you fetch the value from the order number field
- Use Explicit Wait to wait for the existence of the Logout Button. The Logout button only appears after the order number has been generated.

<u>**Solution – 1 – Using Static Wait**</u>

Use the following steps to get the order number using a Static Wait:

1. **Right click** on your existing **ParameterizationTest.java** script, select **Copy**
2. Select the **tests** package folder, **right click** and select **Paste**
3. In the Name Conflict dialog box, enter name as **SynchronizationTest.java** and click **OK**
4. Double click on the newly created "**SynchronizationTest.java**" script to see the script

Note that after we click on the Book Now button we will need to add a Static Wait statement using the Thread.sleep(timeout) method and use GetAttribute to fetch the value from the Order No. field.

Before we do this, we need to add Locator details for the Order No. field and Logout button in our **SharedUIMap** file.

5. Use the **ChroPath** add-on to get the locator value for the Order No. field

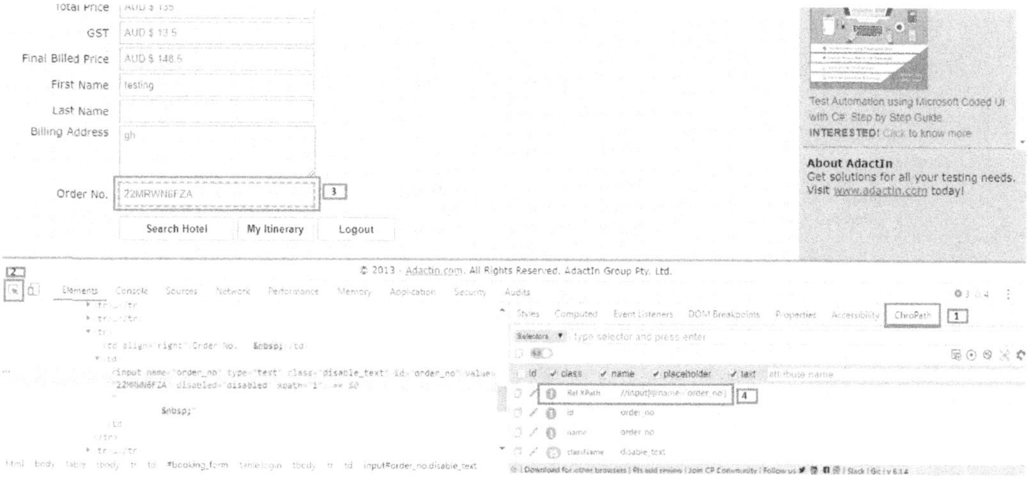

Figure-17-4– Locator value for Order No. field

6. Use the ChroPath add-on to get the locator value for the Logout button

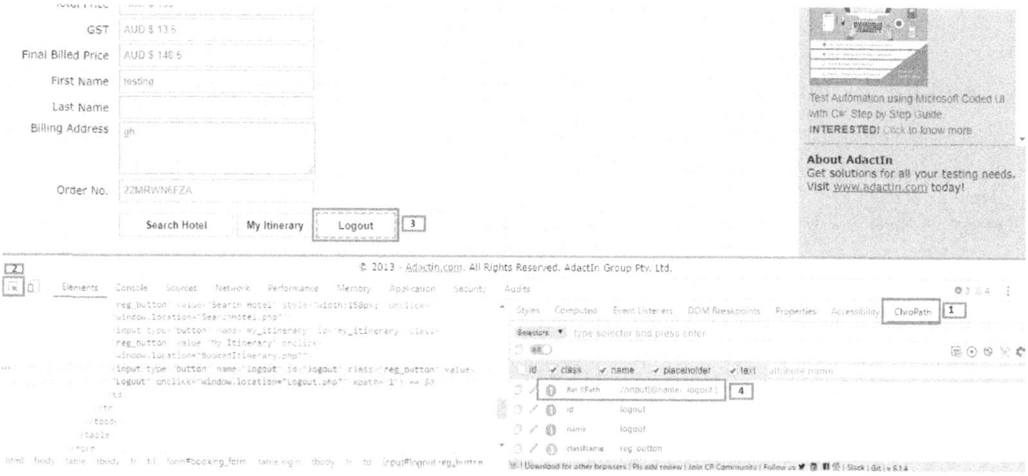

Figure-17-5– Locator value for Logout button

7. Add these locator values to the SharedUIMap **SharedUIMap.properties** file

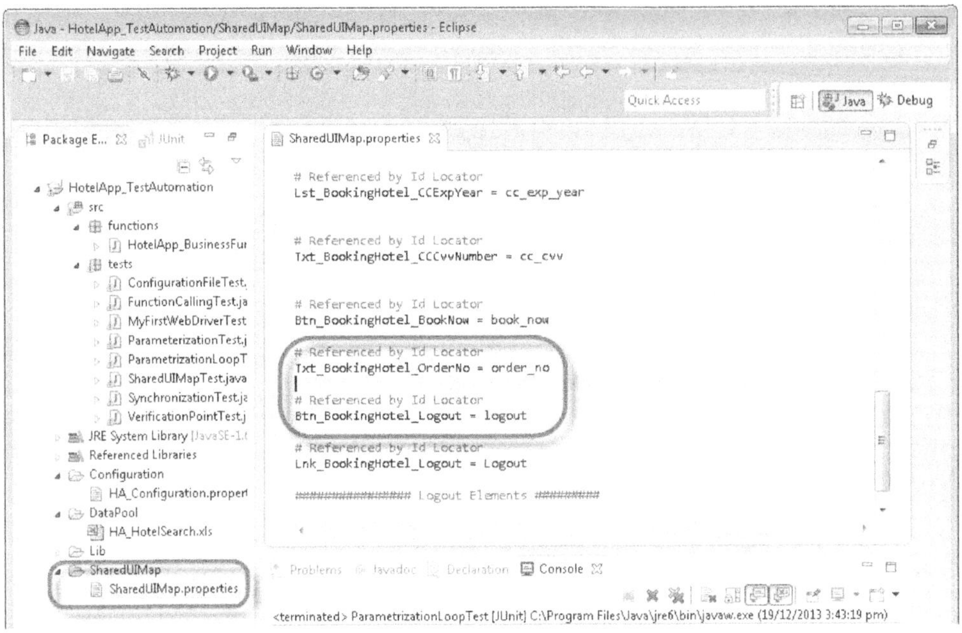

Figure-17-6– Web Elements values added to SharedUIMap

8. Add this locator to the SharedUIMap SharedUIMap.properties file

```
//Step added after Book Now button click
Thread.sleep(10000);
String strOrderNo = driver.findElement(By.id(prop.getProperty("Txt_BookingHotel_OrderNo"))).getAttribute("value");

System.out.println("Order Number generated is "+ strOrderNo);
```

Figure-17-7– Script Code Added to script

```
@Test
public void testMyFirstWebDriver() throws Exception {

  driver.get(sAppURL);
  //Comment - Call to Login Function
  HA_BF_Login(driver, "adactin123", "          ");

  String strFile = "./DataPool/HA_HotelSearch.xls";
  String strLocation =  HA_GF_readXL(1,"Location",strFile);

  new Select(driver.findElement(By.id(prop.getProperty("Lst_SearchHotel_Location")))).selectByVisibleText(strLocation);
  driver.findElement(By.id(prop.getProperty("Btn_SearchHotel_Search"))).click();
  driver.findElement(By.id(prop.getProperty("Rad_SelectHotel_RadioButton_1"))).click();
  driver.findElement(By.id(prop.getProperty("Btn_SelectHotel_Continue"))).click();
  driver.findElement(By.id(prop.getProperty("Txt_BookingHotel_FirstName"))).clear();
  driver.findElement(By.id(prop.getProperty("Txt_BookingHotel_FirstName"))).sendKeys("test");
  driver.findElement(By.id(prop.getProperty("Txt_BookingHotel_LastName"))).clear();
  driver.findElement(By.id(prop.getProperty("Txt_BookingHotel_LastName"))).sendKeys("test");
  driver.findElement(By.id(prop.getProperty("Txt_BookingHotel_Address"))).clear();
  driver.findElement(By.id(prop.getProperty("Txt_BookingHotel_Address"))).sendKeys("test");
  driver.findElement(By.id(prop.getProperty("Txt_BookingHotel_CCNumber"))).clear();
  driver.findElement(By.id(prop.getProperty("Txt_BookingHotel_CCNumber"))).sendKeys("1212121212121212");
  new Select(driver.findElement(By.id(prop.getProperty("Lst_BookingHotel_CCType")))).selectByVisibleText("American Express");
  new Select(driver.findElement(By.id(prop.getProperty("Lst_BookingHotel_CCExpMonth")))).selectByVisibleText("March");
  new Select(driver.findElement(By.id(prop.getProperty("Lst_BookingHotel_CCExpYear")))).selectByVisibleText("2015");
  driver.findElement(By.id(prop.getProperty("Txt_BookingHotel_CCCvvNumber"))).clear();
  driver.findElement(By.id(prop.getProperty("Txt_BookingHotel_CCCvvNumber"))).sendKeys("111");
  driver.findElement(By.id(prop.getProperty("Btn_BookingHotel_BookNow"))).click();

  Thread.sleep(10);
  String strOrderNo = driver.findElement(By.id(prop.getProperty("Txt_BookingHotel_OrderNo"))).getAttribute("value");
  System.out.println("Order Number Generate is "+ strOrderNo);

  driver.findElement(By.linkText(prop.getProperty("Lnk_BookingHotel_Logout"))).click();
  driver.findElement(By.linkText(prop.getProperty("Lnk_Logout_ClickHeretoLoginAgain"))).click();
}
```

```
@Test
public void testMyFirstWebDriver() throws Exception {

    String strFile = "./DataPool/HA_HotelSearch.xls";
    String strLocation =  HA_GF_readXL(1,"Location",strFile);

    HA_BF_Login(driver, "adactin123", "adactin123");
    new Select(driver.findElement(By.id(props.getProperty("Lst_SearchHotel_Location")))).selectByVisibleText(strLc
    driver.findElement(By.id(props.getProperty("Btn_SearchHotel_Search"))).click();
    driver.findElement(By.id(props.getProperty("Rad_SelectHotel_RadioButton_1"))).click();
    driver.findElement(By.id(props.getProperty("Btn_SelectHotel_Continue"))).click();
    driver.findElement(By.id(props.getProperty("Txt_BookingHotel_FirstName"))).clear();
    driver.findElement(By.id(props.getProperty("Txt_BookingHotel_FirstName"))).sendKeys("test");
    driver.findElement(By.id(props.getProperty("Txt_BookingHotel_LastName"))).clear();
    driver.findElement(By.id(props.getProperty("Txt_BookingHotel_LastName"))).sendKeys("test");
    driver.findElement(By.id(props.getProperty("Txt_BookingHotel_Address"))).clear();
    driver.findElement(By.id(props.getProperty("Txt_BookingHotel_Address"))).sendKeys("test");
    driver.findElement(By.id(props.getProperty("Txt_BookingHotel_CCNumber"))).clear();
    driver.findElement(By.id(props.getProperty("Txt_BookingHotel_CCNumber"))).sendKeys("1212121212121212");
    new Select(driver.findElement(By.id(props.getProperty("Lst_BookingHotel_CCType")))).selectByVisibleText("Ameri
    new Select(driver.findElement(By.id(props.getProperty("Lst_BookingHotel_CCExpMonth")))).selectByVisibleText("M
    new Select(driver.findElement(By.id(props.getProperty("Lst_BookingHotel_CCExpYear")))).selectByVisibleText("20
    driver.findElement(By.id(props.getProperty("Txt_BookingHotel_CCCvvNumber"))).clear();
    driver.findElement(By.id(props.getProperty("Txt_BookingHotel_CCCvvNumber"))).sendKeys("111");
    driver.findElement(By.id(props.getProperty("Btn_BookingHotel_BookNow"))).click();

    //Step added after Book Now button click

    Thread.sleep(10000);
    String strOrderNo =   driver.findElement(By.id(props.getProperty("Txt_BookingHotel_OrderNo"))).getAttribute("
    System.out.println("Order Number generated is "+ strOrderNo);

    driver.findElement(By.linkText(props.getProperty("Lnk_BookingHotel_Logout"))).click();
    driver.findElement(By.linkText(props.getProperty("Lnk_Logout_ClickHeretoLoginAgain"))).click();
}
```

Figure-17-8– Script Code in Eclipse for static wait

In the above script we are using Thread.sleep to wait for 10 seconds for the order to be inserted. After the order is inserted we use the getAttribute method to retrieve the value from the Order No. field.

Simple and easy!

The only drawback of the above process is that the system will wait all 10 seconds even though actual order insertion might take 6-7 seconds.

9. **Save** the test and perform Project clean-up using **Project → Clean Up...** and **clean all selected** projects.

10. **Run** the updated script and see if you can see the Order Number in the Output Console

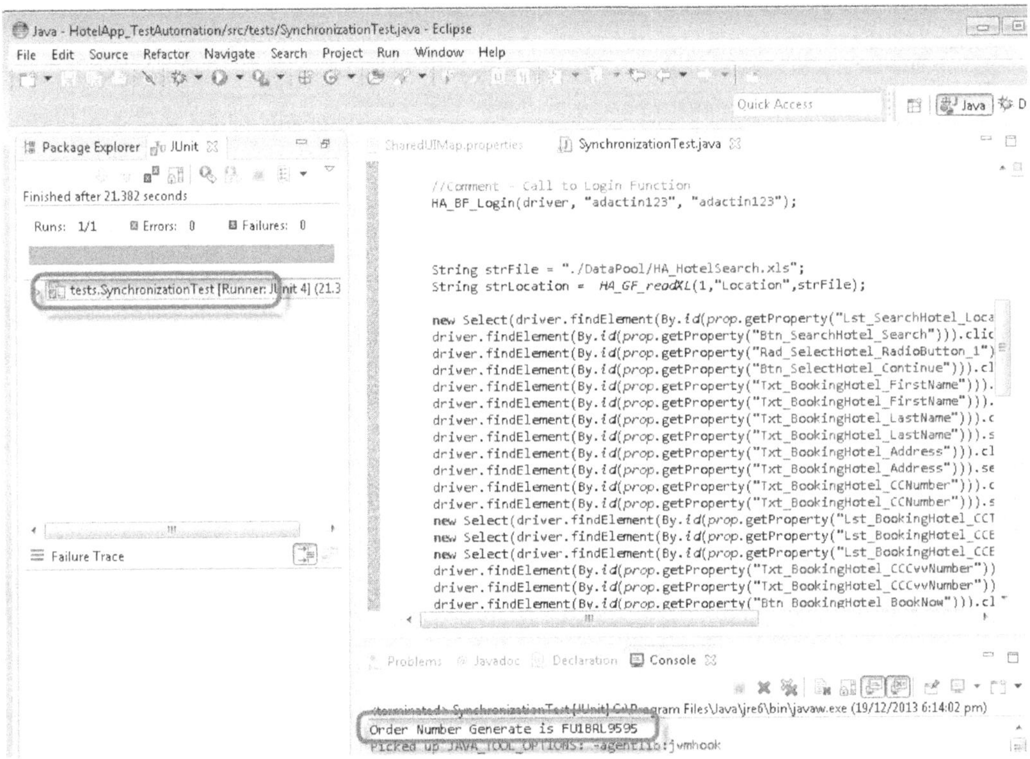

Figure-17-9 - Test Results

Solution – 2– Using Explicit Wait

Let us see how to get the order number using the Static Wait-

1. **Right click** on your existing SynchronizationTest.java script, select **Copy**
2. Select the **tests** package folder, **right click** and select **Paste**
3. In the Name Conflict dialog box, enter the name of the script as SynchronizationExplicitWaitTest.java and click **OK**
4. Double click on the newly created "**SynchronizationExplicitWaitTest.java**" script so that you can see the script

Note that after we click the "Book Now" button we will need to to use the Explicit Custom Wait function ("HA_GF_WaitForElementPresent" as discussed in the previous section) to wait until the order is inserted and then use GetAttribute to fetch the value from the OrderNo field.

Before we do this we need to add Locator details for the Order No. Field and the Logout Button in our SharedUIMap file.

5. Go to Step 9 if you have already added OrderNo and Logout Button in the previous solution
6. Use ChroPath add-ons to get the locator value for the Order No. field.

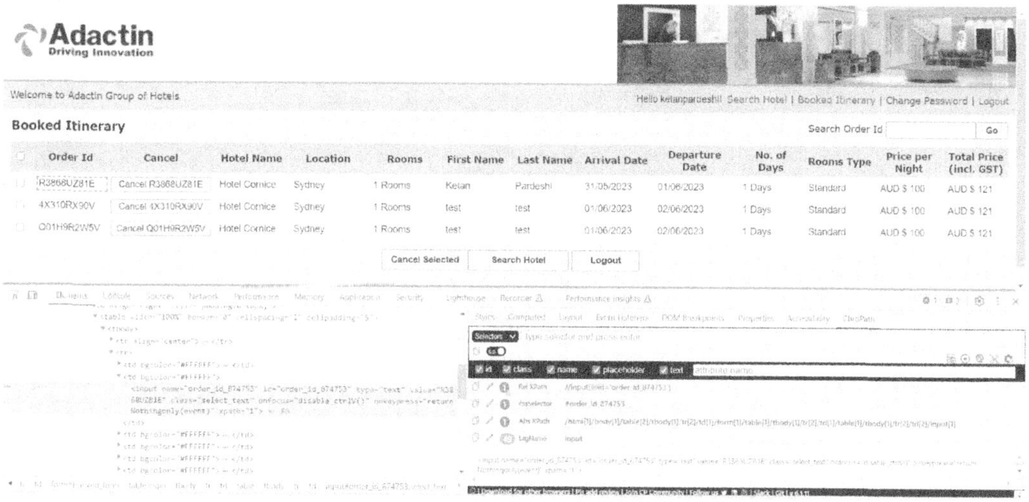

Figure-17-10– Locator value for Order No. Field

7. Use ChroPath add-ons to get the locator value for the Logout Button

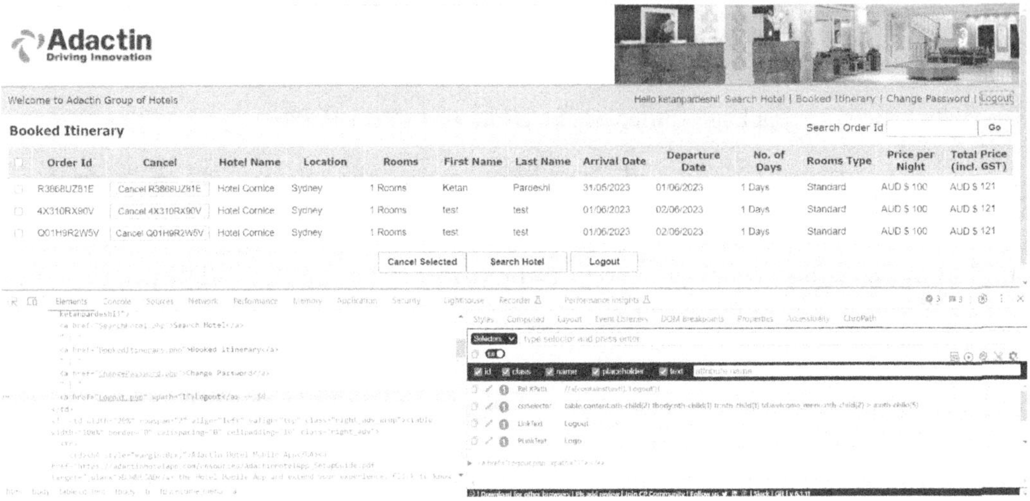

Figure-17-11– Locator value for Logout Button

8. Add the locator values to the SharedUIMap file SharedUIMap.properties file

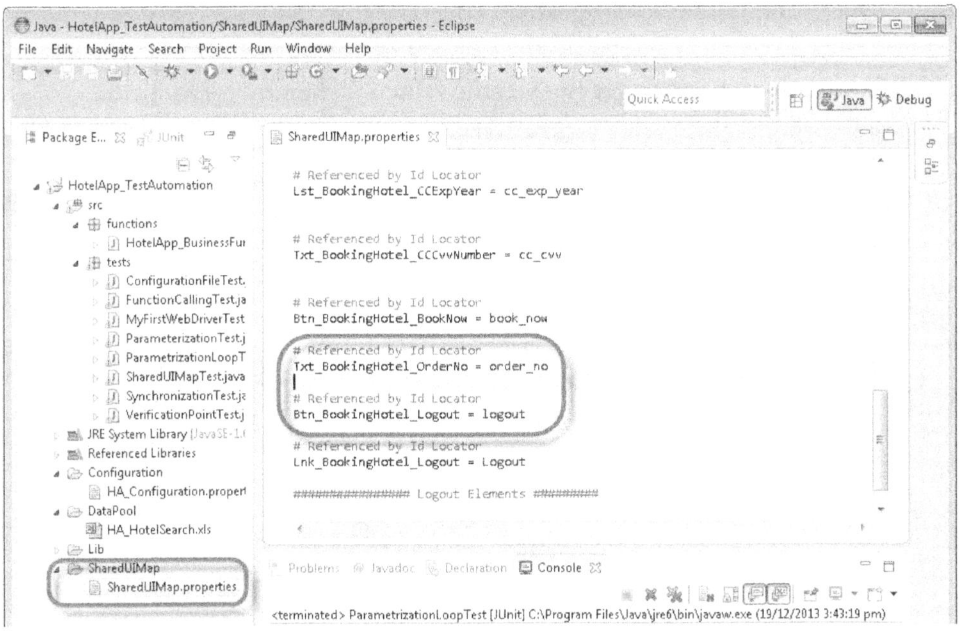

Figure-17-12– Web Elements values added to SharedUIMap

9. Add below code to the script after the step where the user clicks the Book Now button

```
//Step added after Book Now button click
  HA_GF_WaitForElementPresent(driver, By.id(prop.getProperty("Btn_BookingHotel_Logout")), 10000);
  String strOrderNo = driver.findElement(By.id(prop.getProperty("Txt_BookingHotel_OrderNo"))).getAttribute("value");

      System.out.println("Order Number Generate is "+ strOrderNo);
```

Figure-17-13– Script Code Added to script for Explicit Wait

Note: This code assumes we have added the explicit function HA_GF_WaitForElementPresent as part of the previous section in our function library file

```
@Test
public void testMyFirstWebDriver() throws Exception {

  driver.get(sAppURL);
  //Comment - Call to Login Function
  HA_BF_Login(driver, "adactin123", "adactin123");

  String strFile = "./DataPool/HA_HotelSearch.xls";
  String strLocation =  HA_GF_readXL(1,"Location",strFile);

  new Select(driver.findElement(By.id(prop.getProperty("Lst_SearchHotel_Location")))).selectByVisibleText(strLocation);
  driver.findElement(By.id(prop.getProperty("Btn_SearchHotel_Search"))).click();
  driver.findElement(By.id(prop.getProperty("Rad_SelectHotel_RadioButton_1"))).click();
  driver.findElement(By.id(prop.getProperty("Btn_SelectHotel_Continue"))).click();
  driver.findElement(By.id(prop.getProperty("Txt_BookingHotel_FirstName"))).clear();
  driver.findElement(By.id(prop.getProperty("Txt_BookingHotel_FirstName"))).sendKeys("test");
  driver.findElement(By.id(prop.getProperty("Txt_BookingHotel_LastName"))).clear();
  driver.findElement(By.id(prop.getProperty("Txt_BookingHotel_LastName"))).sendKeys("test");
  driver.findElement(By.id(prop.getProperty("Txt_BookingHotel_Address"))).clear();
  driver.findElement(By.id(prop.getProperty("Txt_BookingHotel_Address"))).sendKeys("test");
  driver.findElement(By.id(prop.getProperty("Txt_BookingHotel_CCNumber"))).clear();
  driver.findElement(By.id(prop.getProperty("Txt_BookingHotel_CCNumber"))).sendKeys("1212121212121212");
  new Select(driver.findElement(By.id(prop.getProperty("Lst_BookingHotel_CCType")))).selectByVisibleText("American Express");
  new Select(driver.findElement(By.id(prop.getProperty("Lst_BookingHotel_CCExpMonth")))).selectByVisibleText("March");
  new Select(driver.findElement(By.id(prop.getProperty("Lst_BookingHotel_CCExpYear")))).selectByVisibleText("2015");
  driver.findElement(By.id(prop.getProperty("Txt_BookingHotel_CCCvvNumber"))).clear();
  driver.findElement(By.id(prop.getProperty("Txt_BookingHotel_CCCvvNumber"))).sendKeys("111");
  driver.findElement(By.id(prop.getProperty("Btn_BookingHotel_BookNow"))).click();

  HA_GF_WaitForElementPresent(driver, By.id(prop.getProperty("Btn_BookingHotel_Logout")), 10);
  String strOrderNo = driver.findElement(By.id(prop.getProperty("Txt_BookingHotel_OrderNo"))).getAttribute("value");
  System.out.println("Order Number Generate is "+ strOrderNo);

  driver.findElement(By.linkText(prop.getProperty("Lnk_BookingHotel_Logout"))).click();
  driver.findElement(By.linkText(prop.getProperty("Lnk_Logout_ClickHeretoLoginAgain"))).click();
}
```

```
HA_BF_Login(driver, "adactin", "Adactin@123");
new Select(driver.findElement(By.id(props.getProperty("Lst_SearchHotel_Location")))).selectByVisibleText(strLocation);
driver.findElement(By.id(props.getProperty("Btn_SearchHotel_Search"))).click();
driver.findElement(By.id(props.getProperty("Rad_SelectHotel_RadioButton_1"))).click();
driver.findElement(By.id(props.getProperty("Btn_SelectHotel_Continue"))).click();
driver.findElement(By.id(props.getProperty("Txt_BookingHotel_FirstName"))).clear();
driver.findElement(By.id(props.getProperty("Txt_BookingHotel_FirstName"))).sendKeys("test");
driver.findElement(By.id(props.getProperty("Txt_BookingHotel_LastName"))).clear();
driver.findElement(By.id(props.getProperty("Txt_BookingHotel_LastName"))).sendKeys("test");
driver.findElement(By.id(props.getProperty("Txt_BookingHotel_Address"))).clear();
driver.findElement(By.id(props.getProperty("Txt_BookingHotel_Address"))).sendKeys("test");
driver.findElement(By.id(props.getProperty("Txt_BookingHotel_CCNumber"))).clear();
driver.findElement(By.id(props.getProperty("Txt_BookingHotel_CCNumber"))).sendKeys("1212121212121212");
new Select(driver.findElement(By.id(props.getProperty("Lst_BookingHotel_CCType")))).selectByVisibleText("American Express");
new Select(driver.findElement(By.id(props.getProperty("Lst_BookingHotel_CCExpMonth")))).selectByVisibleText("March");
new Select(driver.findElement(By.id(props.getProperty("Lst_BookingHotel_CCExpYear")))).selectByVisibleText("2030");
driver.findElement(By.id(props.getProperty("Txt_BookingHotel_CCCvvNumber"))).clear();
driver.findElement(By.id(props.getProperty("Txt_BookingHotel_CCCvvNumber"))).sendKeys("111");
driver.findElement(By.id(props.getProperty("Btn_BookingHotel_BookNow"))).click();

//Step added after Book Now button click

HA_GF_WaitForElementPresent(driver, By.id(props.getProperty("Btn_BookingHotel_Logout")), 10000);
String strOrderNo = driver.findElement(By.id(props.getProperty("Txt_BookingHotel_OrderNo"))).getAttribute("value");
System.out.println("Order Number Generate is "+ strOrderNo);

driver.findElement(By.linkText(props.getProperty("Lnk_BookingHotel_Logout"))).click();
driver.findElement(By.linkText(props.getProperty("Lnk_Logout_ClickHeretoLoginAgain"))).click();
```

Figure-17-14– Script Code in Eclipse for explicit wait

In the above script we are not using Thread.sleep to wait for the order to be inserted. Instead we perform a dynamic wait and as soon as the order is inserted the script moves to the next step to fetch the order number value.

10. Save the test and perform Project clean-up using Project → Clean Up… and clean all selected projects.

11. Run the updated script and see if you can see the Order Number printed in the Output Console

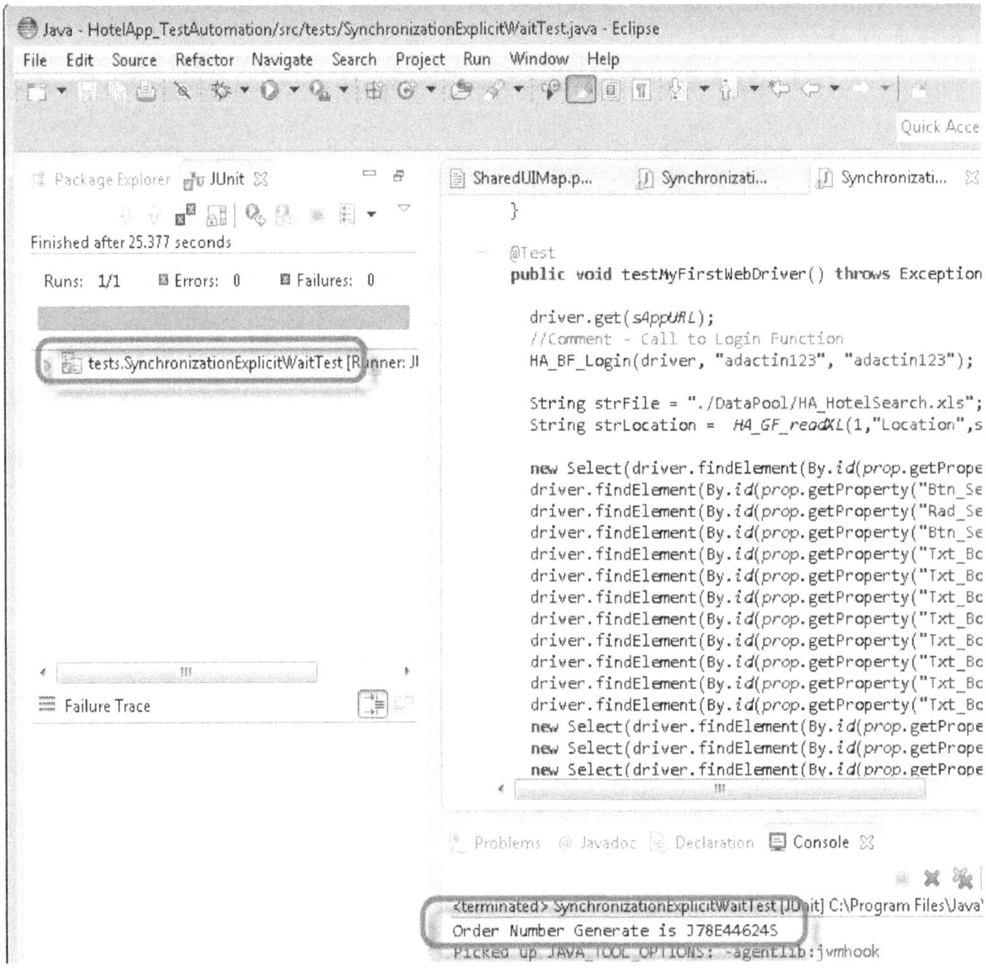

Figure-17-15- Test Results

EXERCISE

1. Use the knowledge gained in this chapter and try automating the workflow below
   a) Book a Hotel
   b) Grab Order number from Booking page
   c) Click on Book Itinerary link
   d) Search for your newly created Order number
   e) Logout

Hint

a) Do not forget to add Web element locators in SharedUI map file in booked Itinerary page
b) Copy/paste previous step and modify them to perform actions as required in Booked Itinerary page

§

18

Handling Pop-up Dialogs and Multiple Windows

Most of the modern-day applications come with various pop-up messages and multiple windows.

For instance, if you would like to delete a record in your application many applications **will throw a javascript based confirmation pop-up dialog before deletion.**

Also applications these days have child pop-up windows.

example We were once automating a mortgage based Web application, which had search customer functionality. When a user clicks on the search link it opens a new pop-up window in which the user could search and select any customer. After the customer was selected the pop-up window would close and the user returns back to the main Web page.

How would Selenium work in these scenarios?

In this chapter we will see

- How WebDriver works with Alert dialogs
- How WebDriver works with multiple windows

18.1. Handle Alerts or Prompts

Test Scenario – Follow the steps below in the Hotel Application

- Login
- Search for the hotel
- Select Hotel
- In the Booking form enter required details but enter Credit Card expiry year as 2011
- Click on Book Now. You will see a pop-up window

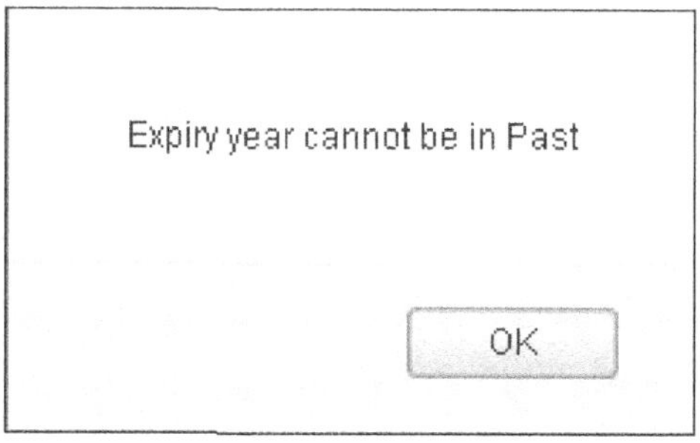

Figure 18-1 – Pop-up dialog

- While the pop-up dialog is open, try to click on Logout link or any other link. You would notice that all other links are disabled since focus is on pop-up dialog

Now we will see how to handle this pop-up dialog using Selenium

1. **Right click** on your existing **ParameterizationTest.java** script, select **Copy**
2. Select the **tests** package folder, **right click** and select **Paste**
3. In the Name Conflict dialog box, enter name of the script as **PopDialogTest.java** and click **OK**
4. Double click on the newly created "**PopDialogTest.java**" script to see the script
5. Modify the CreditCard Expiry Year from "2027" to "2018" to simulate the scenario

```
Test
ublic void testMyFirstWebDriver() throws Exception {

   String strFile = "./DataPool/HA_HotelSearch.xls";
   String strLocation =  HA_GF_readXL(1,"Location",strFile);

   HA_BF_Login(driver, "Adactin", "Adactin@123");
   new Select(driver.findElement(By.id(props.getProperty("Lst_SearchHotel_Location")))).selectByVisibleText(strLocation);
   driver.findElement(By.id(props.getProperty("Btn_SearchHotel_Search"))).click();
   driver.findElement(By.id(props.getProperty("Rad_SelectHotel_RadioButton_1"))).click();
   driver.findElement(By.id(props.getProperty("Btn_SelectHotel_Continue"))).click();
   driver.findElement(By.id(props.getProperty("Txt_BookingHotel_FirstName"))).clear();
   driver.findElement(By.id(props.getProperty("Txt_BookingHotel_FirstName"))).sendKeys("test");
   driver.findElement(By.id(props.getProperty("Txt_BookingHotel_LastName"))).clear();
   driver.findElement(By.id(props.getProperty("Txt_BookingHotel_LastName"))).sendKeys("test");
   driver.findElement(By.id(props.getProperty("Txt_BookingHotel_Address"))).clear();
   driver.findElement(By.id(props.getProperty("Txt_BookingHotel_Address"))).sendKeys("test");
   driver.findElement(By.id(props.getProperty("Txt_BookingHotel_CCNumber"))).clear();
   driver.findElement(By.id(props.getProperty("Txt_BookingHotel_CCNumber"))).sendKeys("1212121212121212");
   new Select(driver.findElement(By.id(props.getProperty("Lst_BookingHotel_CCType")))).selectByVisibleText("American Expr
   new Select(driver.findElement(By.id(props.getProperty("Lst_BookingHotel_CCExpMonth")))).selectByVisibleText("March");
   new Select(driver.findElement(By.id(props.getProperty("Lst_BookingHotel_CCExpYear")))).selectByVisibleText("2027");
   driver.findElement(By.id(props.getProperty("Txt_BookingHotel_CCCvvNumber"))).clear();
   driver.findElement(By.id(props.getProperty("Txt_BookingHotel_CCCvvNumber"))).sendKeys("111");
   driver.findElement(By.id(props.getProperty("Btn_BookingHotel_BookNow"))).click();
```

Figure 18-2– Credit Card Expiry Year – Previous

```
@Test
public void testMyFirstWebDriver() throws Exception {

    String strFile = "./DataPool/HA_HotelSearch.xls";
    String strLocation =  HA_GF_readXL(1,"Location",strFile);

    HA_BF_Login(driver, "Adactin", "Adactin@123");
    new Select(driver.findElement(By.id(props.getProperty("Lst_SearchHotel_Location")))).selectByVisibleText(strLocation
    driver.findElement(By.id(props.getProperty("Btn_SearchHotel_Search"))).click();
    driver.findElement(By.id(props.getProperty("Rad_SelectHotel_RadioButton_1"))).click();
    driver.findElement(By.id(props.getProperty("Btn_SelectHotel_Continue"))).click();
    driver.findElement(By.id(props.getProperty("Txt_BookingHotel_FirstName"))).clear();
    driver.findElement(By.id(props.getProperty("Txt_BookingHotel_FirstName"))).sendKeys("test");
    driver.findElement(By.id(props.getProperty("Txt_BookingHotel_LastName"))).clear();
    driver.findElement(By.id(props.getProperty("Txt_BookingHotel_LastName"))).sendKeys("test");
    driver.findElement(By.id(props.getProperty("Txt_BookingHotel_Address"))).clear();
    driver.findElement(By.id(props.getProperty("Txt_BookingHotel_Address"))).sendKeys("test");
    driver.findElement(By.id(props.getProperty("Txt_BookingHotel_CCNumber"))).clear();
    driver.findElement(By.id(props.getProperty("Txt_BookingHotel_CCNumber"))).sendKeys("1212121212121212");
    new Select(driver.findElement(By.id(props.getProperty("Lst_BookingHotel_CCType")))).selectByVisibleText("American Ex
    new Select(driver.findElement(By.id(props.getProperty("Lst_BookingHotel_CCExpMonth")))).selectByVisibleText("March")
    new Select(driver.findElement(By.id(props.getProperty("Lst_BookingHotel_CCExpYear")))).selectByVisibleText("2018");
    driver.findElement(By.id(props.getProperty("Txt_BookingHotel_CCCvvNumber"))).clear();
    driver.findElement(By.id(props.getProperty("Txt_BookingHotel_CCCvvNumber"))).sendKeys("111");
    driver.findElement(By.id(props.getProperty("Btn_BookingHotel_BookNow"))).click();
```

Figure 18-3– Credit Card Expiry Year – Updated

6. Save the test and perform Project clean-up using **Project → Clean Up…** and **clean all selected projects**.
7. Select and Run the test using **Run → Run As JUnit Test**. Check that the script fails with the error "Modal dialog present"

Figure 18-4– Script Failed

8. In order to fix the script let us add these lines of code in the script after the user clicks the "Book Now" Button

```
//After user clicks on Book Now
Alert javascriptAlert = driver.switchTo().alert();
        System.out.println(javascriptAlert.getText()); // Get text on alert box

        javascriptAlert.accept();
```

Figure 18-5– Alert handler script

In the above script-

"Alert javascriptAlert = driver.switchTo().alert();" - We are creating an instance of an Alert object by switching the focus from the driver to the alert pop-up

javascriptAlert.accept(); - statement is used to Accept the Alert Pop-up or Click Ok on the Alert Pop.

Note: Alternatively we could use the functions below on the alert dialog

javascriptAlert.sendKeys("Name");	To enter data into the alert pop-up (incase it has an edit field)
javascriptAlert.dismiss();	To click on Cancel in the alert pop-up

9. We will see the below script in Eclipse view

```
@Test
public void testMyFirstWebDriver() throws Exception {

    String strFile = "./DataPool/HA_HotelSearch.xls";
    String strLocation =  HA_GF_readXL(1,"Location",strFile);

    HA_BF_Login(driver, "adactin", "adactin@123");
    new Select(driver.findElement(By.id(props.getProperty("Lst_SearchHotel_Location")))).selectByVisibleText(strLocation);
    driver.findElement(By.id(props.getProperty("Btn_SearchHotel_Search"))).click();
    driver.findElement(By.id(props.getProperty("Rad_SelectHotel_RadioButton_1"))).click();
    driver.findElement(By.id(props.getProperty("Btn_SelectHotel_Continue"))).click();
    driver.findElement(By.id(props.getProperty("Txt_BookingHotel_FirstName"))).clear();
    driver.findElement(By.id(props.getProperty("Txt_BookingHotel_FirstName"))).sendKeys("test");
    driver.findElement(By.id(props.getProperty("Txt_BookingHotel_LastName"))).clear();
    driver.findElement(By.id(props.getProperty("Txt_BookingHotel_LastName"))).sendKeys("test");
    driver.findElement(By.id(props.getProperty("Txt_BookingHotel_Address"))).clear();
    driver.findElement(By.id(props.getProperty("Txt_BookingHotel_Address"))).sendKeys("test");
    driver.findElement(By.id(props.getProperty("Txt_BookingHotel_CCNumber"))).clear();
    driver.findElement(By.id(props.getProperty("Txt_BookingHotel_CCNumber"))).sendKeys("1212121212121212");
    new Select(driver.findElement(By.id(props.getProperty("Lst_BookingHotel_CCType")))).selectByVisibleText("American Express");
    new Select(driver.findElement(By.id(props.getProperty("Lst_BookingHotel_CCExpMonth")))).selectByVisibleText("March");
    new Select(driver.findElement(By.id(props.getProperty("Lst_BookingHotel_CCExpYear")))).selectByVisibleText("2011");
    driver.findElement(By.id(props.getProperty("Txt_BookingHotel_CCCvvNumber"))).clear();
    driver.findElement(By.id(props.getProperty("Txt_BookingHotel_CCCvvNumber"))).sendKeys("111");
    driver.findElement(By.id(props.getProperty("Btn_BookingHotel_BookNow"))).click();

    Alert javascriptAlert = driver.switchTo().alert();
    System.out.println(javascriptAlert.getText()); // Get text on alert box
    javascriptAlert.accept();

    driver.findElement(By.linkText(props.getProperty("Lnk_BookingHotel_Logout"))).click();
    driver.findElement(By.linkText(props.getProperty("Lnk_Logout_ClickHeretoLoginAgain"))).click();
}
```

Figure 18-6– Alert handler script in eclipse

10. Save the test and perform Project clean-up using Project → Clean Up... and clean all selected projects.
11. Select and Run the test using Run → Run As JUnit Test. Check that the script runs, passed and prompts the alert text.

Figure 18-7– Test Results

Note: Another traditional way of handing pop-ups in Selenium is using AutoIT script. You can find more details on AutoIT at www.autoitscript.com/

Steps used to handle pop-up with AutoIT

a. Install AutoIT script editor tool
b. Create an AutoIT script to handle the pop-up
c. Compile AutoIT script to create a .exe file
d. Run the .exe file in Selenium using Runtime.getRuntime().exec command when required

18.2. Working with Multiple Windows

An important part when working with multiple windows using WebDriver is to keep switching the focus between parent and child Web pop-up windows.

Key steps to working with multiple windows are:

1. Keep a handle to the existing parent window

Before we switch to the child window it is always appropriate to keep a handle to the existing parent window. We use the method driver.getWindowHandle() to keep a handle to the existing parent window.

String parentWindowHandle = driver.getWindowHandle();

2. Switch to the child or alternate window

WebDriver supports moving between named windows using the "switchTo" method: driver.switchTo().window("windowName");

Note - But how do you know the window's name? Take a look at the javascript or link that opened it:

<a href="somewhere.html" target="windowName">Click here to open a new window</a>

Alternatively if the name is not known, you can use WebDriver.getWindowHandles( ) to obtain a list of known windows. You can pass a "window handle" to the switchTo ( ).window( ) method. Knowing this, it's possible to iterate over every open window like this:

```
for (String handle : driver.getWindowHandles())
{
  driver.switchTo().window(handle);
}
```

3. Perform operations on the child/alternate window
4. Switch the focus back to the Parent window

driver.switchTo().window(parentWindowHandle);

See below a sample script to work with multiple windows using window handles

```
Boolean bfound = false;
String strExpectedWindowtitle = "AdactIn Group;

//save the window handle of parent window for future reference

String parentWindowHandle = driver.getWindowHandle();

//Use for loop to iterate through all the window handles found

                    for (String handle : driver.getWindowHandles())
                    {
                              //Switch focus to window with Handle handle
                              driver.switchTo().window(handle);

                    // use gettitle method to get title of window
                       String stractualtitle = driver.getTitle().trim();

                    //check if the current window has title match

                       if (stractualtitle.equals(strExpectedWindowtitle))
                       {
                              System.out.println(" Window with Title Match found: " + stractualtitle);
                              bfound = true;
                         //close the driver window if needed
                             driver.close();
                              Thread.sleep(2000);

                           //break from for loop since we have found the window
                              break;
                       }

                    }
//giving focus back to parent window
                     driver.switchTo().window(parentWindowHandle);
```

Figure 18-8– Sample of Multiple Windows script

Explanation – The above code looks into the list of existing opened windows and matches the title of those windows with the expected Title.

It starts with the keep handle of the parent window for reference

```
String parentWindowHandle = driver.getWindowHandle();
```

Using the for loop it gets the handle to all of the open windows

```
for (String handle : driver.getWindowHandles())
```

Using switchto statement it switches focus to the current window given by the for loop

```
driver.switchTo().window(handle);
```

Uses getwindowtitle statement to get the window title and if the title matches with the expected string then the user continues the for loop or break

In the end, the focus is switched back to the Parent window handle

Note: You can also switch from frame to frame
driver.switchTo().frame("frameName");

It's possible to access subframes by separating the path with a dot, and you can specify the frame by its index too. See below

driver.switchTo().frame("frameName.0.child");

§

19

Working with Dynamic UI Objects

In most real-life applications, we find objects whose properties change at runtime. For instance, a Web browser will have a variable date/time stamp in the title or the window title will contain the id number of the latest order you have booked. These types of objects, whose properties change at run-time are called dynamic objects.

Since we are using a Shared UI Map and save the object properties statically in our map file, the WebDriver script will fail since the objects' properties are being generated at run time.

In this chapter we will see how WebDriver can handle these dynamic objects.

example

While testing a banking application which had account numbers generated at runtime, hyperlinks to those newly generated account numbers were created. As part of the test case steps, we had to click on the newly created account number link and modify the account details. So how do we deal with such objects like account number links which are dynamic and whose property values are generated at run time?

We will learn how to work with objects whose properties change dynamically in this chapter.

Key objectives:

- Understand Dynamic Objects: Create a Test Scenario
- Handling Dynamic Objects using programming
- Regular Expression alternative to handling Dynamic Objects

19.1. Understanding Dynamic UI Objects

Dynamic objects are objects whose property values change at run time. For example, there are some Web based applications, whose browser title includes the date and time stamp indicating when the browser window was opened. The title property of the browser will always be dependent on current date and time. How do we handle such objects using WebDriver?

Let us take an example from our application:

Test Scenario – Cancel a booked order.

Manual Steps

1. Login (Use the username/password with which you have registered earlier)

2. Search for a Hotel

    i. Select a location, e.g., Sydney
    ii. Select number of rooms, e.g., 2-Two
    iii. Select adults per room, e.g., 2-Two
    iv. Click on Search button

3. Select a Hotel

    v. Select one of the hotel Radio Buttons, e.g., select radio button next to Hotel Creek

4. Book a Hotel

    vi. Enter First Name - Test
    vii. Enter Last Name - Test
    viii. Enter Address - Test
    ix. Enter 16-digit Credit Card number -1212121212121212
    x. Enter Credit Card type – MasterCard
    xi. Enter Expiry Month - October
    xii. Enter Expiry Year - 2018
    xiii. Enter CVV number -111
    xiv. Click on Book Now

5. **In the Booking confirmation page, get the new Order No. generated**

6. Click on the Booked Itinerary link and go to row where you can see your booking order number or alternatively search for your Order No.

7. You should see the Cancel <Order Number> button. Use the ChroPath add-on to inspect the property value of this button.

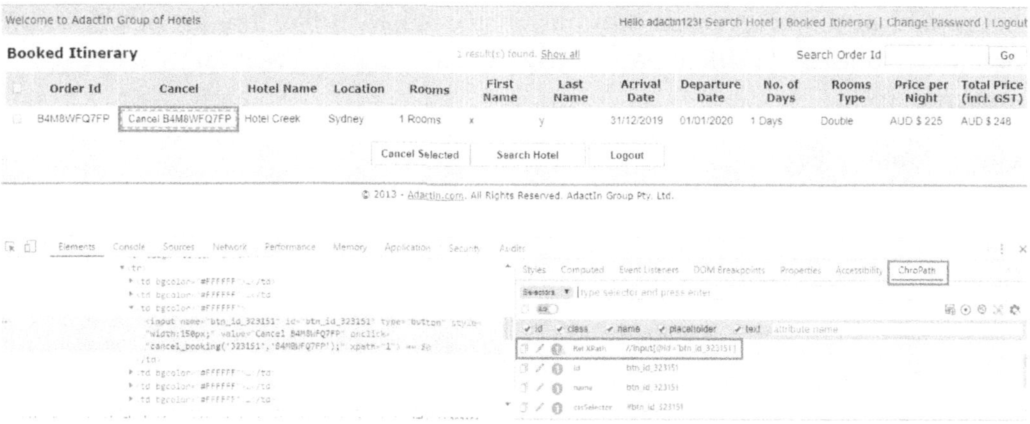

Figure 19- 1 – Cancel Order Button

You will notice that the id corresponding to this button (id='btn_id_323151']) is dynamic and will change for every order we submit.

Also look at the value = Cancel B4M8WFQ7FP which will change automatically with every new order that is inserted.

These types of objects are called dynamic objects because their properties (e.g. Id or Value) change based on when the test is executed.

How will WebDriver work with these objects? We will see this in the next section.

8. Click on the Cancel <Order Number> button and click OK on the Confirmation Prompt

9. Click on Logout and then on link "Click here to login again"

19.2. Handling Dynamic Objects using Programming

WebDriver Script for Cancelling a Booked Order

Let us enhance our existing WebDriver script to code for the steps for Cancelling a Booked order

1. Right click on your existing **SynchronizationTest.java** script, select **Copy**

2. Select the **tests** package folder, right click and select **Paste**

3. In the Name Conflict dialog box, enter name of the script as **DynamicUIObjectTest.java** and click **OK**

4. Double click on SharedUI map file **SharedUIPMap.properties** to see the object properties

Let us add new WebElements into our Shared UI Map file. The list of objects we need to add include (assuming we already have existing objects in our SharedUIMap. properties file):

Hotel Booking Page

1. My Itinerary Button

Booked Itinerary Page

1. Search Order ID text field
2. Go Button

See below snapshot of updated Shared UI Map file

```
# Referenced by Id Locator
Btn_BookingHotel_MyItinerary = my_itinerary

################## Booked Itinerary Elements ##########

# Referenced by Id Locator
Txt_BookedItinerary_SearchOrderid= order_id_text

# Referenced by Id Locator
Btn_BookedItinerary_Go = search_hotel_id

################## Logout Elements ##########
```

Figure 19- 2– Updated Shared UI Map

5. **Save** the Shared UI Map file
6. Double click on newly created "**DynamicObjectTest.java**" script to see the script

Note: Verify that your script has several seconds to get the order number value from the booking page.

7. We need to insert the below additional steps in the script -
   - Click on the **My Itinerary** link

```
driver.findElement(By.id(prop.getProperty("Btn_BookingHotel_MyItinerary"))).click();
```

   - Enter the value into the **Search Order No** field

```
driver.findElement(By.id(prop.getProperty("Txt_BookedItinerary_SearchOrderid"))).sendKeys(strOrderNo);
```

- Click on the **Go** button

```
driver.findElement(By.id(prop.getProperty("Btn_BookedItinerary_Go"))).click();
```

8. Now we need to click the **Cancel <Order No>** button. Using ChroPath we found that

Xpath for Cancel <Order No> = .//*[@id='btn_id_71825']

Figure 19-3– XPath for Cancel order Button

Notice that the id value 71825 is dynamic and we do not have any control on it. But property "value" has the value "Cancel JDJ22MCUPI" which we can use to uniquely identify the object.

So the Xpath value that we will use is .//*[@value= 'Cancel JDJ22MCUPI']

Note: You can confirm that the Xpath value is correct by going to Chropath interface and adding the generated Xpath in the Xpath field and then Press Enter. If you can see the object highlighted it means that your Xpath expression is correct. If you see document Interface shown without any specific object highlighted it means that the XPath is incorrect and does not matches any objects in the application.

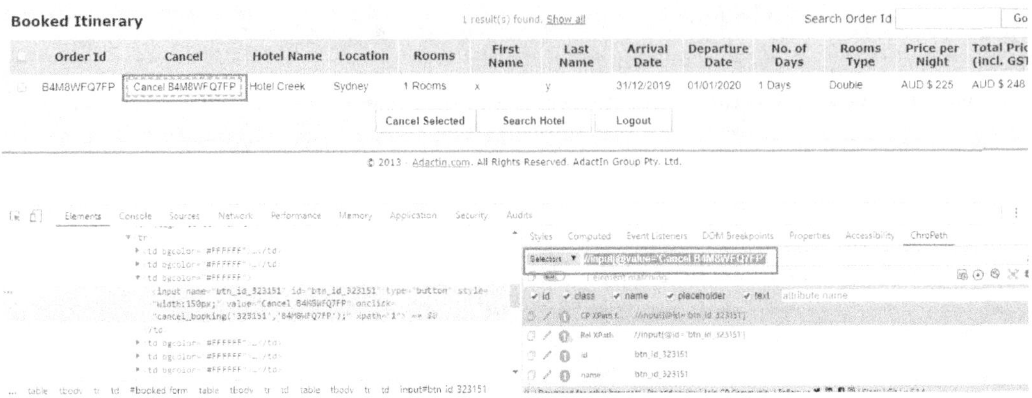

Figure 19-4– Correct XPath

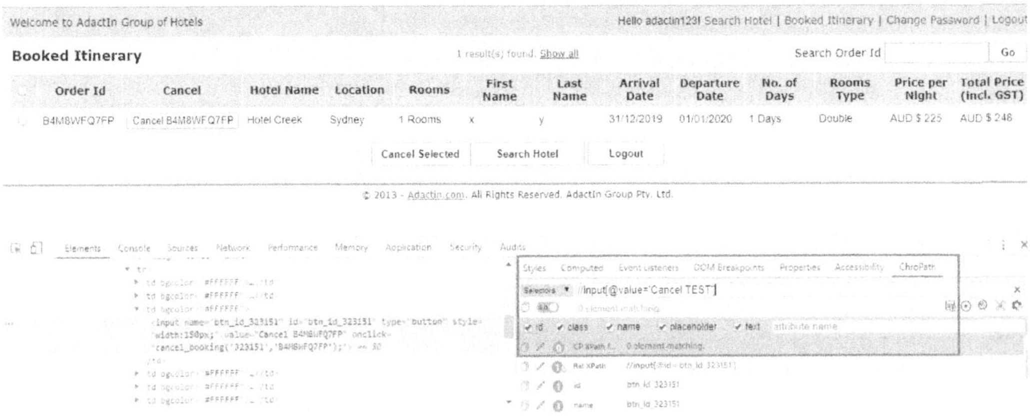

Figure 19-5– XPath Mismatch

Now as you know the Order No value is changing constantly so we will replace the hardcoded Order No with the value fetched using GetAttribute method from the Order No.

The statement we will use in the script will be

```
driver.findElement(By.xpath(".//*[@value='Cancel "+strOrderNo+"']")).click();
```

9. Press OK on the confirmation Prompt

```
Alert javascriptAlert = driver.switchTo().alert();
System.out.println(javascriptAlert.getText()); // Get text on alert box

javascriptAlert.accept();
```

The final script will look like this:

```
new Select(driver.findElement(By.id(props.getProperty("Lst_BookingHotel_CCType")))).selectByVisibleText("American Express");
new Select(driver.findElement(By.id(props.getProperty("Lst_BookingHotel_CCExpMonth")))).selectByVisibleText("March");
new Select(driver.findElement(By.id(props.getProperty("Lst_BookingHotel_CCExpYear")))).selectByVisibleText("2030");
driver.findElement(By.id(props.getProperty("Txt_BookingHotel_CCCvvNumber"))).clear();
driver.findElement(By.id(props.getProperty("Txt_BookingHotel_CCCvvNumber"))).sendKeys("111");
driver.findElement(By.id(props.getProperty("Btn_BookingHotel_BookNow"))).click();

Thread.sleep(10000);
String strOrderNo = driver.findElement(By.id(props.getProperty("Txt_BookingHotel_OrderNo"))).getAttribute("value");
System.out.println("Order Number Generate is "+ strOrderNo);

driver.findElement(By.id(props.getProperty("Btn_BookingHotel_MyItinerary"))).click();

driver.findElement(By.id(props.getProperty("Txt_BookedItinerary_SearchOrder_id"))).sendKeys(strOrderNo);

driver.findElement(By.id(props.getProperty("Btn_BookedItinerary_Go"))).click();

driver.findElement(By.xpath(".//*[@value='Cancel "+strOrderNo+"']")).click();

Alert javascriptAlert = driver.switchTo().alert();
System.out.println(javascriptAlert.getText()); // Get text on alert box
javascriptAlert.accept();

driver.findElement(By.linkText(props.getProperty("Lnk_BookingHotel_Logout"))).click();
driver.findElement(By.linkText(props.getProperty("Lnk_Logout_ClickHeretoLoginAgain"))).click();
```

Figure 19-6– Final Script

10. Save the test and perform Project clean-up using **Project → Clean Up…** and clean all selected projects.Select and Run the test using **Run → Run As JUnit Test**. Check that the script run passes

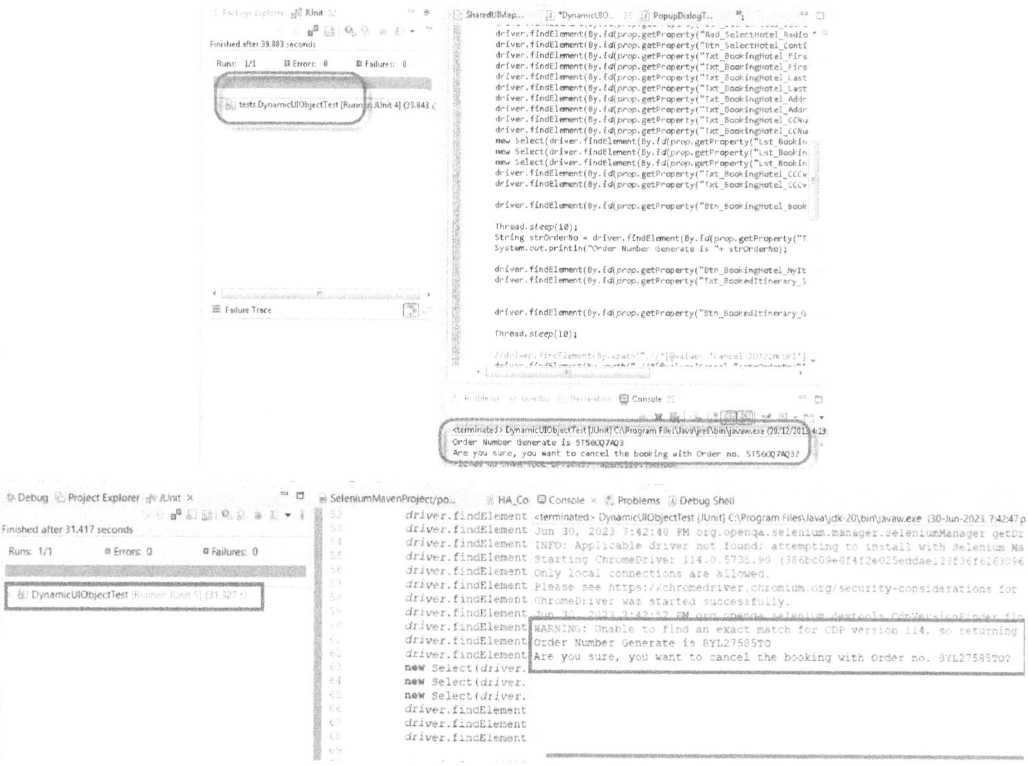

Figure 19-7– Test Results

19.3. Handling Dynamic Objects using Partial Match

Another way of handling dynamic objects is using Partial Match. A partial match describes or matches a set of strings. It is often called a pattern match.

We can use the statement below for a partial match in Xpath

Syntax	**Example**	*Description*
startswith()	.//[starts-with(@id,'btn_id_')]	For example, if the ID of an element is btn_id_786, this will locate and return elements with btn_id_at the beginning of the ID
endswith()	.// [ends-with(@id,'_btn_id')]	For example, if the ID of an element is 786_btn_id, this will locate and return elements with _btn_id at the end of the ID.
contains()	.//[contains(@id, 'btn_id_)]	For example, if the ID for an element is Test_ btn_id_786, this will use the btn_id_part in the middle to match and locate the element.

Figure 19-8– Partial Match Patterns

In our Scenario to match the Xpath statement

Xpath for Cancel <Order No> = .//*[@id='btn_id_323151'] or .//*[@value= 'Cancel B4M8WFQ7FP']

We could use any of the following partial matches to match the dynamic object

.//*[@id='btn_id_323151']	.//*[contains(@id,'btn_id_')]
	.//[starts-with(@id,'btn_id_')]
.//*[@value= 'Cancel B4M8WFQ7FP']	.//*[contains(@value,'Cancel')]
	.//*[starts-with(@value,'Cancel')]

Figure 19-9– Examples of Partial match

Let us modify our existing script to use Regular Expression

1. Double click on the newly created "**DynamicObjectTest.java**" script so that you can see the script

2. Comment the existing statement when we click on Cancel <Order No> button and copy/paste it as a separate statement
3. Modify the statement to show as below based on options highlighted above

   driver.findElement(By.xpath(".//[contains(@id,'btn_id_')]")).click();*

```
driver.findElement(By.id(prop.getProperty("Btn_BookingHotel_BookNow"))).click();

Thread.sleep(10);
String strOrderNo = driver.findElement(By.id(prop.getProperty("Txt_BookingHotel_OrderNo"))).getAttribute("value");
System.out.println("Order Number Generate is "+ strOrderNo);

driver.findElement(By.id(prop.getProperty("Btn_BookingHotel_MyItinerary"))).click();
driver.findElement(By.id(prop.getProperty("Txt_BookedItinerary_SearchOrderid"))).sendKeys(strOrderNo);


driver.findElement(By.id(prop.getProperty("Btn_BookedItinerary_Go"))).click();

Thread.sleep(10);

//driver.findElement(By.xpath(".//*[@value='Cancel JD122MCUPI']")).click();
//driver.findElement(By.xpath(".//*[@value='Cancel "+strOrderNo+"']")).click();
driver.findElement(By.xpath(".//*[contains(@id,'btn_id_')]")).click();

Alert javascriptAlert = driver.switchTo().alert();
System.out.println(javascriptAlert.getText()); // Get text on alert box
javascriptAlert.accept();
```

Figure 19-10– Partial match used in script

11. Perform **Project → Clean**... and **Save** the test, Select and run the test using **Run → Run As JUnit Test.** Check that the script runs and passes.

Note: Partial match should be used after generic recognition of an object WebDriver finds only one instance of the object. If WebDriver finds more than one object matching the string specified by WebDriver partial match, then we cannot use partial match. In that case we would have to use programming. For example, in our above exercise to delete an order, if we use partial match to identify the cancel order No. button (e.g., Cancel.*), it will recognize all the Cancel <Order No> buttons, not just the one which we have booked (assuming that our Itinerary screen will show multiple orders). So we cannot use regular expression in the above exercise as WebDriver will not be able to identify the unique object. The best option in the above scenario would be to use programming to perform an exact match.

§

20

Multiple Choice Questions Set-2

1. Select the method used to Assert and compare two strings in a JUnit WebDriver script
   A. assertTrue
   B. assertNotNull
   C. assertEquals
   D. assertSame

2. Select the method used to read data from the Properties file
   A. getProperty
   B. getAttribute
   C. getVal
   D. getSeleniumVal

3. Which is the preferred method for using objects in WebDriver?
   A. Local Web elements
   B. Shared Web elements

4. Which statement is used to return data from a function?
   A. Function name
   B. Return
   C. Returnval
   D. FunctionReturn

5. Which of the statements is used for static wait?
   A. Thread.sleep
   B. wait
   C. sleep
   D. All of the above

6. Which class is used for Explicit Wait in WebDriver?
   A. Wait
   B. WebDriverSleep
   C. Thread.sleep
   D. WebDriverWait

7. What class of object is used to handle a confirmation pop-up?
   A. Pop-up
   B. Alert
   C. Alertwindow
   D. Childwindow

8. Which method in WebDriver is used to move the focus to a Child window of the browser?
   A. MoveTo
   B. SwitchTo
   C. FocusTo
   D. SetFocus

9. Which is a valid Regular Expression statement to handle a dynamic id with format "txt.Book.4345"?
   A. .//*[contains(@id,'Book.4345')]
   B. .//[starts-with(@id,'Book.')]
   C. .//*[contains(@id,'txt.Book.')]
   D. .//[starts-with(@id,'txt.4345')]

10. Which method is used by Selenium WebDriver to get the Handle of a Parent window?
   A. driver.getHandle
   B. driver.getParentWindowHandle
   C. driver.getWindowHandle
   D. driver.Browserhandle

Answers

Q1. Answer: C
Explanation – assertEquals compares that two string values are the same.

Q2. Answer: A
Explanation – getProperty method is used to get the property value from the Properties file. We specifically use this method when we use SharedUIMap.properties or Configuration.Properties files.

Q3. Answer: B
Explanation – Shared Web elements is a better way to store object information instead of Local map. Shared Web elements present one central location for saving object properties making automation scripts more maintainable.

Q4. Answer: B
Explanation – Return statement is used to return values from the function.

Q5. Answer: A
Explanation – Thread.sleep(waittime) is used for static wait in WebDriver.

Q6. Answer: D
Explanation – WebDriverWait class is used for Explicit Wait in WebDriver. Alternatively you could also use custom wait methods to wait for objects' property values.

Q7. Answer: B
Explanation – Alert Class is used to handle windows driven confirmation pop-ups.

Q8. Answer: B
Explanation – SwitchTo method is used to move focus to Child windows in WebDriver.
Q9. Answer: C

Explanation – Since value 4345 will keep varying and only part "txt.Book." is constant statement .//*[contains(@id,'txt.Book.')] is the correct answer.

Q10. Answer: C
Explanation – getWindowHandle is used to get handle of the Parent or any other window. Before switching focus to the Child window, we use getWindowHandle to store the handle id for the Parent window so that once an action on the Child window is finished the focus can be shifted back to the Parent window.

§

21

Debugging Scripts

I believe one cannot be an expert in any tool if he or she does not know how to debug issues, or does not know how to troubleshoot problems encountered while using the tool.

Debugging is an integrated feature of any automation tool. Automation scripts do fail, and we should be able to pinpoint exactly where the issue is, so that it can be fixed.

WebDriver within Eclipse IDE comes with some nifty debugging features, which should be used while isolating reasons for failure of scripts. We would want to have the ability to execute one step at a time, or pause at a particular step or be able to peek at values of variables at runtime. All these features can be found within Eclipse.

example

At one of our telecom clients, they had around 150 automation scripts, and they were not executable after a new build was released to the test team. The core automation team had left after the release and nobody knew how to fix the scripts. The only way we could figure out what went wrong with the script - and understand the application workflow - was by using the debugging features. This proved invaluable to us in getting the scripts up and running again!

In this chapter we will learn how to debug WebDriver scripts.

Key objectives:

- Debugging features
- Execute tests in debug mode
- Step commands, Variables and Watches

21.1. Debugging Features

Debugging allows you to run a program interactively while watching the source code and the variables during the execution. Basically, debugging is a way to pause the execution of a program so that we can examine its internals at that point in time to deduce what is going wrong. To that end, we have a few basic notions:

Breakpoints: These are locations in the code that we can specify where code execution will pause and we can examine the execution environment and gain manual control over the execution process. Setting a breakpoint in Eclipse is as simple as right-clicking on the desired line of code and selecting Toggle Breakpoint. We will see how to insert a Breakpoint in the next section. In support of breakpoints are two main control buttons:

- **Run**-- execute the code as normal until the next breakpoint is encountered
- **Stop**-- terminate the program execution completely

Execution Stepping: This manual control over the execution process allows us to advance the execution one line of code at a time. Eclipse provides convenient buttons for stepping control. There are generally 3 ways of stepping that one uses most often:

- **Step Into**-- Advance to the next line of code, following the execution path into every method that is executed
- **Step Over**-- Advance to the next line of code, but do not go into any methods that are encountered
- **Step Out**-- Advance to the next line of code following the end of the current method in the calling method, i.e., finish the current method

Variables Inspection: Once execution has stopped by a breakpoint, we can examine all the variables, fields and objects that are in scope (visible) at that moment. Eclipse has a very nice "object inspector" for this job.

Expressions Inspection – Apart from variables you might also want to see results of expressions. For e.g., the Variables tab will show you the value of variables 'a' and 'b', but if you want to evaluate and view the value of expression "a+b", we can see this using Expressions features in Eclipse. Expressions are also referenced using the term "Watch" in programming tools.

Basic Debugging Procedure

There is really no "one right way" to debug. Debugging is as much an art as it is a skill. One has to think like a detective, looking for clues and applying deductive reasoning to explain what you've found. Always remember that the location, at which bugs manifest themselves as some sort of visible effect, is often not where the problem actually occurs. You must always keep an understanding of how all the pieces in your system relate to each other.

Most debugging scenarios can be broken down into these steps:

1. Set breakpoints at key locations in your code. Typical places to put breakpoints include:
    - Wherever the problem clearly manifests itself. If possible, put the breakpoint a few lines before the problematic line so you can step up to it in a controlled manner.

- Wherever the key objects are constructed or key relationships are being established. For instance, null pointer errors are always due to calling a method on an object reference that is null, i.e., it was never assigned or instantiated properly.

2. Examine all variables that are visible at the breakpoint location and make sure that their values are what you expect them to be.
3. Step slowly through your code, checking all your variables as you go. Watch for the unexpected! Remember, if your code did what you expected, it would have run without any errors.
4. Use Step Over and Step Return (Step Out Of) only when you are positive, i.e., you already did a Step Into at least once already, that everything that is being skipped works properly. It is very common to get over-confident and miss where an error occurs because one has skipped right over the critical code.
5. Add more breakpoints when you see something amiss and you deduce that an error might be due to potential problems in another section of the code. You can use the "Resume" button to quickly advance to the next breakpoint.

Using breakpoints in the source code, you specify where the execution of the program should pause. Breakpoints and watch points can be summarized as pause points. Once the program is paused you can investigate variables, change their content, etc.

21.2. Run Tests in Debug mode with Breakpoints

In this section we will execute one of our scripts in Debug mode with Breakpoints.

1. **Right click** on your existing **SharedUIMapTest.java** script, select **Copy**
2. Select the **tests** package folder, **right click** and select **Paste**
3. In the Name Conflict dialog box, enter the name of the script as **DebugModeTest.java** and click **OK**
4. Double click on **DebugModeTest.java** to see the script

Breakpoints

5. At the step where the user selects location ("Sydney") insert a breakpoint. To insert a breakpoint
   a. Double click on the leftmost vertical bar next to the line of code at which you want to insert the Breakpoint

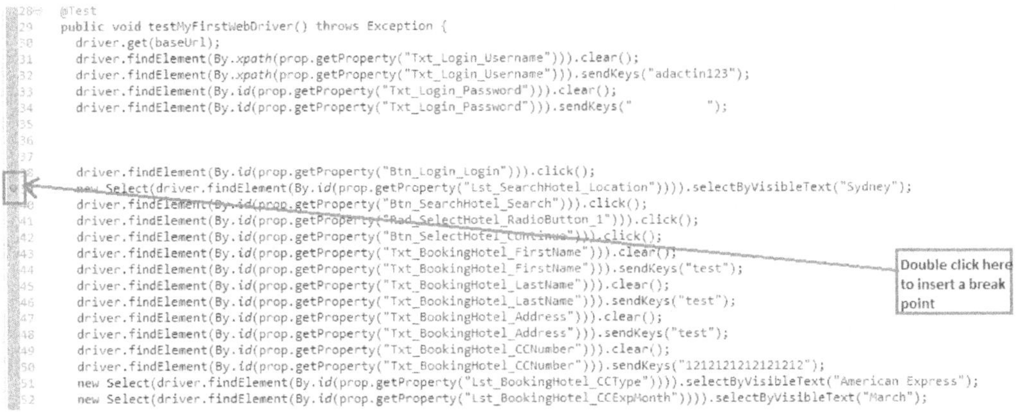

Figure 21-1 – Breakpoint Insertion

b. Alternatively, select the step at which you would want to insert the Breakpoint and select **Run → Toggle Breakpoint**

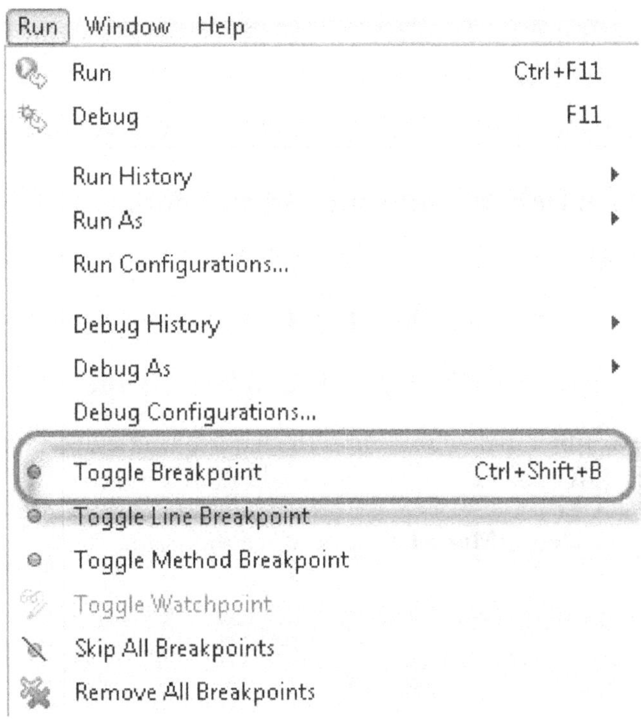

Figure 21-2 – Toggle Breakpoint

Debug Mode Run

6. Once the Breakpoint is inserted, let us execute the script in Debug mode. Select your script, right click and select **Debug As → JUnit Test.**

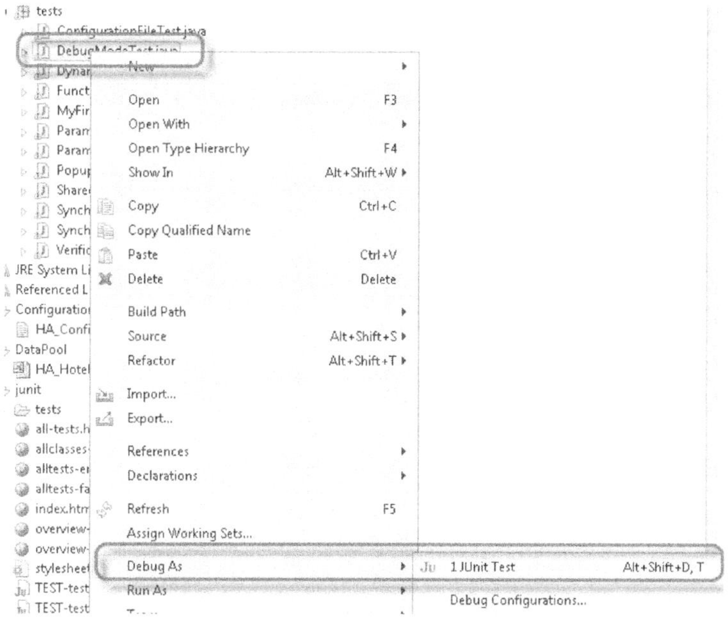

Figure 21-3 – Debug Mode Run

Note: If we run the script in normal Run mode (**Run As → JUnit Test**) our script would not pause at the breakpoint. It is only in Debug mode run that the script will pause at the breakpoint.

7. Once the test starts running you will notice that the script will pause before executing the Breakpoint step and will throw a "**Confirm Perspective Switch dialog**".

8. Press **Switch** on Confirm Perspective Switch dialog.

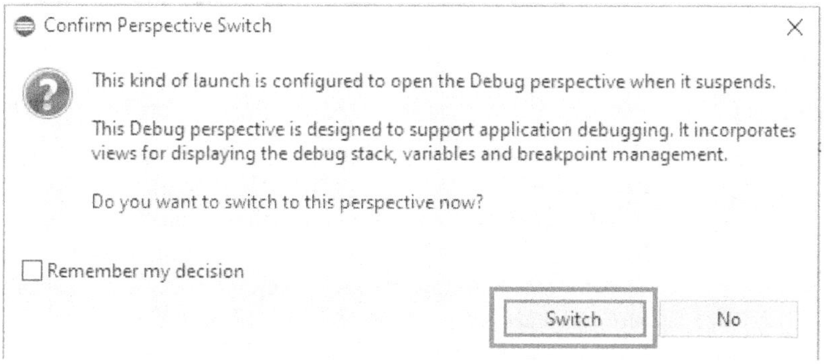

Figure 21-4 – Perspective Switch dialog

9. You will notice Eclipse switches to debug perspective as in the snapshot below.

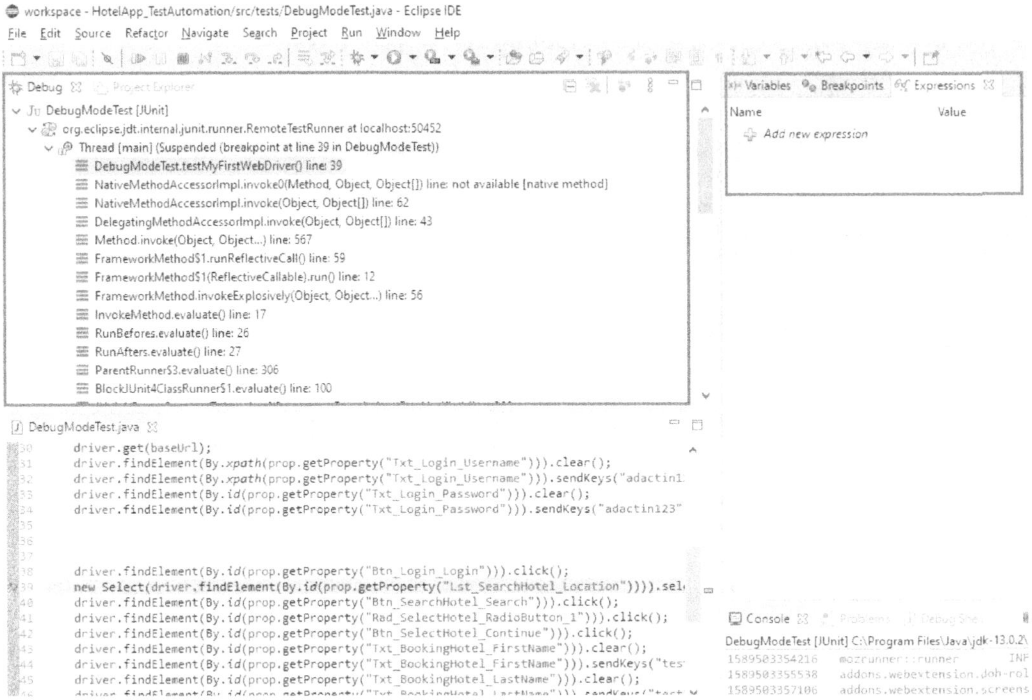

Figure 21-5 – Debug Mode Perspective

10. Go to your sample application and you will notice that execution is halted and the application is on the Search Hotel Page after login

11. Click on the **Run icon** at the top left of the window

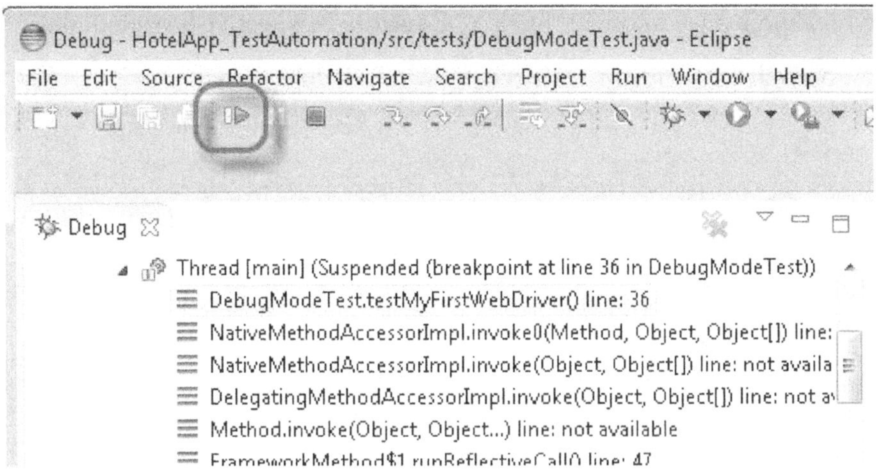

Figure 21-6 – Run Icon

12. You will notice that the test runs until completion

Note: Breakpoint insertion is useful when you are not sure of the state of your application at a particular step of your test run. After the script pauses at the breakpoint, the user can either choose to run the script until completion or run one step at a time.

13. To go back to the main view, click on the **Java** button in the top right corner

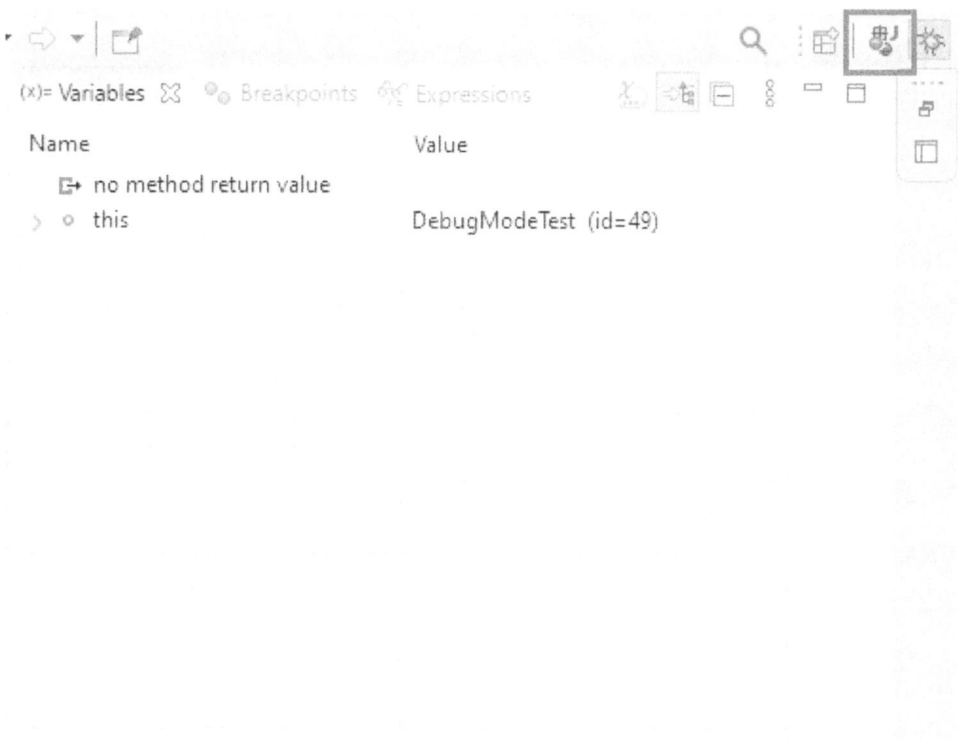

Figure 21-7- Java button to go to Main view

14. Click **Package Explorer**

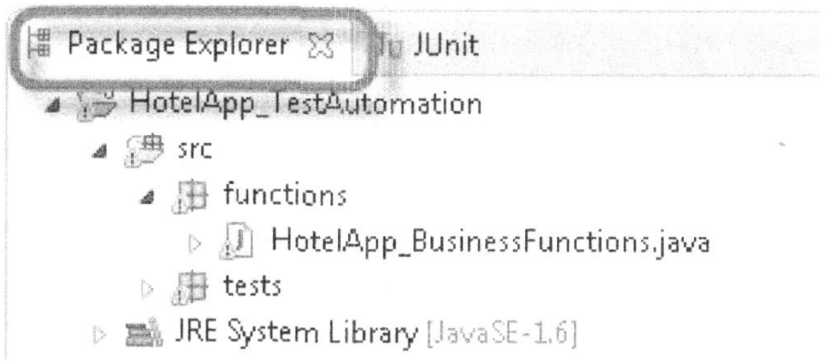

Figure 21-8- Package Explorer View

21.3. Step Commands, Variables and Watch

Another benefit of running a script in debug mode is that the user can run a single step at a time which helps him to debug a specific part of the code which is causing the issue. We can achieve this using Step Into, Step Over and Step Out commands.

Also, as we are running our scripts at each step we would like to see values stored for variables used within the script or alternatively values evaluated for Expressions. We can use the Variables and Expressions tabs for that.

In the following steps we will see how to use Step commands and also how to see variable values.

1. Double click on **DebugModeTest.java** to see the script
2. Verify that there is a breakpoint at the step where the user selects location ("Sydney"). If not, insert a breakpoint as mentioned in previous section.
3. Once Breakpoint is inserted, let us execute the script in Debug mode. Select your script , right click and select **Debug As → JUnit Test**
4. Once the test starts running you will notice that the script will pause before executing the Breakpoint step and throws a "**Confirm Perspective Switch dialog**". Press **Yes** on **Confirm Perspective Switch** dialog.
5. You will notice Eclipse switches to debug perspective. Now let's run one step at a time using Step Into, Step Over and Step Out option.

   Step Into (F5) -- Advance to the next line of code, following the execution path into every method that is executed

   Step Over (F6) -- Advance to the next line of code, but do not go into any methods that are encountered

   Step Return(F7) -- Advance to the next line of code following the end of the current method in the calling method, i.e., finish the current method

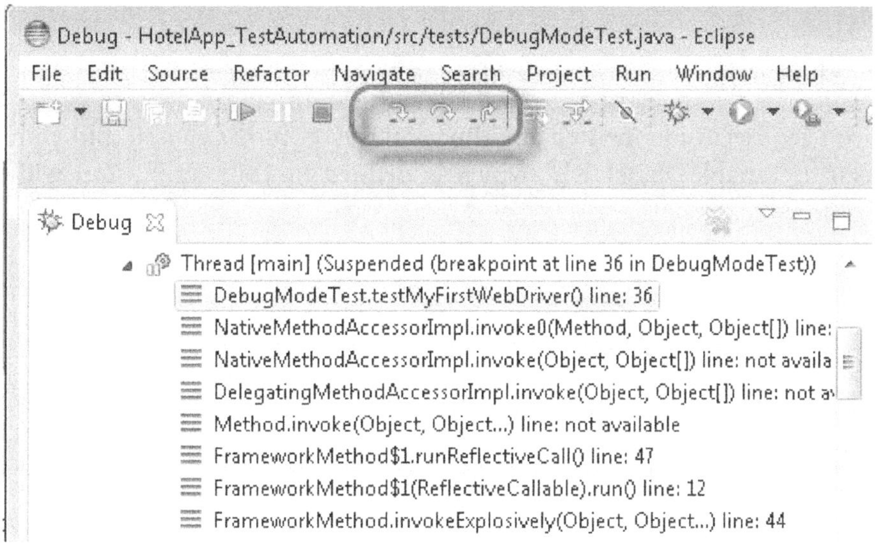

Figure 21-9- Step Options

6. Let us use Step Over (F6) command to run over each step without getting into underlying methods. Click **Step Over icon** or press **F6**.
7. You would notice that the script has moved on to the next step and has again paused waiting for user action

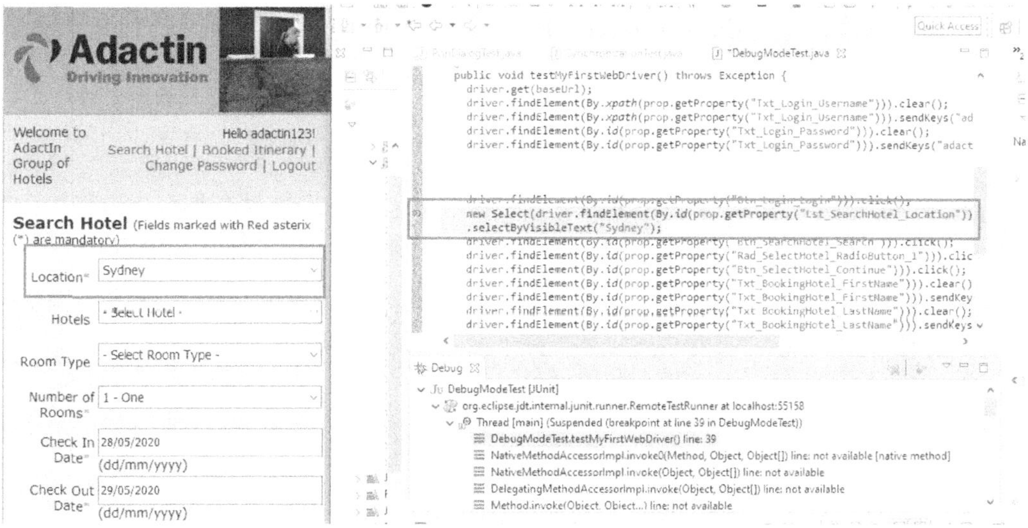

Figure 21-10- Step Over

8. You can continue to press F6 (Step Over) and move the script one step at a time.

Variables

Let us see how to look for variable values

9. As you are executing one step at a time and the script has paused, put your mouse cursor at variable "baseURL" which is defined in your script. You will see the variable value.

Figure 21-11- Variable Value

10. Alternatively, select the **Variables** tab in the top left pane

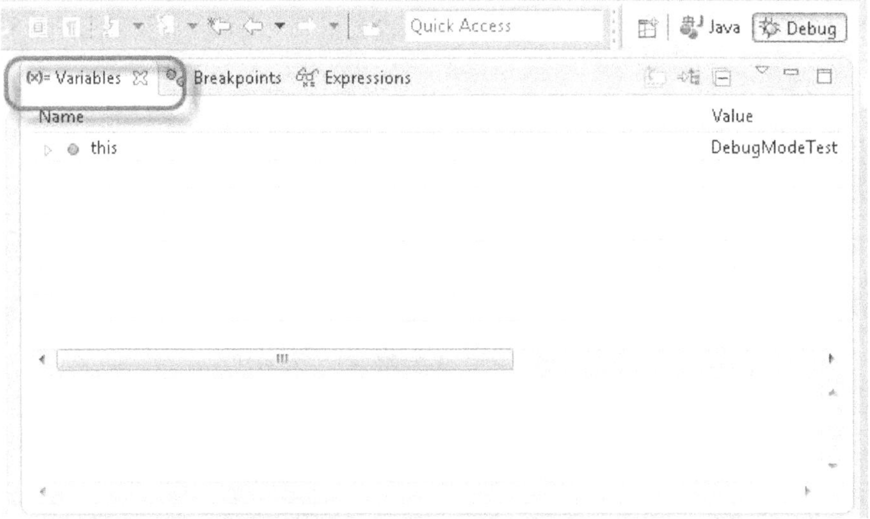

Figure 21-12- Variables Tab

11. Expand **this** node and select **baseUrl** variable

Expressions

Let us see below how to look for Expression values

12. As you are executing one step at a time and the script has paused, select the expression **baseUrl** which is defined in your script
13. Right click and select **Watch**

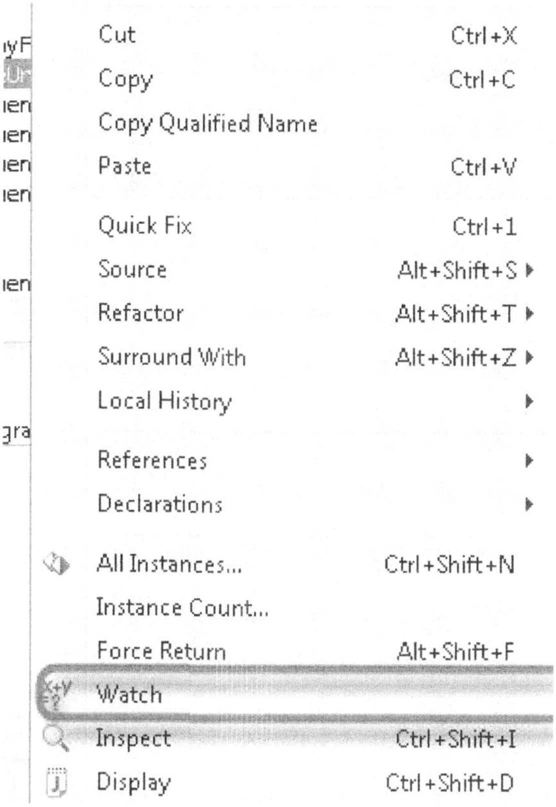

Figure 21-14- Select Watch

14. Click on the Expressions tab in the top right pane and notice that the expression has been added to the list

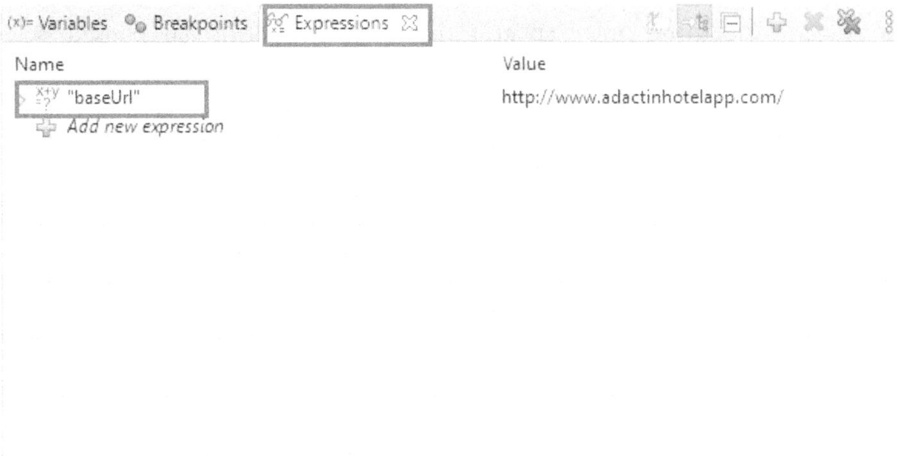

Figure 21-15- Expressions Tab

15. Expand the expression and note the expression value

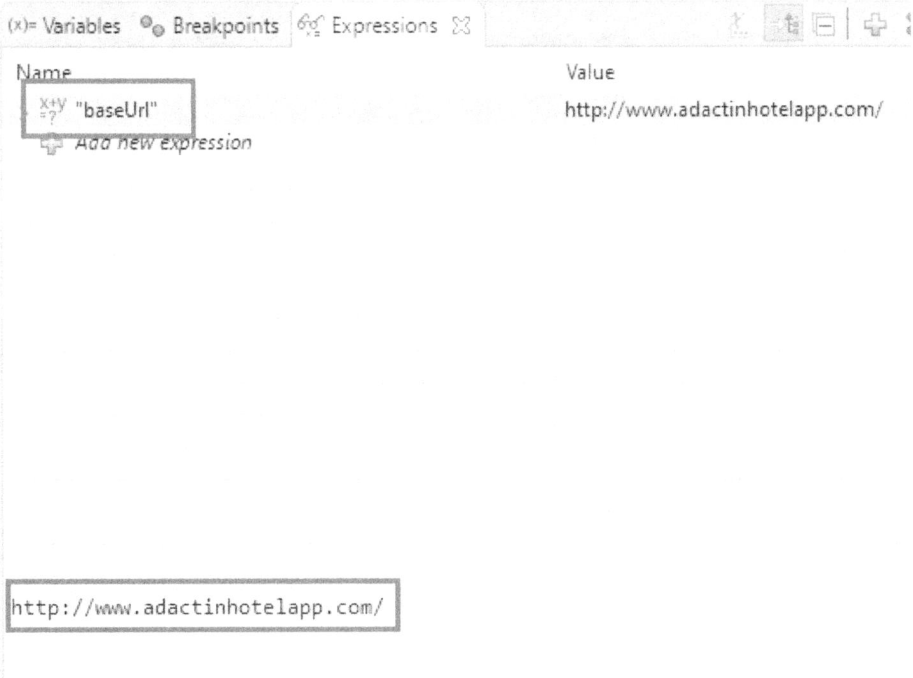

Figure 21-16- Evaluated Expression value

16. Finish the script by clicking on the **Run** icon. Save the script.

EXERCISE

1. Try running the ConfigurationFileTest.java script in step by step mode. Insert breakpoints and verify the value of variables returned and stored in the variable sAppURL.
2. Also look for values returned for expressions "prop.getProperty("Lst_SearchHotel_Location")"

§

22

Exception Handling in WebDriver

We have already learned about exceptions and their handling techniques in the chapter on Basics of Java. An exception is an event which occurs during the execution of a program that disrupts the normal flow of the program's instructions.

Exception handling in Selenium is also a crucial exercise. Most of the time when a selenium script fails, it is because it has landed into an exception. The cause could be anything like:

- Element not found
- Couldn't click the element
- Element not visible

The moment the driver comes across an exception it will halt the test. So it's important for a tester to foresee these exception conditions and handle them according to the script or test requirements. This way the script failures are contributed to failures of test conditions and not to unhandled code exceptions. So, we have a bug corresponding to every test failure- which is our ultimate goal.

To catch an exception we first put the code which we suspect will throw an error into a try block like

```
WebElement txtbox_username = driver.findElement(By.id("username"));
try{
if(txtbox_username.isEnabled()){
txtbox_username.sendKeys("adactin123");}
 }
 catch(NoSuchElementException e){
System.out.println(e.toString());}
```

This is followed by a catch block of code where we tell the system what should be done when the exception occurs. Generally this is where we display the message of the exception object so that we know which exception has occurred and why.

22.1. Handling WebDriver Exceptions

In WebDriver we can use **try-catch** blocks or the **throws** statement with the purpose of handling the exceptions. The key point is that the exceptions we are catching here are Selenium exceptions rather than Java exceptions.

Test Scenario – If we provide an invalid username and password to the Login function, then the script should exit gracefully with a message.

Let us follow the steps to implement the above scenario.

1. Now what visual cue tells us that a user is logged in to the application? There can be multiple visual cues but let us pick one of them being a welcome message.

Figure 22-1 – User Welcome message

2. Let's use **ChroPath** to get its locator value

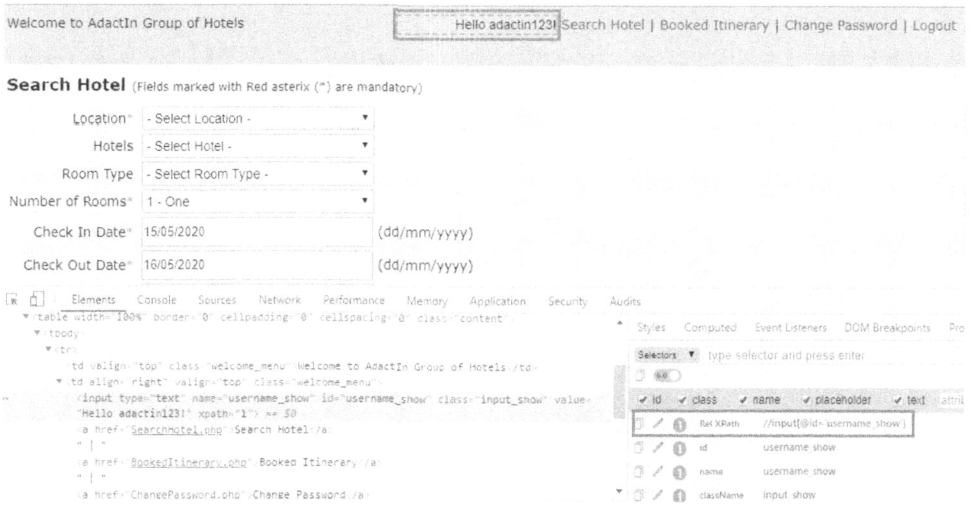

Figure 22-2 – Locator value for Welcome message

You see in the above snapshot the value for locator id is **username_show**

3. Add this to our SharedUIMap.properties file for further use

Figure 22-3 – Welcome message added to Shared UI Map

4. Double click and open our HotelApp_BusinessFunctions.java script to view our existing login function

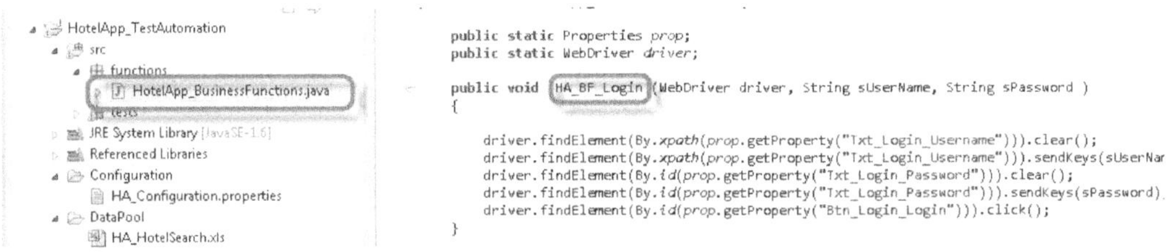

Figure 22-4 – Existing Login Function

5. Modify the function as given below to handle a successful or unsuccessful login

```
public void HA_BF_Login (WebDriver driver, String sUserName, String sPassword ) throws
InterruptedException{

    // Provide user name.
        driver.findElement(By.xpath(prop.getProperty("Txt_Login_Username"))).clear();
    driver.findElement(By.xpath(prop.getProperty("Txt_Login_Username"))).sendKeys(sUserName);

    // Provide Password.
    driver.findElement(By.id(prop.getProperty("Txt_Login_Password"))).clear();
    diver.findElement(By.id(prop.getProperty("Txt_Login_Password"))).sendKeys(sPassword);

    // Click on Login button.
    driver.findElement(By.id(prop.getProperty("Btn_Login_Login"))).click();
    Thread.sleep(4000);

    // Verify for welcome message.
    WebElement welcomeTxt = driver.findElement(By.id(prop.getProperty("Lbl_SearchHotel_
    WelcomeMessage")));
    String text = welcomeTxt.getAttribute("value");

    if(text.contains("Hello "+ sUserName))
        System.out.println("Login Test Pass for: "+ sUserName);

    else
        System.out.println("Login Test Fail for: "+sUserName);

}
```

Figure 22-5 – Updated Login Function

6. **Right click** on your existing **FunctionCallingTest.java** script, select **Copy**
7. Select the **tests** package folder, right click and select **Paste**
8. In the Name Conflict dialog box, enter the name of the script as **ExceptionHandingTest.java** and click **OK**
9. Double click on **ExceptionHandingTest.java** to see the script
10. Modify the Function call in the test to call the function with an invalid username and invalid password

```
//Comment - Call to Login Function
HA_BF_Login(driver, "InvaldUser", "InvalidPassword");
```

Figure 22-6 – Call to Login Function

11. Save the test and perform Project clean-up using **Project → Clean Up… and clean all** selected projects

12. Run the Test by right clicking on the test and **selecting Run As → JUnit Test**

Our expectation would be that the system should print " **Login test Fail for: InvalidUser**"

13. But if you notice the results script fails with "**NoSuchElementException**" instead of throwing a valid message

Figure 22-7 – Unable to locate element Exception

It is because the control never goes to the else part of the condition. Whenever you try log in using invalid user details the browser will still display the login page only.

Figure 22-8 - Invalid Login error

While the next line of our code is looking for a page element which is displayed on the home page of the application. This leads to a **NoSuchElementException** Exception and any line written after the findElement statement is not executed.

How do we resolve this gracefully?

14. We'll catch and report this exception and end the script gracefully.

Let's see the code for that

```
public void HA_BF_Login (WebDriver driver, String sUserName, String sPassword ) throws InterruptedException{

        // Provide user name.
                driver.findElement(By.xpath(prop.getProperty("Txt_Login_Username"))).clear();
driver.findElement(By.xpath(prop.getProperty("Txt_Login_Username"))).sendKeys(sUserName);

        // Provide Password.
        driver.findElement(By.id(prop.getProperty("Txt_Login_Password"))).clear();
                driver.findElement(By.id(prop.getProperty("Txt_Login_Password"))).sendKeys(sPassword);

        // Click on Login button.
        driver.findElement(By.id(prop.getProperty("Btn_Login_Login"))).click();
        Thread.sleep(4000);

        // Verify for welcome message.
        try
                {

                WebElement welcomeTxt =
driver.findElement(By.id(prop.getProperty("Lbl_SearchHotel_WelcomeMessage")));
                String text = welcomeTxt.getAttribute("value");

                        if(text.contains("Hello "+ sUserName))
                        {
                                System.out.println("Login Test Pass for: "+ sUserName);
                                return;
                        }
        }
        catch( Exception e)
        {
          e.printStackTrace();
          System.out.println("Login Test Fail for: "+sUserName);
          // to show the results as fail.
          assert false;
        }

        }
```

Figure 22-9 – Login function with Try-Catch statements

In the above code we have inserted the exception prone code in a try block followed by a catch block to report proper results and not end poorly.

PrintStacktrace is used to print out the program's execution stack.

When an exception is encountered and the printStackTrace() method is called, the contents of the execution stack are printed out to standard output (such as the command line or screen) so that the developer or user can figure out which class, method and line caused the exception.

Note: As in the above example, we should add Try-Catch blocks for all of our scripts and functions.

15. Now if we see the JUnit report, it comes out with a genuine message that Login Failed.

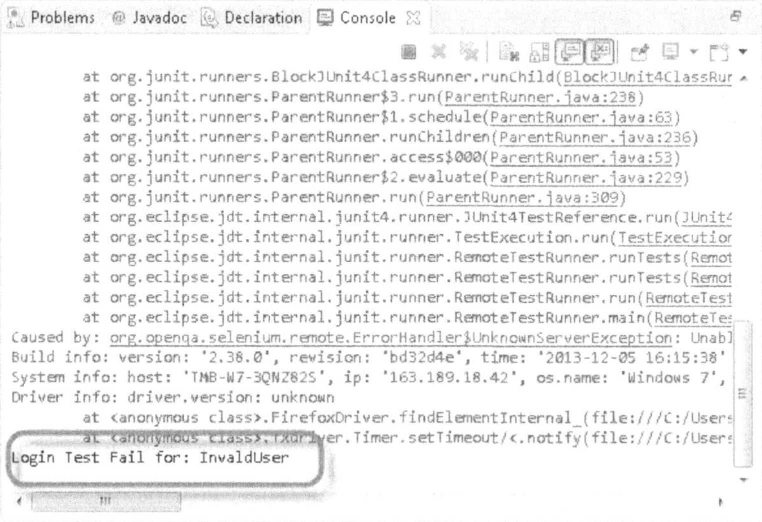

Figure 22-10 – Exception Trace

22.2. Handle Specific Exceptions

Note that apart from the generic way of handling exceptions we can also catch specific exceptions using multiple catch blocks.

```
catch(BiffException be)
   {

         System.out.println("The given file should have .xls extension.");
   }
   catch(Exception e)
   {
         e.printStackTrace();

   }
```

Figure 22 -11 – Multiple Exception Blocks

The above example shows that we could use a specific catch block to handle BiffException and other generic exceptions can be handled in other common Catch blocks.

22.3. Common WebDriver Exceptions

Let us see some of the common WebDriver exceptions.

Common WebDriver Exceptions

Exception Type	Description
ElementNotSelectableException	Thrown when trying to select an unselectable element.
ElementNotVisibleException	Thrown when an element is present on the DOM, but it is not visible, and so cannot be interacted with. Most commonly encountered when trying to click or read text of an element that is hidden from view.
InvalidSwitchToTarget Exception	Thrown when frame or window target to be switched doesn't exist.
NoAlertPresentException	Thrown when switching to no presented alert. This can be caused by calling an operation on the Alert() class when an alert is not yet on the screen.
NoSuchAttributeException	Thrown when the attribute of element could not be found. You may want to check if the attribute exists in the particular browser you are testing against. Some browsers may have different property names for the same property.
NoSuchElementException	Thrown when element could not be found. If you encounter this exception, you may want to check the following: • Check your selector used in your find_by... • Element may not yet be on the screen at the time of the find operation,
TimeoutException	Thrown when a command does not complete in enough time.
UnexpectedAlertPresent Exception	Thrown when an unexpected alert appears.
WebDriverException	Base WebDriver exception.
StaleElementReference Exception	A Stale Element Reference Exception is a runtime error that occurs when the state of an element has changed while your code was in the middle of using it.

Figure 22 -12 – WebDriver Exceptions

EXERCISE

1. Add Try-Catch Blocks for all the functions we have created so far (if they do not have Try-Catch Blocks already), specifically
    - HA_GF_readXL
    - HA_GF_XLRowCount
    - HA_GF_WaitForElementPresent
2. Add a breakpoint at e.printStackTrace() Step and verify that code gets into Catch Block by running script in Debug mode.

§

23

Reporting in Selenium

One of the very important features of a test automation solution is its reporting structure. After test execution we inspect the test report for results and defect detection. Selenium does not have its own mechanism for reporting results. Rather, it allows the automation tester to build their own reporting structure, customized to their needs, using features of the programming language of your choice.

As part of this section, we are going to try to understand Test Framework Reporting tools

Key objectives:

- Test Framework Reporting Tools
- Configuring Junit HTML Report
- Configuring TestNG reports
- Custom Excel or Database reports

23.1. Test Framework Reporting Tools

Building your own reporting structure! It's great! But what if you simply want something quick that's already done for you? Often an existing library or test framework can meet your needs faster than developing your own test reporting code.

Test frameworks are available with all programming languages. Along with their primary function of providing a flexible test engine for executing your tests, they also include library code for reporting results. For example, Java has two commonly used test frameworks, JUnit and TestNG. .NET also has its own, NUnit.

What's The Best Approach?

Most people new to testing frameworks will begin with the framework's built-in reporting features since that's less time consuming than developing your own.

As you begin to use Selenium no doubt you will start putting in your own "print statements" for reporting progress. That may gradually lead to you developing your own reporting, possibly in parallel to using a library or test framework. Regardless, after the initial, but short learning curve, you will naturally develop what works best for your own situation, existing testing frameworks or your custom framework.

Test Reporting Examples in Java

We'll direct you to some specific tools supported by Selenium. The ones listed here are common and have been used extensively (and therefore recommended).

- If Selenium Test cases are developed using JUnit then JUnit Report can be used to generate test reports. To use JUnit Report you would need to integrate Eclipse with ANT.
- If Selenium Test cases are developed using TestNG then no external task is required to generate test reports. The TestNG framework generates an HTML report which list details of tests.

Advantages of using these frameworks:

- Very good reporting structure is available
- Can generate XML, HTML reports
- There are options available to create test methods, test suites, etc.
- Utilizes Selenium IDE or ChroPath to record test scripts

Disadvantages:

- We will not be able to define our own reporting format

23.2. Configuring JUnit HTML Reports

To create a JUnit based HTML report we will be using ANT with Eclipse. We had installed WinANT as part of the earlier setup components chapter.

ANT - Apache Ant is a Java based build tool from Apache whose aim is to drive processes described in build files as targets and extension points dependent upon each other. The main known usage of Ant is the build of Java applications. Ant supplies a number of built-in tasks allowing to compile, assemble, test and run Java applications. Ant can also be used effectively to build non Java applications, for instance C or C++ applications. More generally, Ant can be used to pilot any type of process which can be described in terms of targets and tasks.

Apache Ant's build files are written in XML and take advantage of the open standard, portable and easy to understand nature of XML.

Ant is extremely flexible and does not impose coding conventions or directory layouts to the Java projects which adopt it as a build tool.

Key steps in creation of JUnit HTML report include

- Generate ANT Build
- Run the Build file as Ant Build and Junit report option

Let us follow the below steps

1. Right click on the HotelApp_TestAutomation project and select **Export → General → Ant Buildfiles**

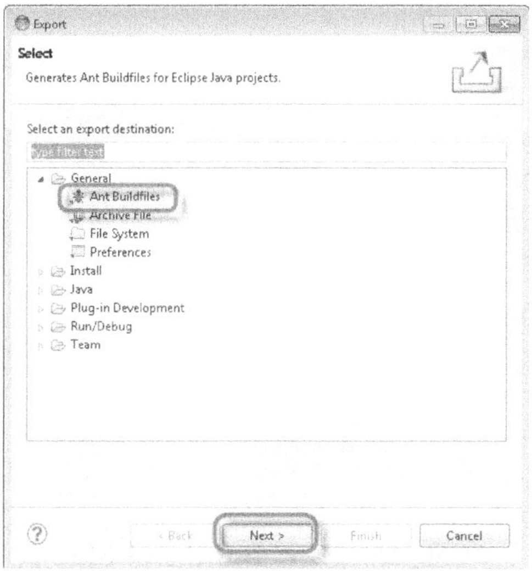

Figure 23-1 – Export AntBuildFiles

2. Click **Next** and **check** the checkbox for your project **HotelApp_TestAutomation** and press **Finish**

The default JUnit output directory is appropriately named "junit".

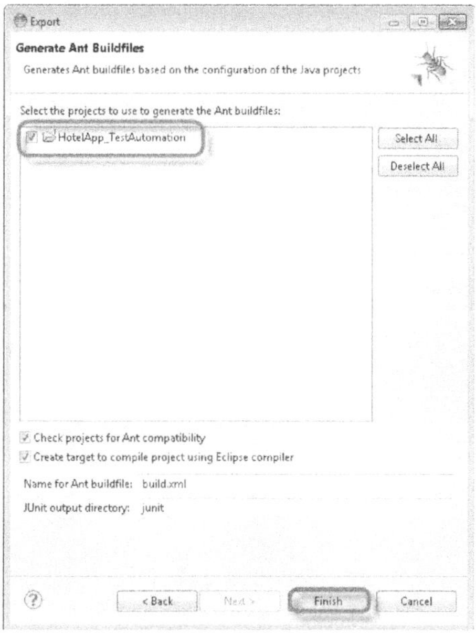

Figure 23-2 –Select Project

3. You will notice a build.xml file generated in package explorer

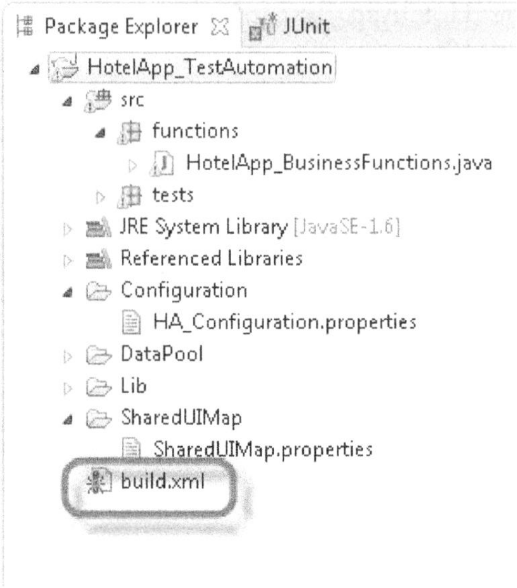

Figure 23-3 –Build.xml file

4. The next thing you have to do is resolve your dependencies by making sure that the junit.jar is added to Ant's "Global Entries". Go to **Window → Preferences → Ant → Runtime → Global Entries**

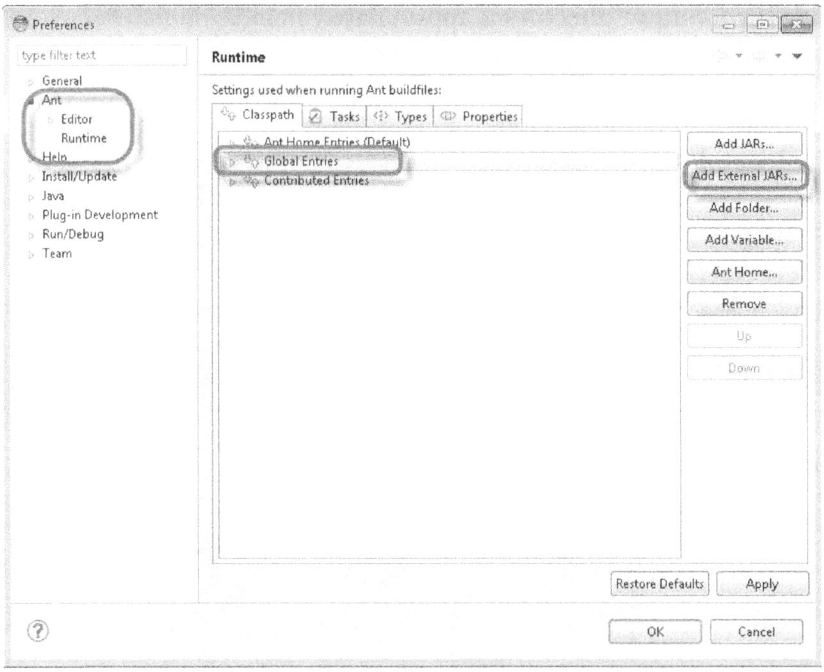

Figure 23-4 –Ant Global Entries

5. Click on **Add External JARs…**

6. Navigate to your "eclipse" directory (where eclipse is installed). (IT is basically inside folder C:\Users\<username>\eclipse\java-2023-033\eclipse\plugins) **Open** Junit.jar which is under the "plugins" directory and presently the junit.jar is in a directory named "org.junit_4.11.0.vxxxx". If this is missing then you will see error messages since junit is an optional Ant task.

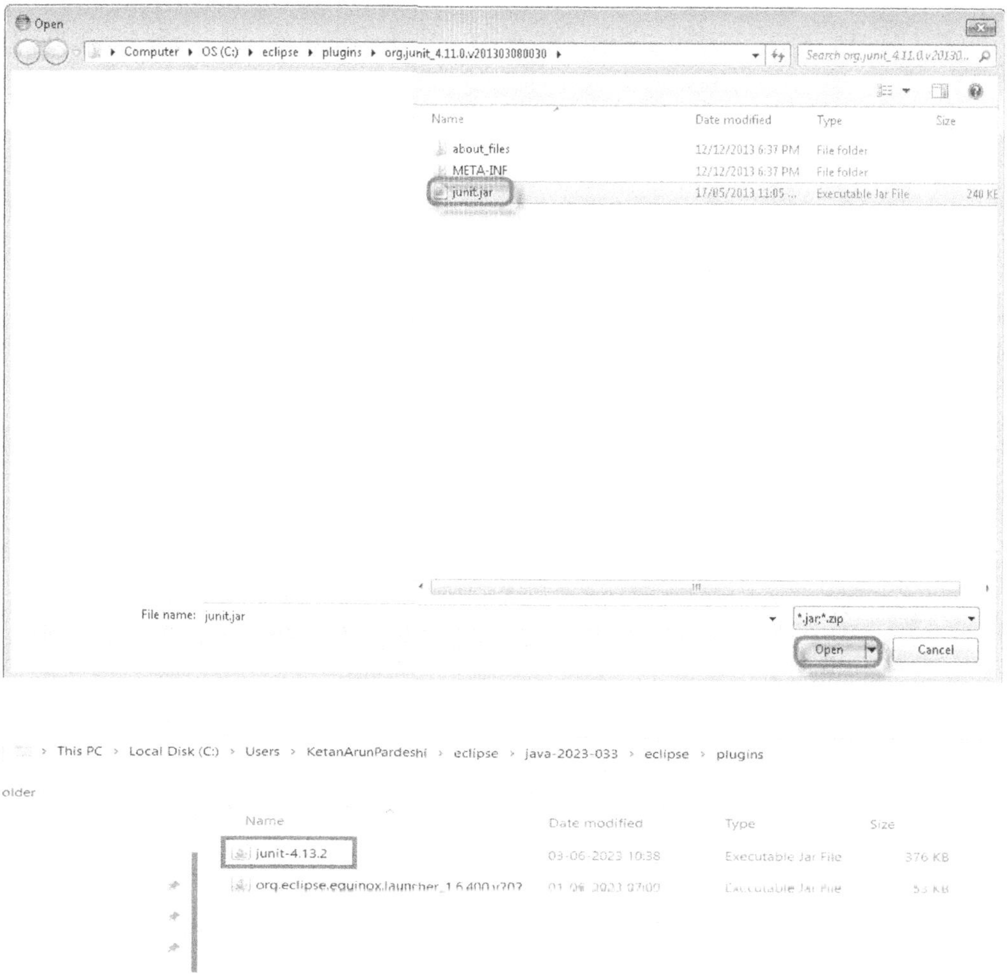

Figure 23-5 –Add External Junit.jar file

7. If the junit.jar file is missing in this directory, navigate to the URL https://mvnrepository.com/artifact/junit/junit and click on latest version in Central Tab

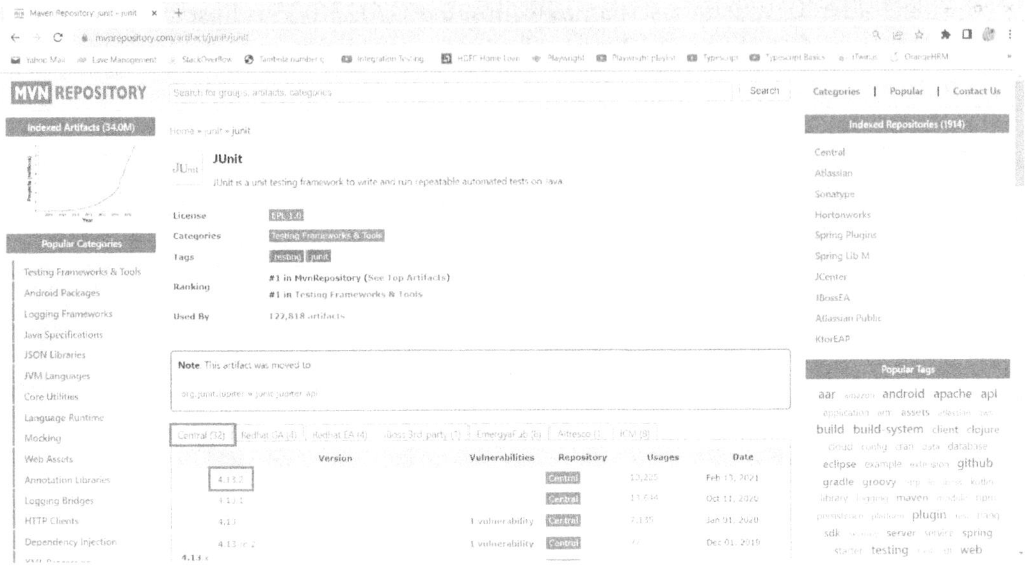

Figure 23-6 –Download junit

8. Now click on Jar link in front of Files.

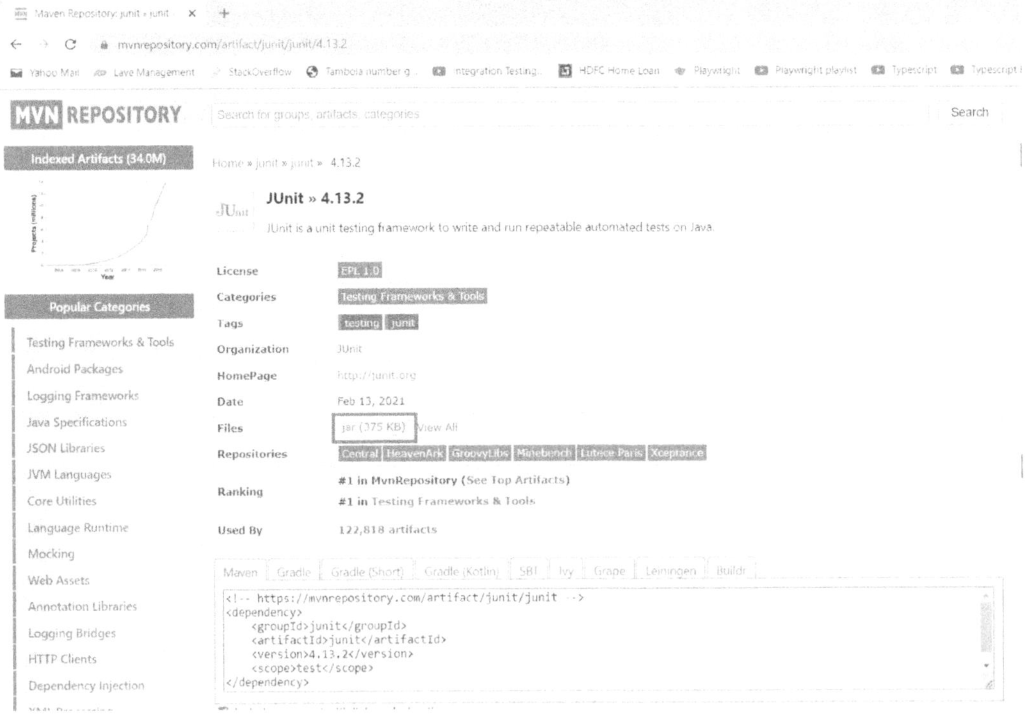

Figure 23-7 – Select Junit

9. Junit jar file gets downloaded in local machine.

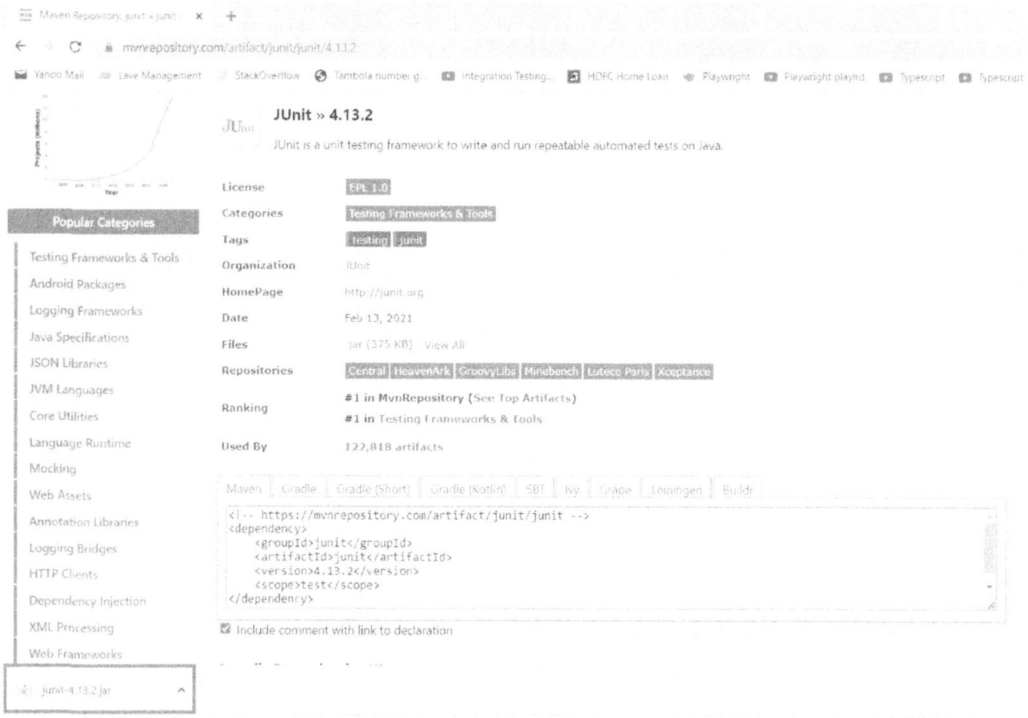

Figure 23-8 – Select Download

10. Now copy paste this jar file in the plugins folder opened and add it to External Jar

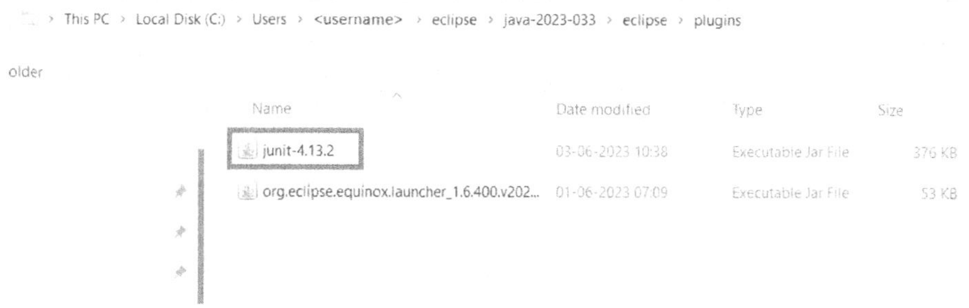

Figure 23-9 –Add External Junit.jar file

11. You will see Junit.jar added to the preferences. Click **OK** to close the Preferences dialog

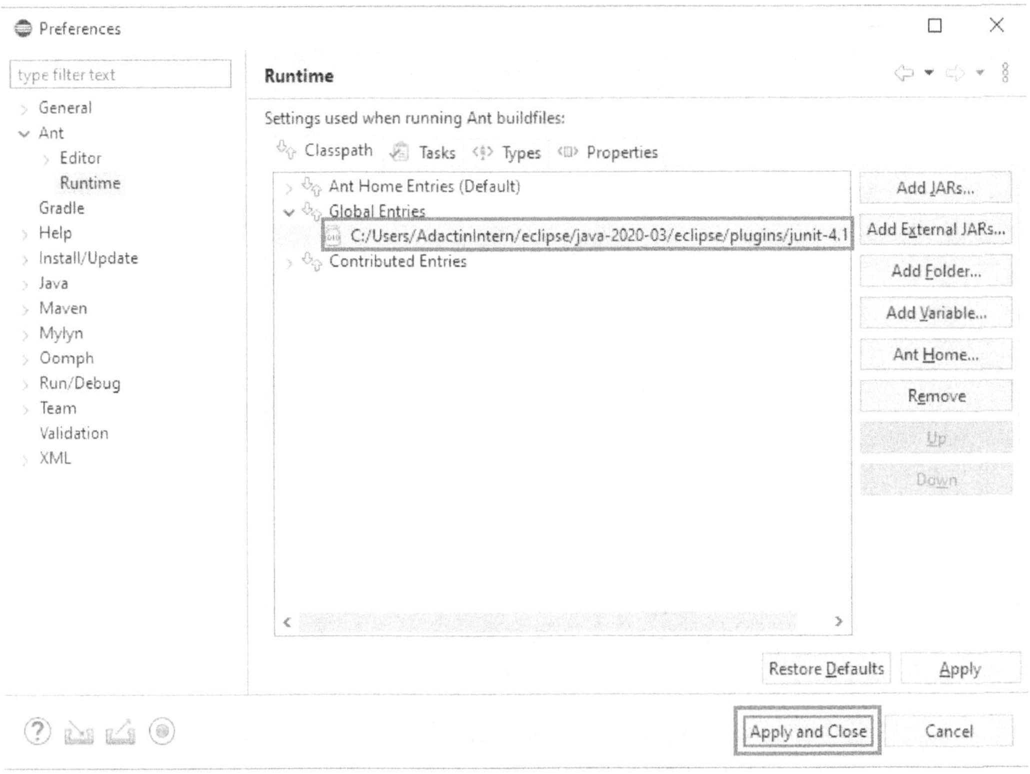

Figure 23-10 – Preferences Dialog

12. Save the test and perform Project clean-up using **Project → Clean Up**... and clean all selected projects

13. Right-click on the Ant build file **build.xml** and **Run As→ 2 Ant Build**

Note – If you select **1 Ant Build** it would not prompt with a configuration dialog and will run the scripts with an already saved configuration. After we have done the configuration we will use 1 Ant Build for speedy execution.

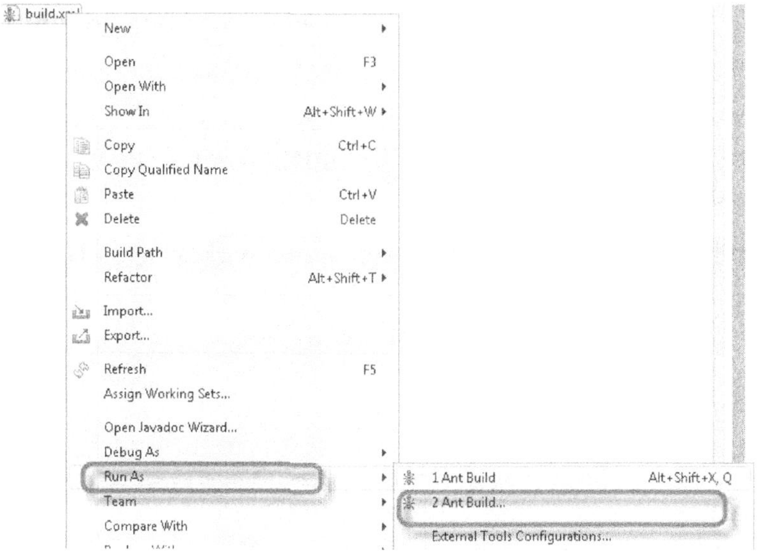

Figure 23-11 – Run As ANT build

14. This will display a list of targets. All of the launch configurations you have previously configured will have a corresponding target in your Ant build file. Select MyFirstWebDriverTest, Verification Test and junitreport (very important).

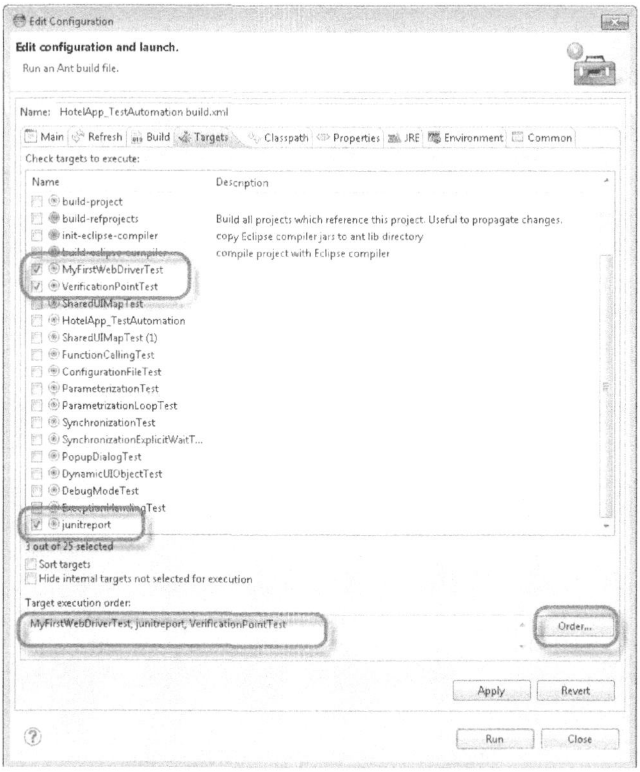

Figure 23-12 – Edit Configuration

15. Check the "**Target execution order**" text field to make sure the **junitreport** is run last

You will notice in the above dialog in "Target Execution order", junitreport is not the last one being executed. Let us fix that. If junitreport is the last in the order, we can ignore the next step.

16. Click on the **Order** Button and move the order by clicking on **Up/Down** buttons to bring junitreport to run last

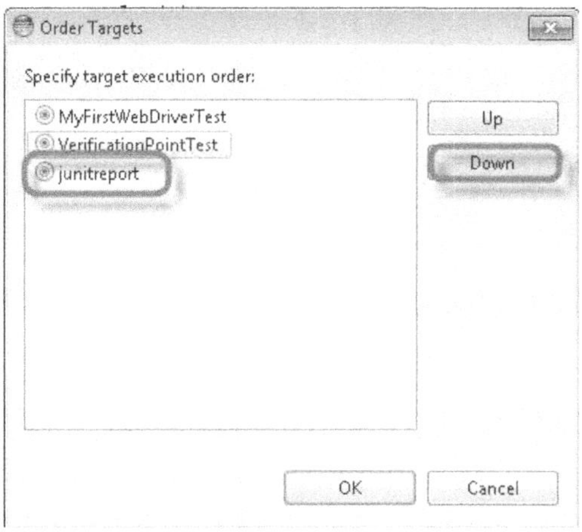

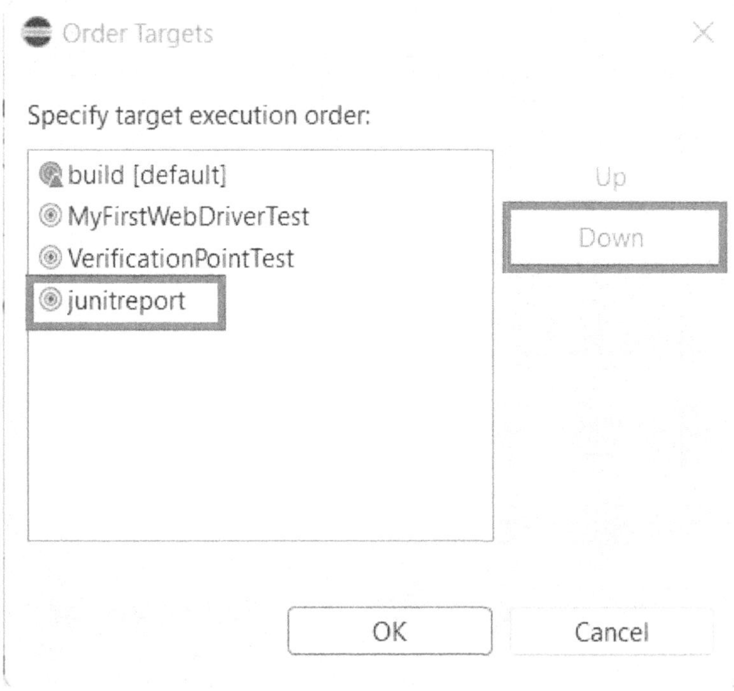

Figure 23-13 – Order Targets

Note - Whenever you run a test case it will generate results and they will be placed in the junit output directory. These results are formatted into an HTML report by the junitreport target and stored in the junit output directory in the project folder. So it is important to select junitreport and run it as a last script.

17. Click on the **Run** Button. This will execute the scripts

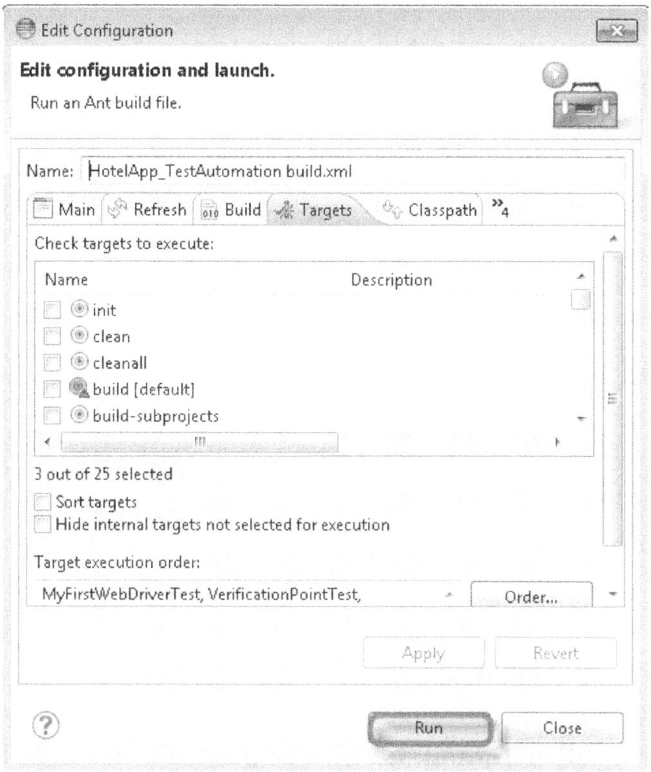

Figure 23-14 – Run Build

18. Once the tests finish running, Select the project, right click and select **Refresh**

19. You will notice a new **JUnit** Folder. **Expand** the **folder**

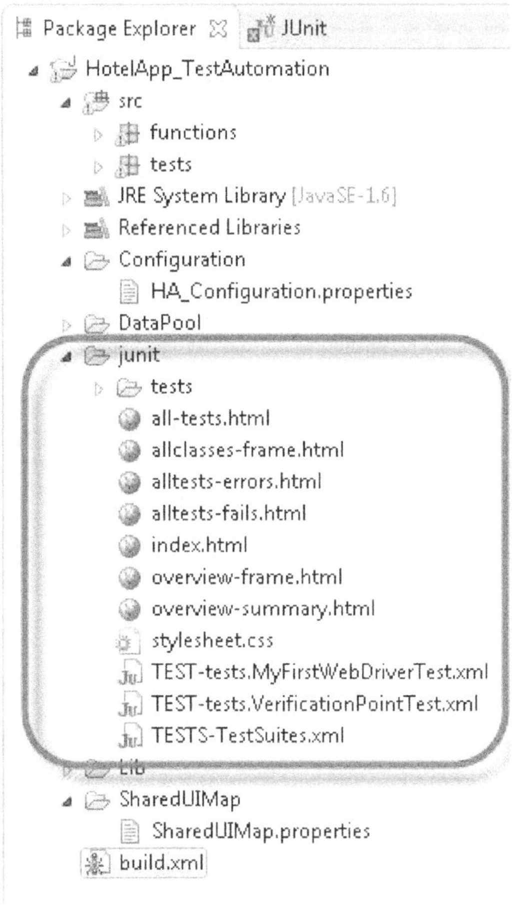

Figure 23-15 – Junit Results folder

20. Double click on **all-tests.html**. It will open the results in the browser.

Figure 23-16 – Results in browser

21. Click on MyFirstWebDriverTest link. This will open up more details of HTML results

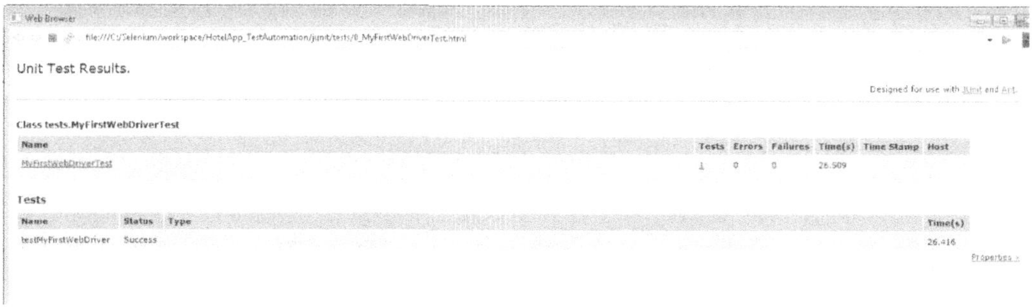

Figure 23-17 – Detailed HTML Results

23.3. Configuring TestNG Report for your Tests

For using **TestNG** reporting framework you need to add the TestNG plugin within your Eclipse setup. Secondly, you would need to modify your script to work as per TestNG Plug-in libraries. Additionally you would need to add the Reporter function for reporting results to TestNG report.

So to summarize we have 4 key steps-

1. Install TestNG Plugin (one time effort)
2. Update script to use TestNG Libraries
3. Add Reporter functions in our script to report results in TestNG results
4. Execute scripts using TestNG

Note:

TestNG is supported for only the following versions of Eclipse: 2018-09 (4.9), Photon (4.8), Oxygen (4.7), Neon (4.6), Mars (4.5), Luna (4.4), Kepler (4.3), Juno (4.2, 3.8), Previous to Juno (<=4.1). If the Eclipse version downloaded is a newer version, it would need to be downgraded. This may also cause an older version of Java to be installed and build path to be re-configured and Java compiler properties to be updated to the older version installed.

Install TestNG Plugin

1. Go to **Help → Eclipse Marketplace**

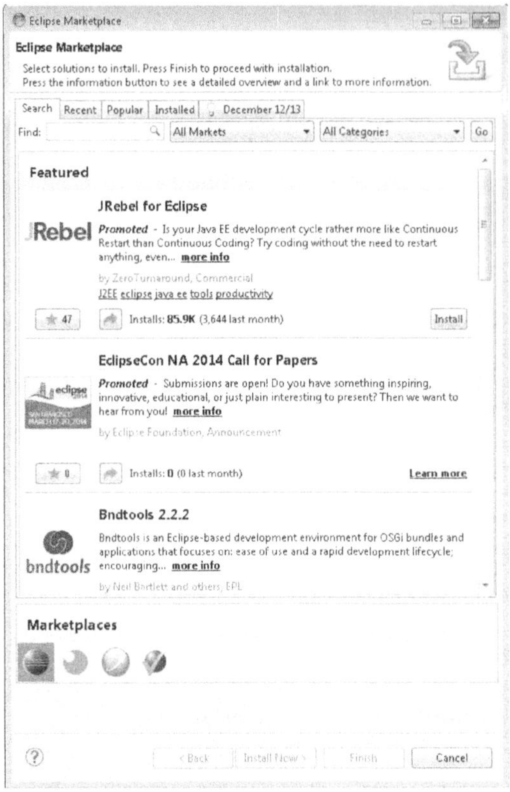

Figure 23-18 – Eclipse Market place

2. If you get an Error, then Eclipse is unable to connect to Marketplace, it is quite possible it could be because of your internal proxy settings. If you do not get this error proceed to Step-3.

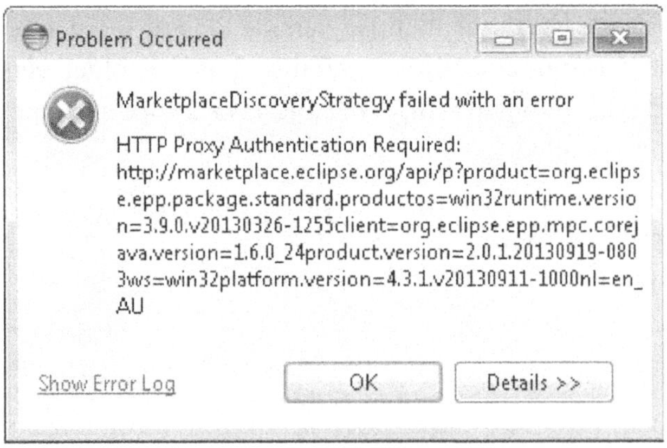

Figure 23-19 – Proxy Error

a. To Resolve the above error, go to **Window → Preferences → General → Network Connections**

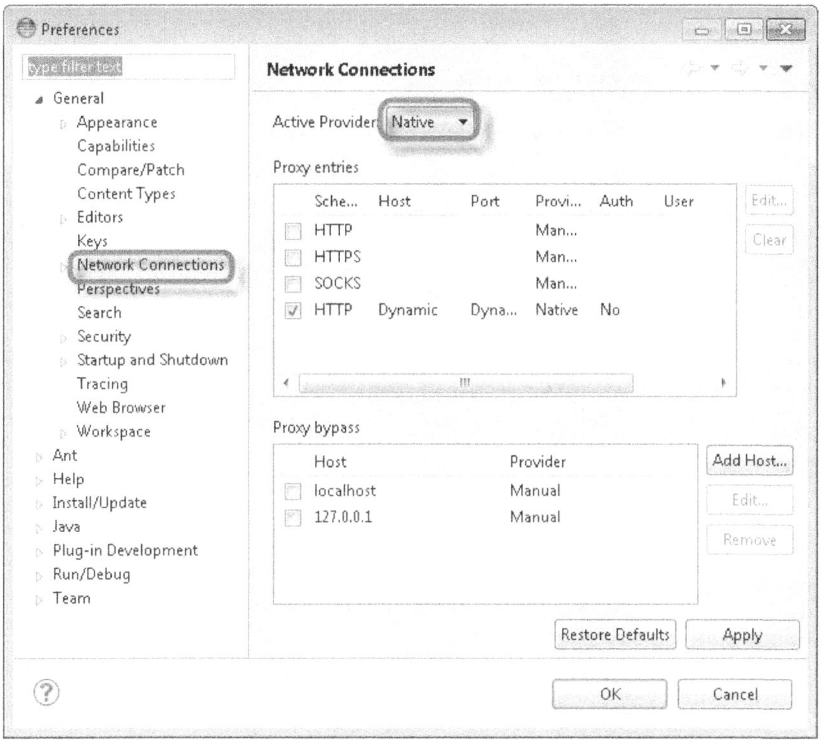

Figure 23-20 – Network Connection Preferences

b. Change **Active Provider** to **Manual**

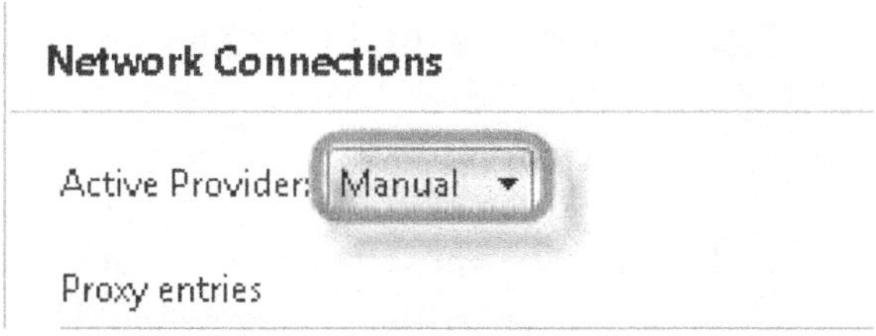

Figure 23-21 – Change Active Provider

c. Select **HTTP** row and click on **Edit** Button and in the Edit Proxy Entry **enter Host Name** and **Authentication details** (if required)

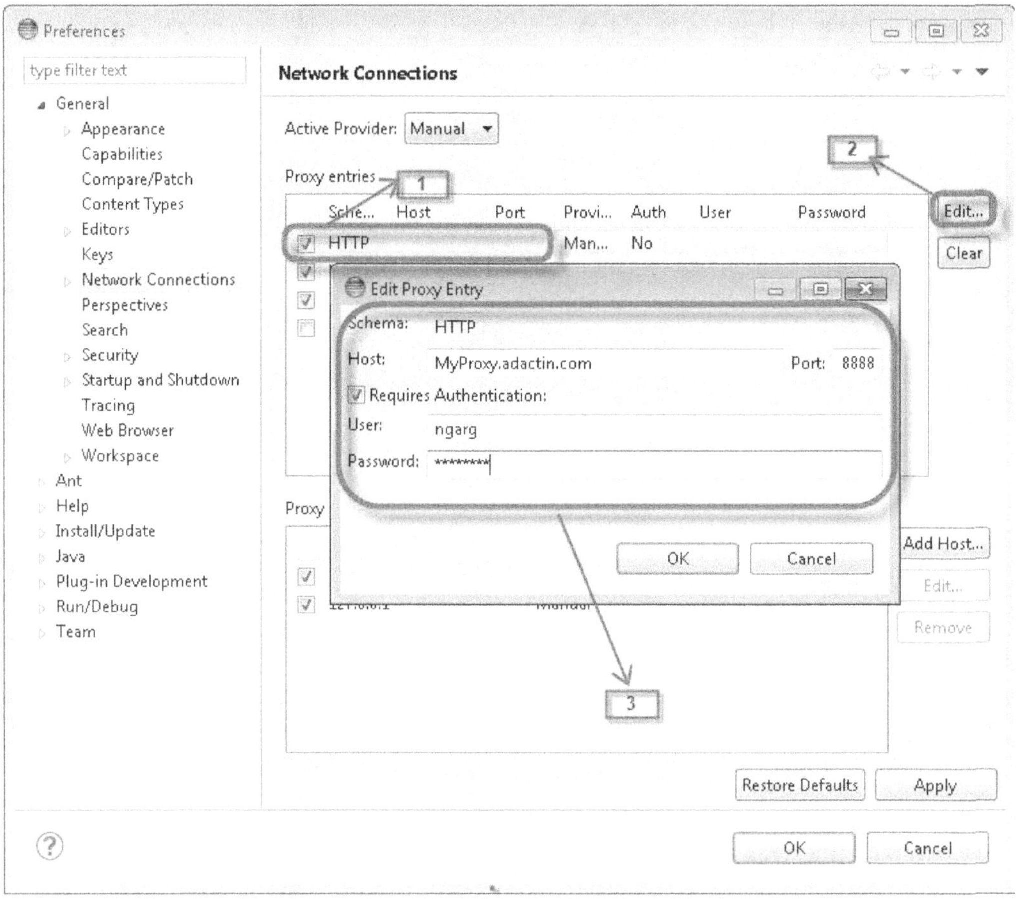

Figure 23-22 –Setup Proxy Details

d. Repeat the same Steps for **HTTPS**

e. Leave SOCKS blank (It is very important to leave SOCKS blank. It is one of the most common issues people struggle with when they try to resolve proxy related errors)

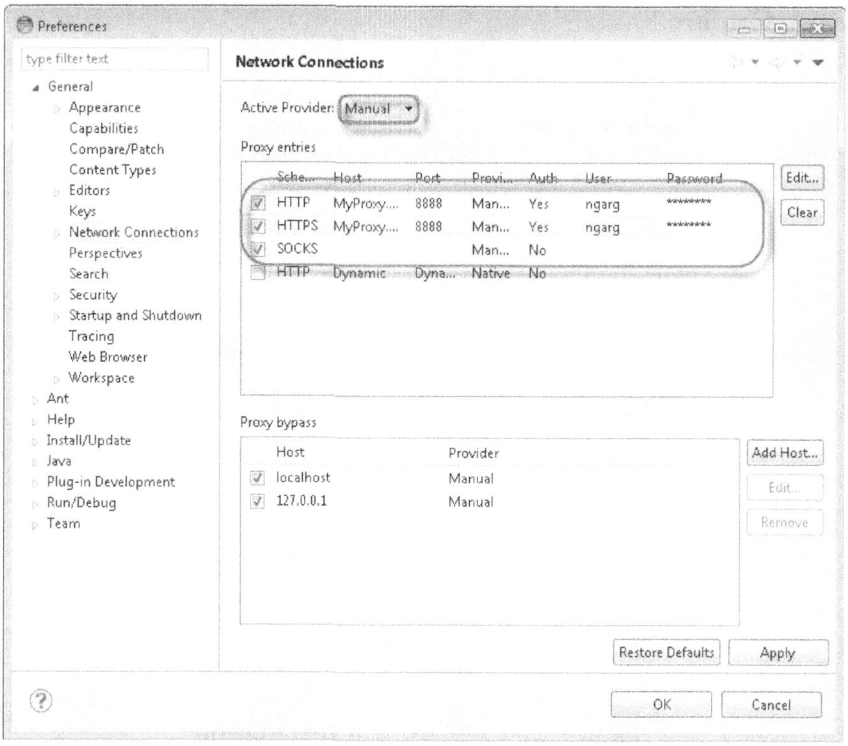

Figure 23-23 –Final Proxy Settings

f. Click **OK**.

g. **Close** and **Re-launch** Eclipse and Open Eclipse Marketplace using **Help → Eclipse Marketplace**

3. Search for **TestNG** and Install it by clicking on **Install** button

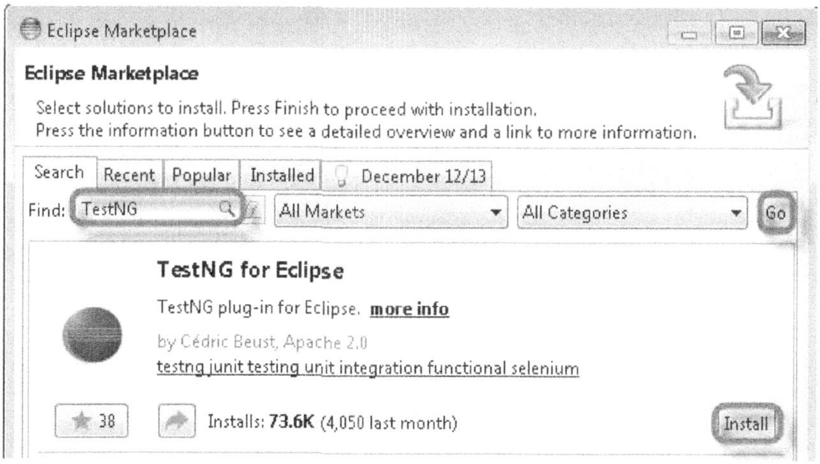

Figure 23-24–Search and Install TestNG

4. Click on **Confirm** Button

Figure 23-25 –Confirm Installation

5. Review License conditions and **Accept** terms and click **Finish**

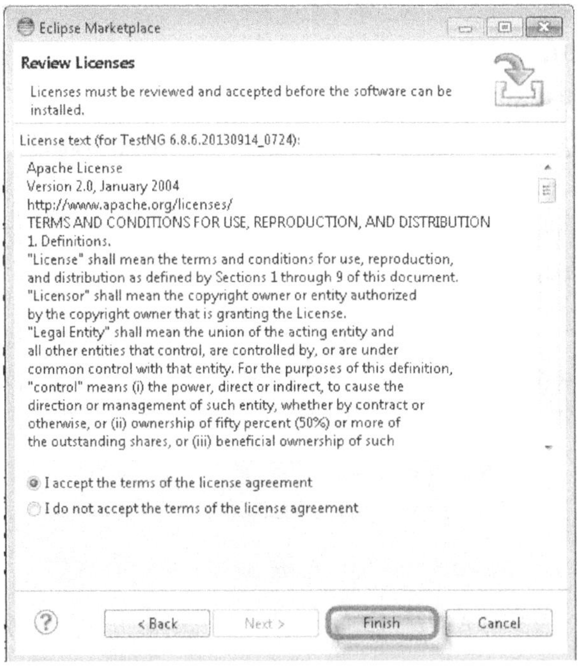

Figure 23-26 – License Review

6. Now TestNG plug-in is installed. You can **accept** if you see any Security warning.

7. **Restart Eclipse**.

8. To confirm that TestNG has been correctly installed, Go to **Window → Show View → Other → Java**. Verify that you can see TestNG in the list.

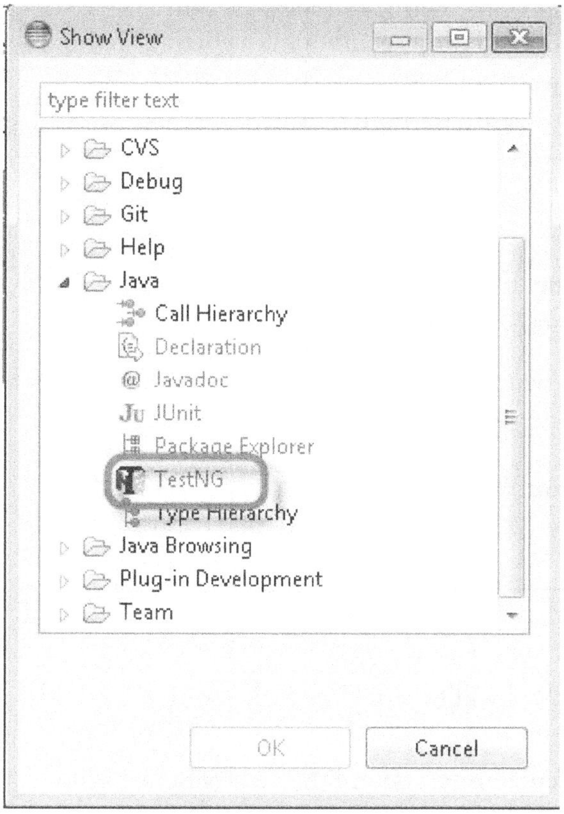

Figure 23-27 – Show View

Create a Test with TestNG Annotations

Note that TestNG test has similar yet different annotations as compared to Junit test.

9. **Right click** on your existing **MyFirstWebDriverTest.java** script, select **Copy**

10. Select the **tests** package folder, **right click** and select **Paste**

11. In the Name Conflict dialog box, enter the name of the script as **MyTestNGTest.java** and click **OK**

12. Double click on **MyTestNGTest.java** to see the **script**

13. Now remove all JUnit imports from your test class (import org.junit.*; import static org.junit.Assert.*;) . This will lead to an error in your script. This should be fixed by the following imports:

```
import org.testng.annotations.AfterMethod;
import org.testng.annotations.Test;
import org.testng.annotations.BeforeMethod;
import org.testng.Assert;
import org.testng.Reporter;
```

Figure 23-28 – TestNG imports

14. Replace after and before annotation with respective TestNG ones which are @ BeforeMethod and @AfterMethod.

See the table below which compares key annotation differences between TestNG and JUnit

Annotation Description	TestNG	JUnit
The @Test annotation identifies a method as a test method.	@Test	@Test
The annotated method will be run before each test method.	@BeforeMethod	@Before
The annotated method will be run after each test method.	@AfterMethod	@After
This method is executed once, before the start of all tests.	@BeforeSuite	@BeforeClass
This method is executed once, after all tests have been finished.	@AfterSuite	@AfterClass

Figure 23-29 – Annotation comparison TestNG vs JUnit

Note: TestNG supports similar Asserts as in JUnit. We just need to import testng. Assert class.

15. Replace **fail** (verificationErrorString) statement with **Assert. fail**(verificationErrorString) wherever you see Asserts in the script.

Also you can replace any assertTrue or assertEquals statement with **Assert.assertTrue** or **Assert.assertEquals** statement as per TestNG syntax.

Note: As an alternative, If you right click on any Junit Test and select **TestNG → Convert to TestNG** , it will display what changes need to be done in the existing JUnit script to make it compatible for a TestNG run. You can understand the changes and manually modify the JUnit script.

Figure 23-30 – Covert to TestNG

We will not be discussing the conversion in detail since most of the conversion can be handled manually.

Add Reporter Method to TestNG test

TestNG has a built in Class Reporter for test methods to log messages that will be included in the HTML reports generated by TestNG.

One of the key methods for the Reporter object is log.

Use Reporter.log () to log your own message to this HTML report. See below

Reporter.log("The Search Results are correct");

It also gives you an argument to log to the standard console apart from logging to the report.

Reporter.log("The Search Results are correct", true);

16. To demonstrate the usage of the Reporter.log method, modify your MyTestNGTest to include the below method after login

```
    driver.findElement(By.id("login")).click();

Reporter.log("Login was successful", true);


new Select(driver.findElement(By.id("location"))).selectByVisibleText("Sydney");
```

Figure 23-31 – Reporter.log method

Note: You would need to import org.testng.Reporter to use the Reporter.log method.

Your final script should look like this:

```
package tests;


import java.time.Duration;
import org.testng.annotations.AfterMethod;
import org.testng.annotations.Test;
import org.testng.annotations.BeforeMethod;
import org.testng.Assert;
import org.testng.Reporter;
import org.openqa.selenium.By;
import org.openqa.selenium.WebDriver;
import org.openqa.selenium.chrome.ChromeDriver;
import org.openqa.selenium.support.ui.Select;


public class MyTestNGTest {


        private WebDriver driver;

        @BeforeMethod
        public void setUp()
```

```
        {
                driver = new ChromeDriver();
                driver.get("https://www.adactinhotelapp.com");
                driver.manage().window().maximize();
                driver.manage().timeouts().implicitlyWait(Duration.ofSec-
onds(20));
        }

        @Test
         public void testMyFirstWebDriver() throws Exception {


         driver.findElement(By.xpath(".//*[@id='username']")).clear();
         driver.findElement(By.xpath(".//*[@id='username']")).sendKeys("Adctin");
         driver.findElement(By.id("password")).clear();
         driver.findElement(By.id("password")).sendKeys("Adactin123");
         driver.findElement(By.id("login")).click();

         Reporter.log("Login was successful", true);
         new
Select(driver.findElement(By.id("location"))).selectByVisibleText("Sydney");
         driver.findElement(By.id("Submit")).click();
         driver.findElement(By.id("radiobutton_1")).click();
         driver.findElement(By.id("continue")).click();
         driver.findElement(By.id("first_name")).clear();
         driver.findElement(By.id("first_name")).sendKeys("test");
         driver.findElement(By.id("last_name")).clear();
         driver.findElement(By.id("last_name")).sendKeys("test");
         driver.findElement(By.id("address")).clear();
         driver.findElement(By.id("address")).sendKeys("test");
         driver.findElement(By.id("cc_num")).clear();
         driver.findElement(By.id("cc_num")).sendKeys("1212121212121212");
         new
Select(driver.findElement(By.id("cc_type"))).selectByVisibleText("American     Ex-
press");
         new
Select(driver.findElement(By.id("cc_exp_month"))).selectByVisibleText("March");
         new
Select(driver.findElement(By.id("cc_exp_year"))).selectByVisibleText("2015");
         driver.findElement(By.id("cc_cvv")).clear();
         driver.findElement(By.id("cc_cvv")).sendKeys("111");
         driver.findElement(By.id("book_now")).click();
         driver.findElement(By.linkText("Logout")).click();
         driver.findElement(By.linkText("Click here to login again")).click();
         }


        @AfterMethod
        public void cleanUp()
        {
                driver.quit();
        }
}
```

Figure 23-32 – Final TestNG script

Run the Test as TestNG

17. Save the test and perform Project clean-up using **Project → Clean Up…** and clean all selected projects
18. Select your **MyTestNGTest**, right click and Select **Run As → TestNG**

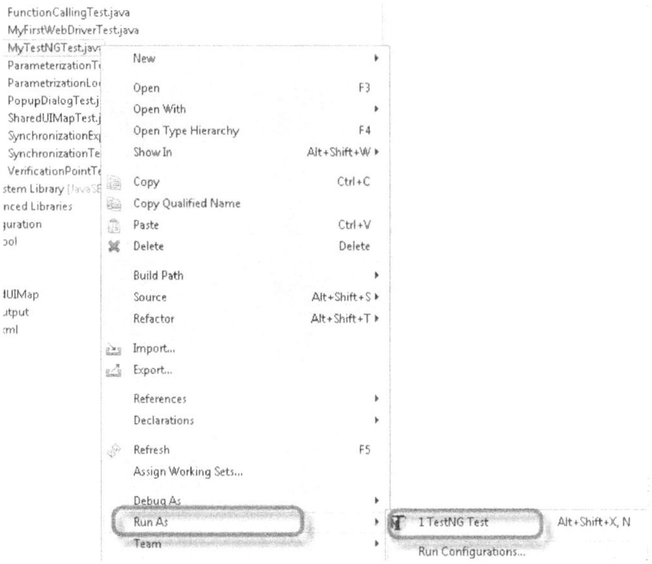

Figure 23-33- Executing TestNG test

19. After Test completion, select your project from the Package Explorer and press **F5**. It'll add a new folder to your project structure called **test-output**

Figure 23-34 Test Results using TestNG

20. Expand **Test-output** and open **index.html** in a browser to see the html report.

Figure 23-35 - TestNG Web Report

21. Click on option "**Reporter output**" to see the reporter results.

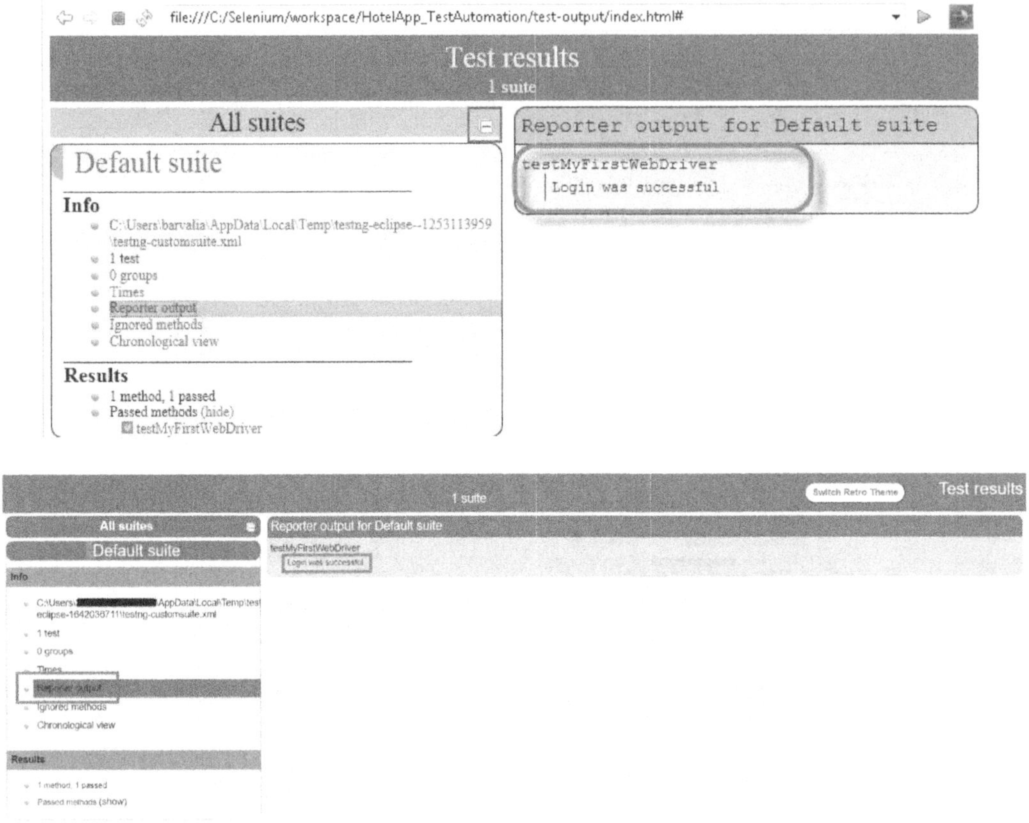

Figure 23-36 Reporter logs

Note: You can create TestNG.xml file to run multiple tests in one go or parameterize tests. For further details on TestNg framework please refer to http://testng.org/doc/documentation-main.html.

A lot of automation testers use TestNG as a way of reporting results. With its built in Reporter class it clearly has an advantage over JUnit HTML reports.

23.4. Custom Reporting in Excel Sheets or Databases

Apart from using JUnit and TestNG reporting options, another option which is widely used is custom reporting in Excel Sheets or in databases.

One of the key needs for reporting is to have all test reports at one central location. Also, automation testers would want to refer back to the report at later stages. This is one of the disadvantages of using the TestNG or JUnit framework.

A possible alternative is to use Excel based reporting. As seen in one of the previous chapters we can use the jxl.jar library to read or write into an Excel sheet. So we can create our own custom functions to write results in the Excel based sheet which can be kept in a central location.

See the snapshot below containing a sample results format.

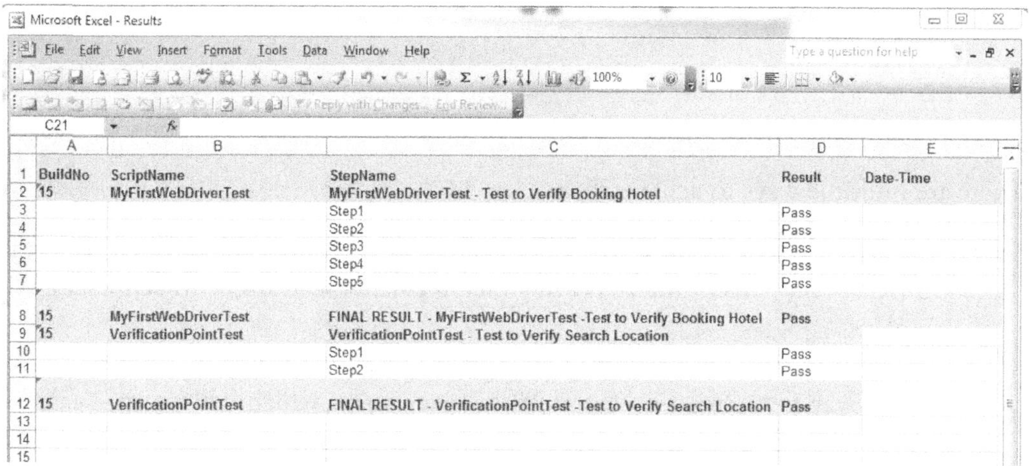

	A	B	C	D	E
1	BuildNo	ScriptName	StepName	Result	Date-Time
2	15	MyFirstWebDriverTest	MyFirstWebDriverTest - Test to Verify Booking Hotel		
3			Step1	Pass	
4			Step2	Pass	
5			Step3	Pass	
6			Step4	Pass	
7			Step5	Pass	
8	15	MyFirstWebDriverTest	FINAL RESULT - MyFirstWebDriverTest -Test to Verify Booking Hotel	Pass	
9	15	VerificationPointTest	VerificationPointTest - Test to Verify Search Location		
10			Step1	Pass	
11			Step2	Pass	
12	15	VerificationPointTest	FINAL RESULT - VerificationPointTest -Test to Verify Search Location	Pass	
13					
14					
15					

Figure 23-37- Excel Results format

We can drive the Result and Step Name based on our custom function which will write into Excel based result sheet.

Note: We are not covering how to report into Excel based sheets, as it would be done purely using Java language and using jxl.jar library. You can find a lot of reference material on the internet on how to write into an Excel sheet.

Apart from Excel based reporting we can also report into Databases. The mysql database is pretty popular along with WebDriver since they are free. You can report results into mysql database and create a small php based Web page to read and show data in that page. This is my personal favourite as we would have results in a central location and anyone can see the results using the Web based URL.

EXERCISE

1. Convert (use copy/paste and create new scripts) your VerificationPointTest and SharedUIMapTest from JUnit test to TestNG test.

Hint – You can use **Convert to TestNG** feature to understand the script changes required.

2. Create a custom function call ExcelReporter (stepname, stepstatus) which will report results to an Excel sheet. Call this custom function from one of your WebDriver scripts.

§

24

Batch Execution

A batch run means executing all the scripts in a suite at the same time, unattended.

There are multiple ways to achieve this:

- Use JUnit's integration with ANT in eclipse to run multiple scripts at the same time (we saw this in the Reporting chapter while working with JUnit HTML reports)
- Use TestNG framework and its testng.xml file to run scripts as a batch
- Third way is to create a Master WebDriver script which will execute calls to other WebDriver scripts as call to classes defined in those scripts

You can use any of the above 3 ways to run your script as a batch suite.

In this chapter, we will see how to run WebDriver scripts as Batch using

- TestNG Framework
- Master WebDriver script

24.1. Batch Execution with TestNG

Now in our project explorer we have different classes for each test. We create an XML file where we call all the classes we want to execute in one go. For this we

1. Select any of your existing Tests in TestNG e.g. MyTestNGTest
2. Right Click on the Test and Select **TestNG → Convert to TestNG**
3. Refactoring window comes up. Click **Next** in this window

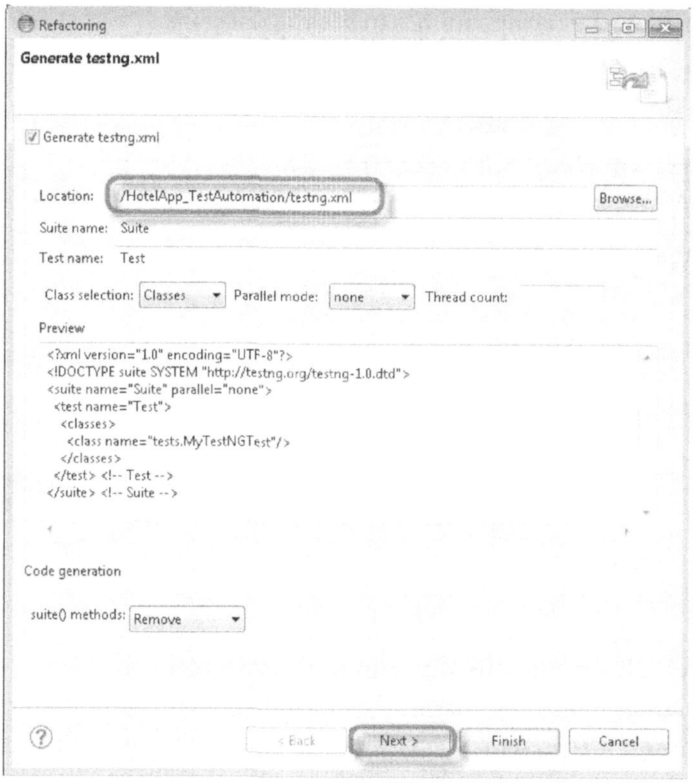

Figure 24-1- Refactoring Window

4. Window will prompt changes to be performed. Click **Finish**
5. A **testng.xml** file will be created

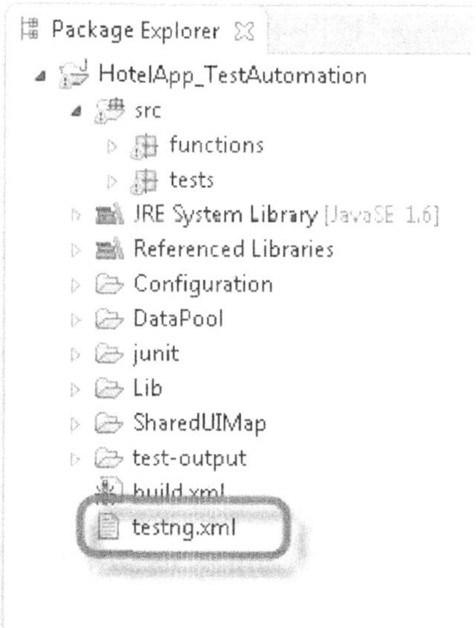

Figure 24-2- TestNG file

6. Double click on the testng.xml file to see the contents

testng.xml   MyTestNGTest.java

```
<?xml version="1.0" encoding="UTF-8"?>
<!DOCTYPE suite SYSTEM "http://testng.org/testng-1.0.dtd">
<suite name="Suite" parallel="none">
  <test name="Test">
    <classes>
      <class name="tests.MyTestNGTest"/>
    </classes>
  </test> <!-- Test -->
</suite> <!-- Suite -->
```

Figure 24-3- TestNG file contents

Now let us understand the tags of testing.xml file

- **Parallel**: This is being used to run tests in parallel.
- **suite name**: This is the name of your test suite.
- **test name** : It represents the type of tests that are being executed. For instance Regression, Sanity or Smoke tests. Note that this is not the actual name of the class file or webdriver tests being executed.
- **classes**: Names of classes (tests) you want to add in this batch execution.

7. To run multiple scripts modify the code by adding more classes

```
<?xml version="1.0" encoding="UTF-8"?>
<!DOCTYPE suite SYSTEM "http://testng.org/testng-1.0.dtd">
<suite name="Suite" parallel="none">
  <test name="Test">
    <classes>
      <class name="tests.MyTestNGTest"/>
      <class name="tests.VerificationPointTest"/>
      <class name="tests.SharedUIMapTest"/>
    </classes>
  </test> <!-- Test -->
</suite> <!-- Suite -->
```

Figure 24-4- Calling Multiple Tests

8. Now if we right click this file and select **Run As → TestNG** Suite this will execute all the given tests one after the other in the given order.
9. Press **F5** to refresh the **Test Output folder**. You can see the results of each individual test in TestNG Web report - **index.html**.
10. If you want to run the tests in parallel you just change the line

    <suite name="E2ESuite" parallel="none">

    With

    <suite thread-count="2" name="Suite" parallel="tests">

    This will run two tests in parallel at the same time.

24.2. Batch Execution with Master WebDriver Script

We can also control the execution of WebDriver scripts using a master script. This means that when we run that file it will internally make calls to other Webdriver scripts defined in that master file. We will run this suite as JUnit Tests

Note: This option is most useful if you are reporting into an Excel sheet or Database.

Let us see how.

Step1 - Changes to Test

1. **Right click** on your existing **MyFirstWebDriverTest**.java script, select **Copy**
2. Select the **tests** package folder, right click and select **Paste**
3. In the Name Conflict dialog box, enter the name of the script as **CalledbyTestSuiteTest** and click **OK**
4. Double click on the newly created "**CalledbyTestSuiteTest.java**" script to see the script
5. You would have noticed that steps to create a new driver and set timeouts are defined in the setup function

```
    driver = new ChromeDriver();
  baseUrl = "http://www.adactinhotelapp.com/";

driver.manage().timeouts().implicitlyWait(Duration.ofSeconds(10));
```

6. Cut and paste these steps at the start of testMyFirstWebDriver function (which is the main test)

7. You would have noticed that step to quit the Driver is in teardown function.

```
driver.quit();
```

8. Cut and Paste this step at the end of the function testMyFirstWebDriver function.
9. So your final **"CalledbyTestSuiteTest.java" script will look like below**

```
@Test
public void testMyFirstWebDriver() throws Exception {

        driver = new ChromeDriver();
        baseUrl = "http://www.adactinhotelapp.com/";
        driver.manage().timeouts().implicitlyWait(Duration.ofSeconds(30));
        driver.get(baseUrl");
  driver.findElement(By.xpath(".//*[@id='username']")).clear();
  driver.findElement(By.xpath(".//*[@id='username']")).sendKeys("adactin123");
  driver.findElement(By.id("password")).clear();
  driver.findElement(By.id("password")).sendKeys("adactin123");


  driver.findElement(By.id("login")).click();


  new Select(driver.findElement(By.id("location"))).selectByVisibleText("Sydney");
  driver.findElement(By.id("Submit")).click();
  driver.findElement(By.id("radiobutton_1")).click();


  driver.findElement(By.id("continue")).click();
  driver.findElement(By.id("first_name")).clear();
  driver.findElement(By.id("first_name")).sendKeys("test");
  driver.findElement(By.id("last_name")).clear();
  driver.findElement(By.id("last_name")).sendKeys("test");
  driver.findElement(By.id("address")).clear();
  driver.findElement(By.id("address")).sendKeys("test");
  driver.findElement(By.id("cc_num")).clear();
  driver.findElement(By.id("cc_num")).sendKeys("1212121212121212");
  new Select(driver.findElement(By.id("cc_type"))).selectByVisibleText("American Express");
  new Select(driver.findElement(By.id("cc_exp_month"))).selectByVisibleText("March");
  new Select(driver.findElement(By.id("cc_exp_year"))).selectByVisibleText("2015");
  driver.findElement(By.id("cc_cvv")).clear();
  driver.findElement(By.id("cc_cvv")).sendKeys("111");
  driver.findElement(By.id("book_now")).click();
  driver.findElement(By.linkText("Logout")).click();
  driver.findElement(By.linkText("Click here to login again")).click();
  driver.quit();
}
```

Figure 24-5- Updated Test Script

Note that the reason we are adding both invoking of the driver and quit on driver in the same method is that it will make it much easier to call just one method from any other external script.

Step2 – Create a Suite Script

10. Again **right click** on your existing **MyFirstWebDriverTest**.java script, select **Copy**
11. Select the **tests** package folder, right click and select **Paste**
12. In the Name Conflict dialog box, enter the name of the script as **HA_TestSuite** and click **OK**
13. Double click on the newly created "**HA_TestSuite.java**" script to see the script
14. Remove the code within the function @Test so we have only the following empty declaration

```
@Test
public void testMyFirstWebDriver() throws Exception {

}
```

15. Add below code

```
@Test
public void testMyFirstWebDriver() throws Exception {

        //Instantiating Object for each of the tests

         CalledbyTestSuiteTest bp1 = new CalledbyTestSuiteTest();

                // More tests can be added in similar fashion
        // VerificationPointTest bp2 = new VerificationPointTest();
         //SynchronizationTest bp3 = new SynchronizationTest();
         //SharedUIMapTest bp4 = new SharedUIMapTest();

        // Defining which tests to run

         Boolean bCalledbyTestSuiteTest = true;

       // More variables can be added in similar fashion for other tests
               // Boolean bVerificationPointTest = false;
        // Boolean bSynchronizationTest = false;
        // Boolean bSharedUIMapTest = false;

  //Call and run tests

              if(bCalledbyTestSuiteTest)
                  bp1.testMyFirstWebDriver();
}
```

Figure 24-6- Test Suite Script

If you look at the above code,

The first section instantiates all the script classes. Note that we first need to add all the scripts and instantiate them if you would want them to be executed as part of the suite.

The second section defines variables which tell us whether we want to run a particular script or not. If a value is true it means that the script will be executed as part of the current run of the suite. We might choose not to run a few scripts if they are not relevant for that run cycle.

Note - Alternatively we can also keep this variable's value in an Excel sheet and read from it using a custom method to read from Excel

The third section is where we call the test method within the script and run the script.

16. Save the test and perform Project clean-up using **Project → Clean Up...** and clean all selected projects
17. Select the test "**HA_TestSuite.java**", right click and **Run As → Junit Test** and verify that the scripts run one after the other as a suite.

EXERCISE

1. Add 2 more scripts (VerificationPointTest and bSynchronizationTest) to the suite file. Modify the above script and confirm all scripts execute as a suite.

§

25

Continuous Integration with Jenkins

Why do we need Continuous Integration tools for test automation?

Continuous Integration (CI) tools assists in creating builds frequently (usually on a daily basis) and running developer driven tests (unit tests) to provide timely feedback on application quality.

We can integrate our Selenium based functional test automation scripts with CI tools to execute our scripts as soon as a new build is created which will provide instant feedback on application issues.

Popular open source tools include Hudson, Jenkins (the fork of Hudson), CruiseControl and CruiseControl.NET.

Commercial tools include ThoughtWorks' Go, Urbancode's Anthill Pro, Jetbrains' Team City and Microsoft's Team Foundation Server.

As part of this chapter we will learn how Selenium WebDriver scripts integrate with Jenkins, one of the popular open source CI tools.

About Jenkins

Jenkins is a popular continuous integration server in the Java development community. It is derived from the Hudson CI server. It supports configuration management tools including CVS, Subversion, Git, Mercurial, Perforce, and ClearCase, and can execute Apache Ant and Apache Maven based projects as well as arbitrary shell scripts and Windows batch commands.

Jenkins can be deployed to set up an automated testing environment where you can runSelenium WebDriver tests unattended based on a defined schedule, or every time changes are submitted in configuration.

Key objectives

- Install Jenkins tool
- Jenkins Configuration
- Run Jenkins with ANT
- Scheduling Auto-Runs

25.1. Installing Jenkins Tool

Let us first install Jenkins.

Note: In a typical software development environment you would already have it installed by the development team.

1. Go to URL - https://www.jenkins.io/download/
2. Download the Jenkins for the correct environment (in our case it is Windows)

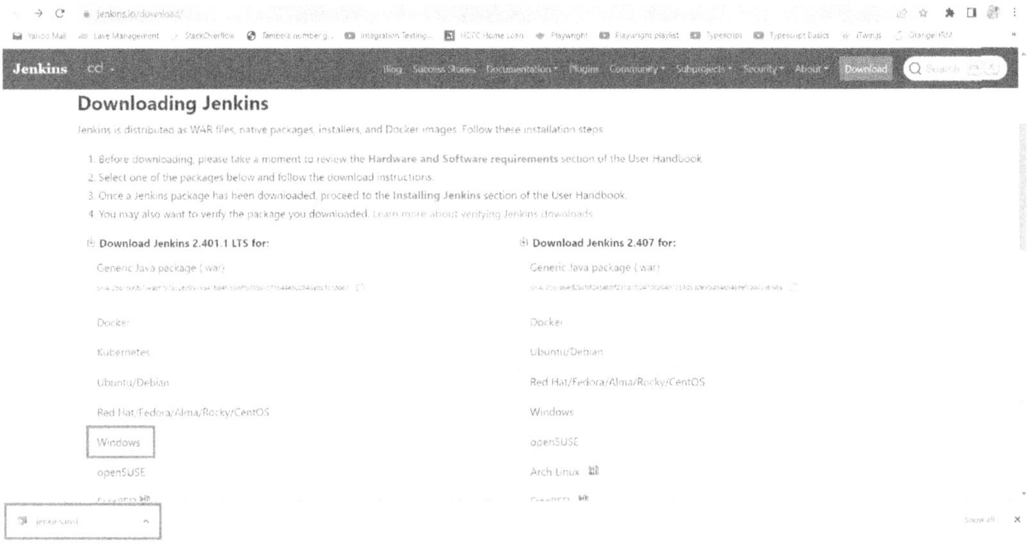

Figure 25-1- Download Jenkins

3. Unzip the install file and click on **jenkins file** (incase of windows)

Figure 25-2- Jenkins Setup

4. Keep clicking **Next** to perform default installation and finally click on **Install** button

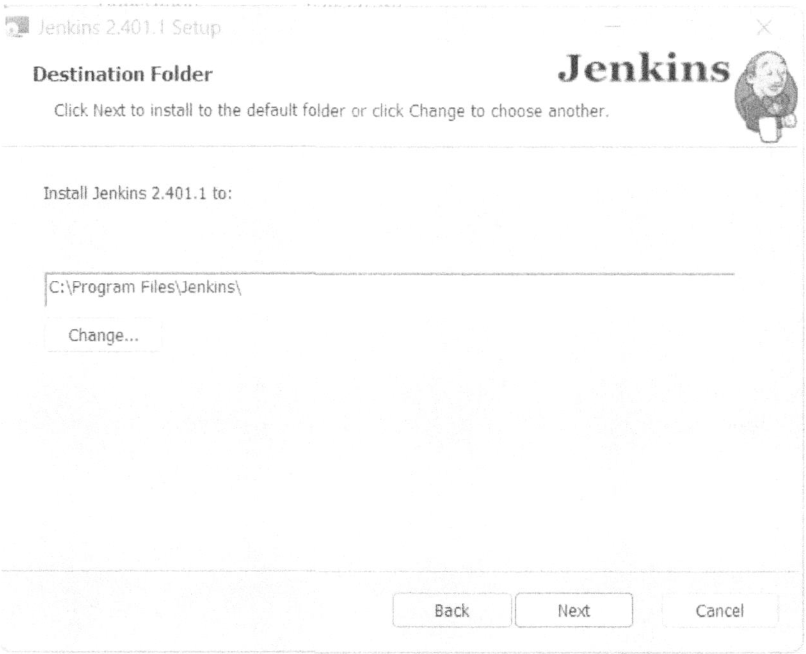

Figure 25-3- Default Jenkins installation

5. Click on 'Allow' when a prompt message is displayed asking "Allow this app to make changes to your system?",
6. Click **Finish** once installation is finished

Figure 25-4- Finish Installation

7. This opens a sign in page in Jenkins asking for Administrator password.

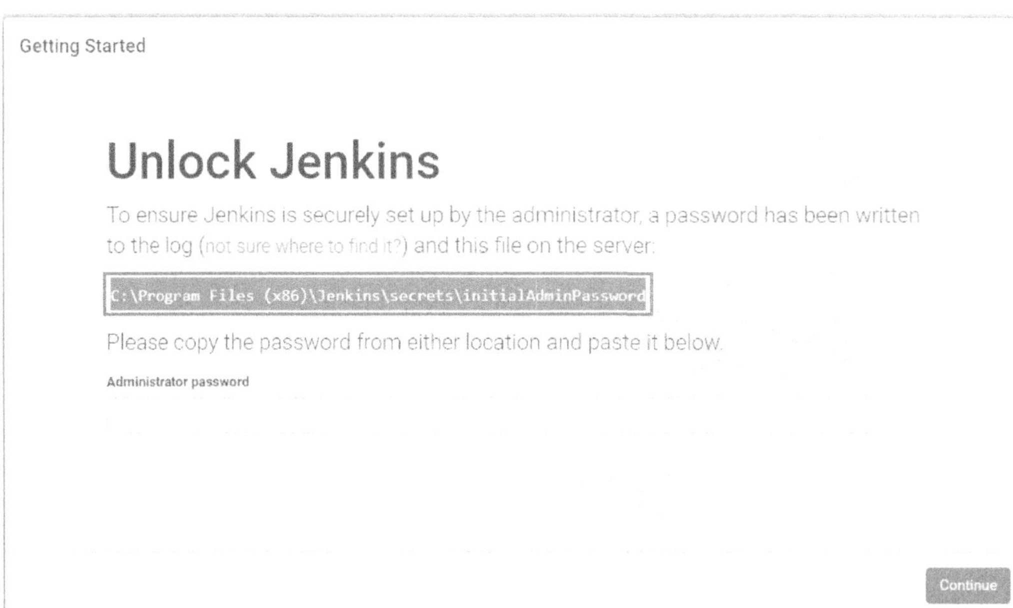

Figure 25-5- Unlock Jenkins

8. Copy paste the path highlighted into the Windows Explorer and open the file. Copy paste the password into the Administrator Password textbox and click on Continue

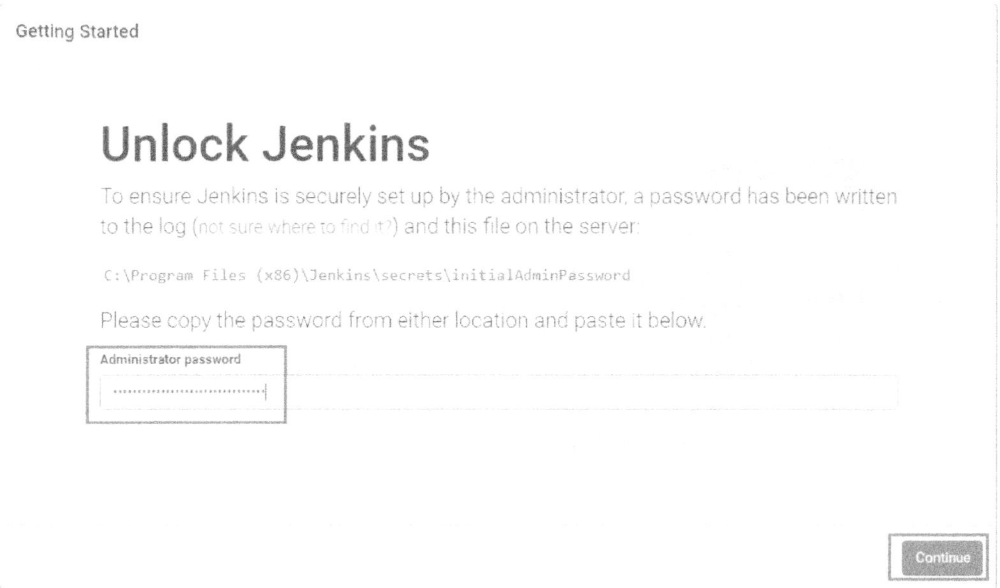

Figure 25-6- Enter Administrator password

9. Page is now open to customize Jenkins. Select the option: Install suggested Plugins. This will install all the required plugins

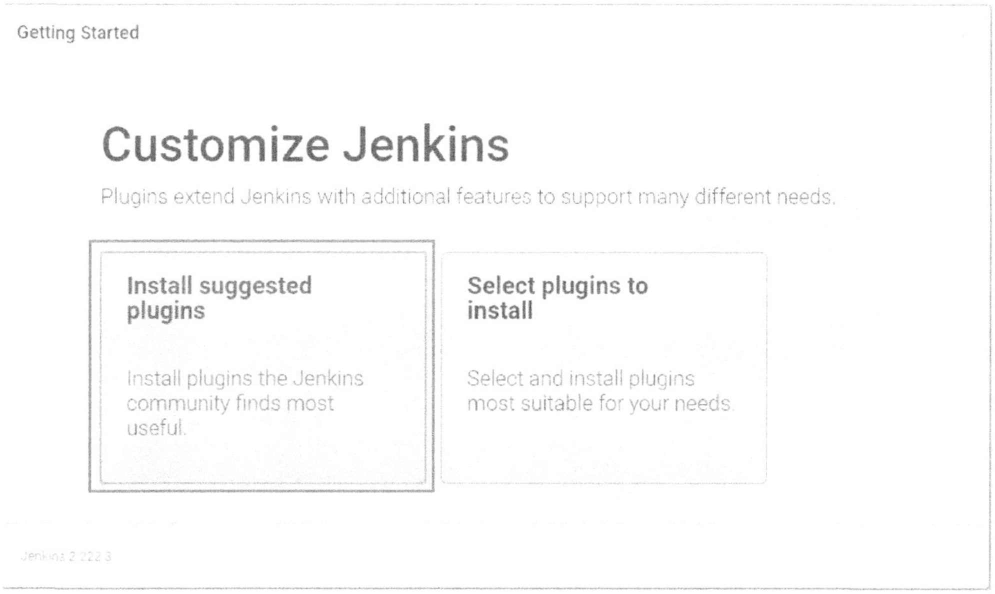

Figure 25-7- Customize Jenkins

10. Now a page is open with the option to create a user. Add all the details and click on Save and Continue.

Figure 25-8- Create First Admin User

11. This opens a page with Instance configuration. Click on '**Save and Finish'** here.

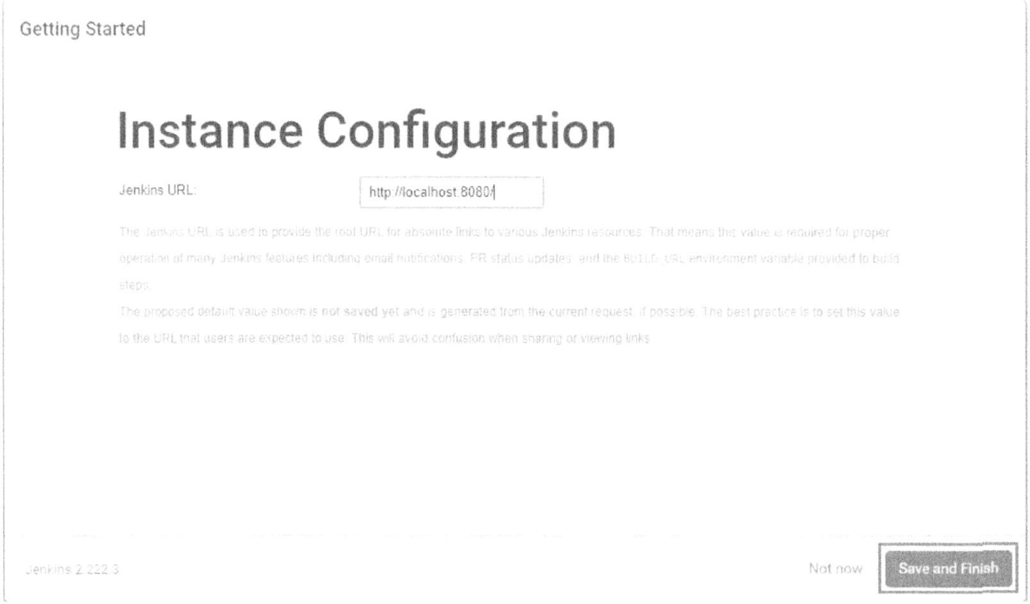

Figure 25-9 - Instance Configuration

12. A page is open with the text: Your Jenkins setup is complete. Click on '**Start Using Jenkins**'

13. This should open Jenkins default Web-page with URL (http://localhost:8080/)

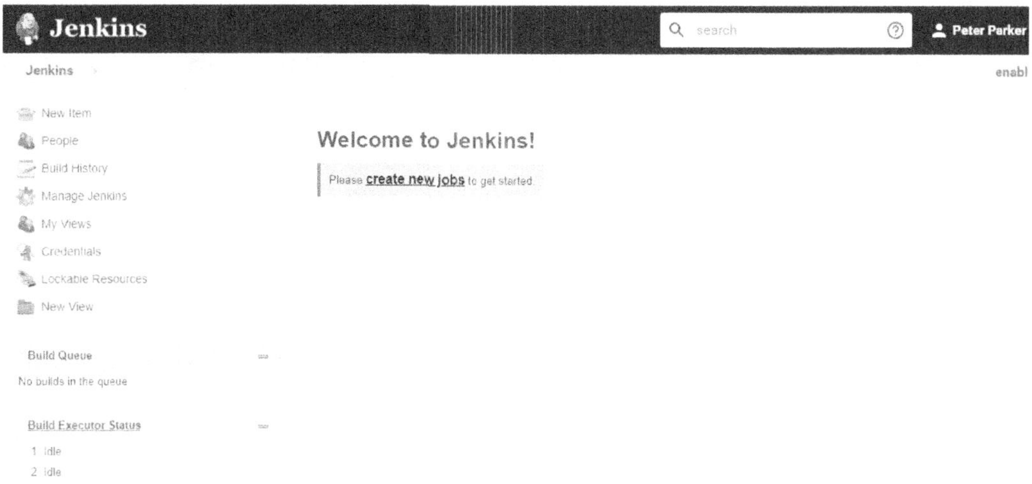

Figure 25-10- Jenkins Dashboard

25.2. Jenkins Configuration

In this section we will configure Jenkins.

1. Navigate to the Jenkins Dashboard (http://localhost:8080) by default
2. On Jenkins Dashboard, click on the **Manage Jenkins** link

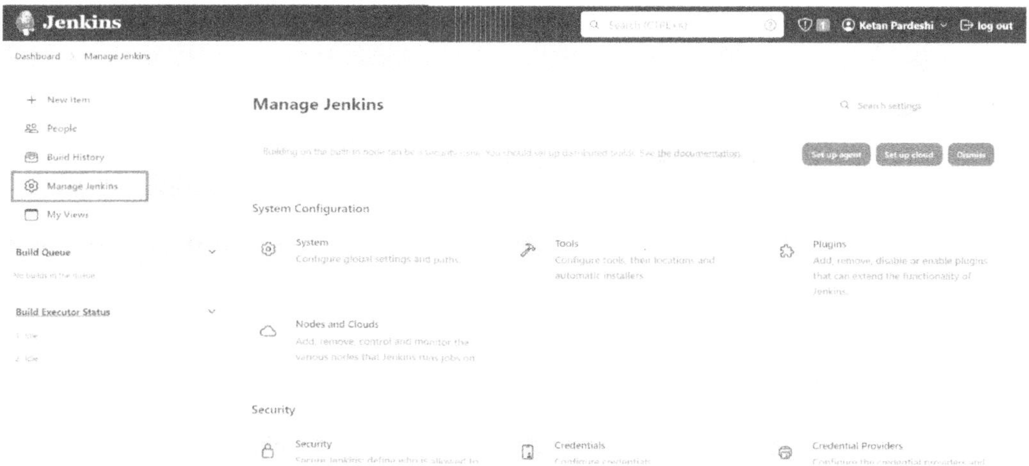

Figure 25-11- Manage Jenkins Link

3. Click on **Global Tool Configuration** link in Manage Jenkins page

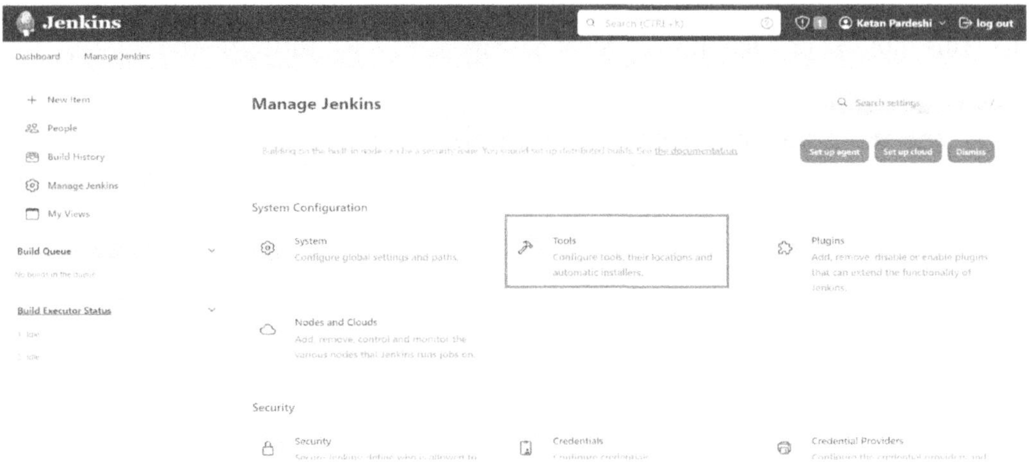

Figure 25-12- Tool Configuration Link

4. You will see Global Tool Configuration Page

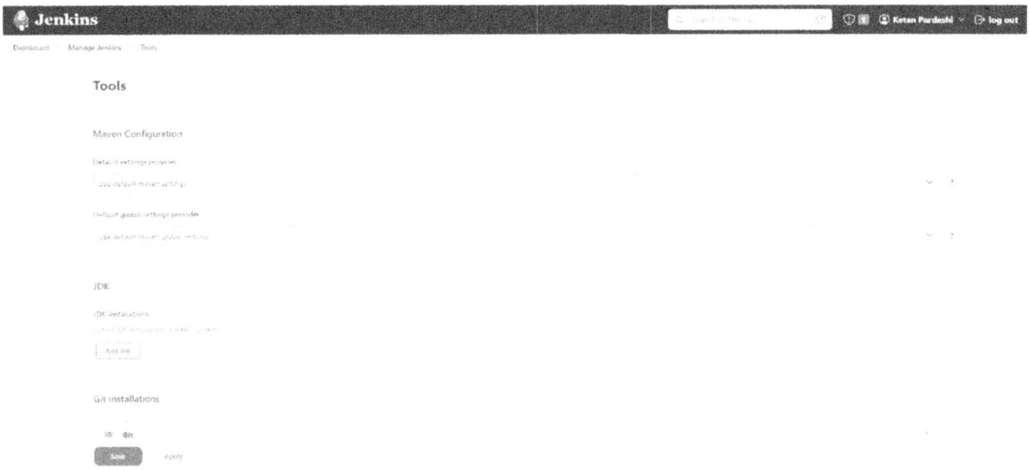

Figure 25-13- Tool Configuration Page

Configuring JDK

5. On the Global Tool Confguration page, locate the **JDK section**
6. Click on the **Add JDK** button in the JDK section
7. Specify JDK1.8 in the **Name** field and unselect the **Install automatically** checkbox

8. In the **JAVA_HOME** textbox, enter the path of the JDK folder from your system. By default it should be **C:\Program Files (x86)\Java\jdk1.8.0**

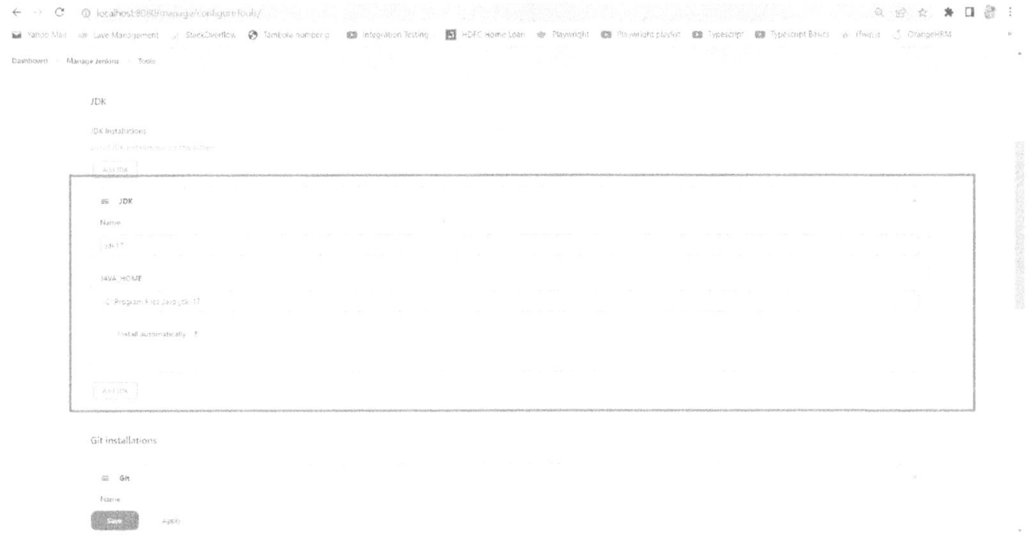

Figure 25-14- JDK Configuration

Configuring ANT

We can run Selenium WebDriver tests in Jenkins using ANT. Make sure ANT is installed and configured on the same machine as the Jenkins server. We had earlier installed ANT as part of "Installing Selenium Components" chapter.

9. On the Configure System page, locate the **Ant** section
10. Click on the **Add Ant** button in the Ant section
11. Specify Ant in the **Name** field and unselect the Install automatically checkbox
12. In the **ANT_HOME** textbox, enter the path of the Ant folder from your system. By default ANT path on window is C:\Program Files (x86)\WinAnt
13. Click on Save after these configurations are complete.

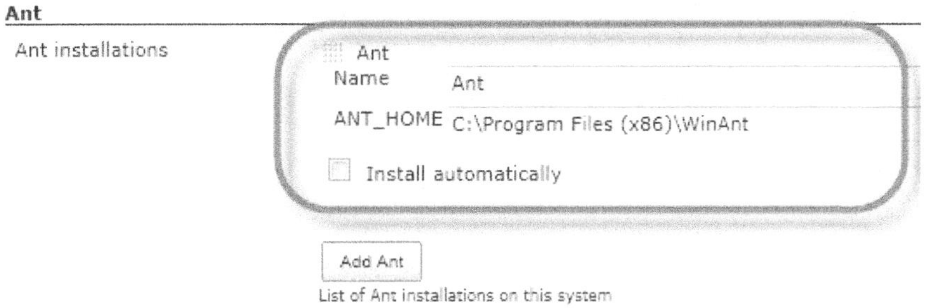

Figure 25-15- ANT Configuration

25.3. Selenium WebDriver Test Execution in Jenkins

1. Navigate to the **Jenkins Dashboard** (http://localhost:8080 by default)
2. On Jenkins Dashboard, click on the **New Item** link to create a new CI item.

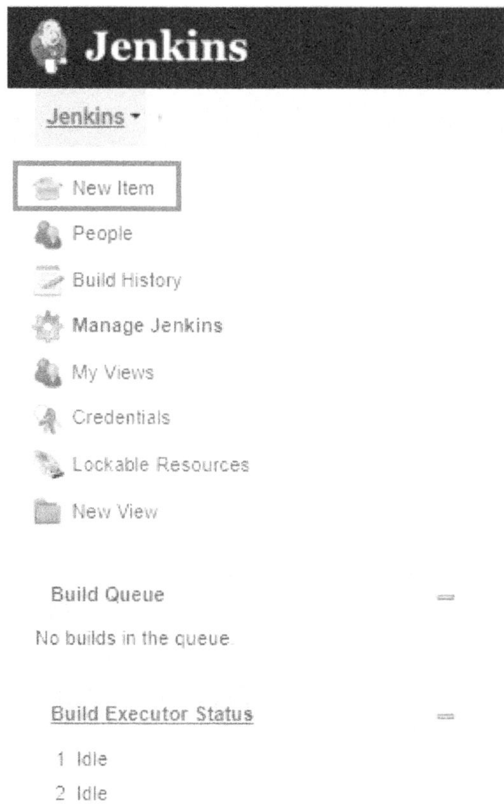

Figure 25-16- New Item Link

3. Enter **HotelApp_TestAutomation** in the **Job name** field
4. Select **Free-style** project option
5. Click **OK**

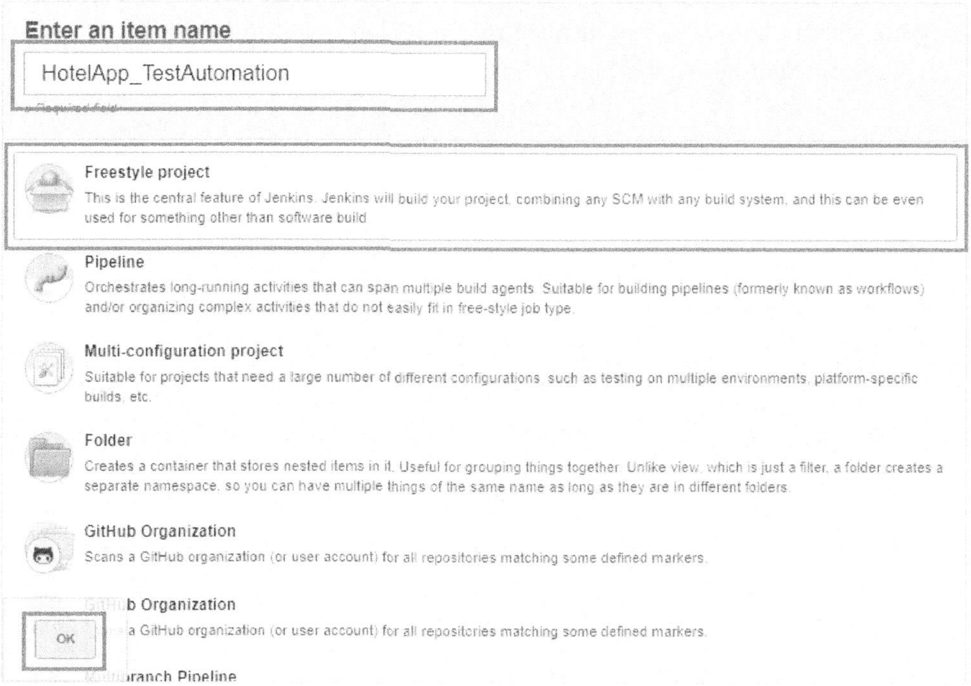

Figure 25-17- Item Name

6. In the configuration screen if you are using any existing Source code management (SCM) system you can select it (Subversion or Git). You would need to provide extra configuration details for that. See the snapshot below for Git settings.

Figure 25-18- Git settings

7. In case you would want to run the script locally without any SCM you can select **None**. In this case you would need to copy your Selenium Project folder (which exists in workspace) to the Jenkins installation directory

Figure 25-19- None setting

8. For the scope of this book we will assume users will have the **None** setting
9. In the Build Section, select **Add build Step → Invoke Ant**

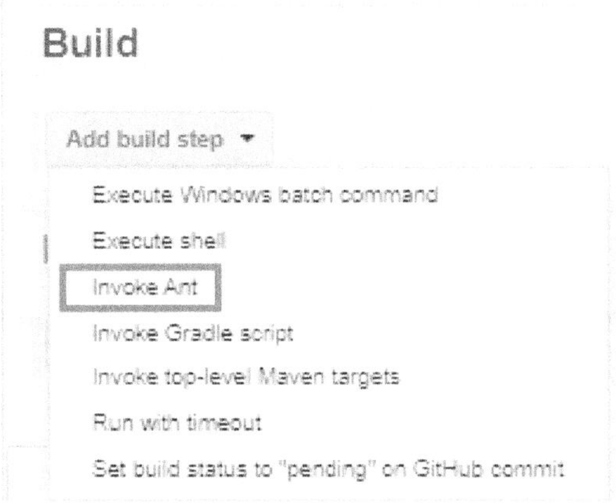

Figure 25-20 – Invoke Ant

10. In Ant version select **ANT** from the Ant Version drop-down

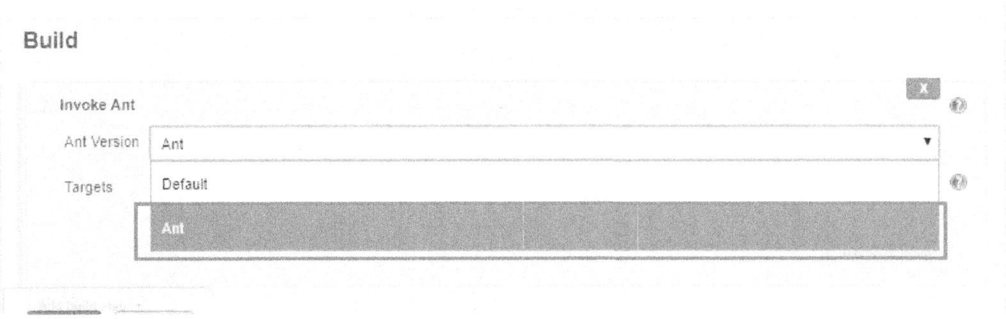

Figure 25-21 – ANT version selection

11. Click on the **Save** button

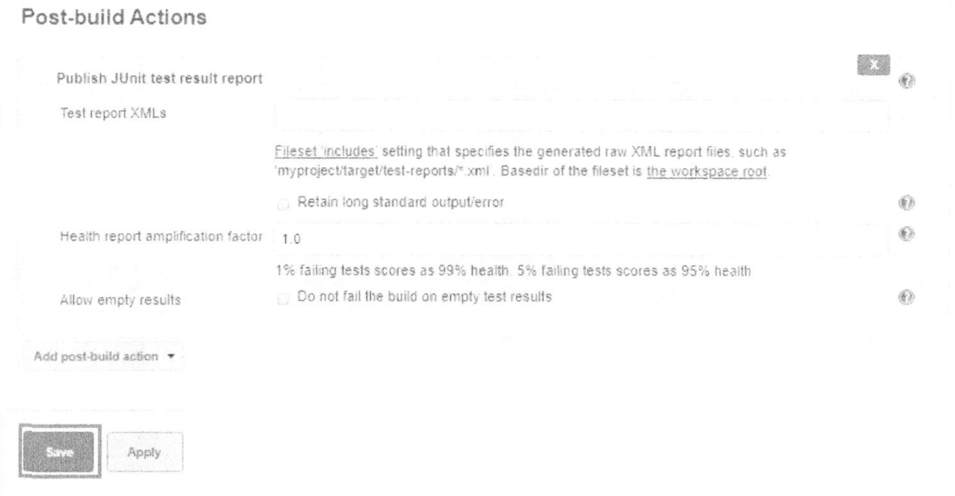

Figure 25-22- Save Settings

12. Since we will be using local Jenkins workspace, we have to run the job once to create a workspace placeholder. So **click on Build Now button.**

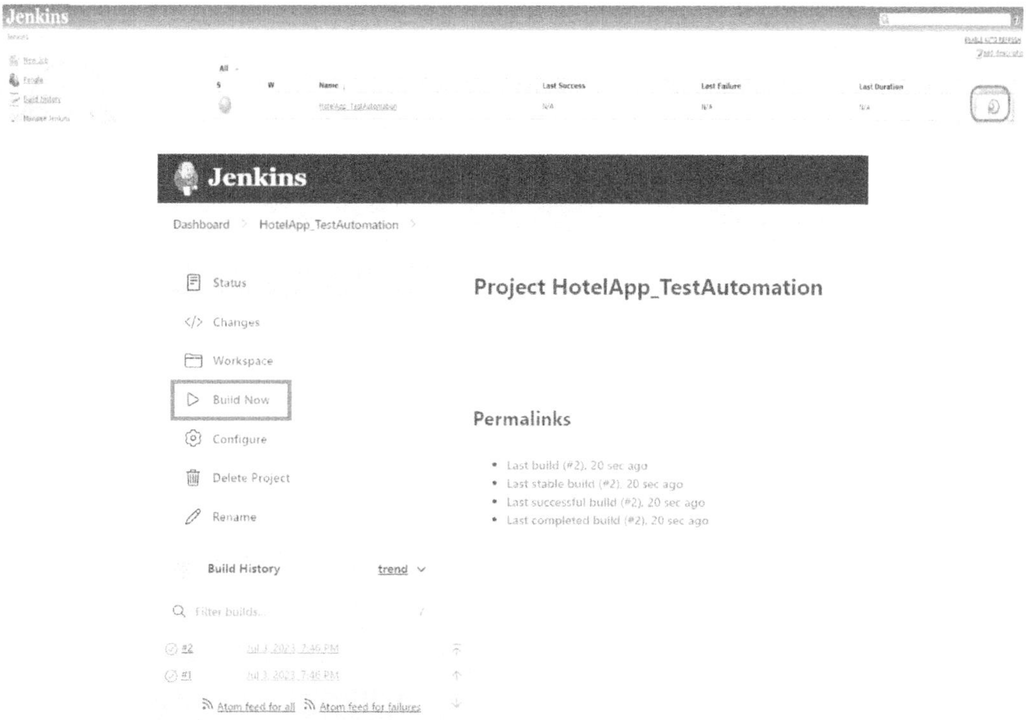

Figure 25-23- Build Now

13. This will run the Job and create a placeholder workspace folder in the Jenkins default directory in our case C:\ProgramData\Jenkins\.jenkins

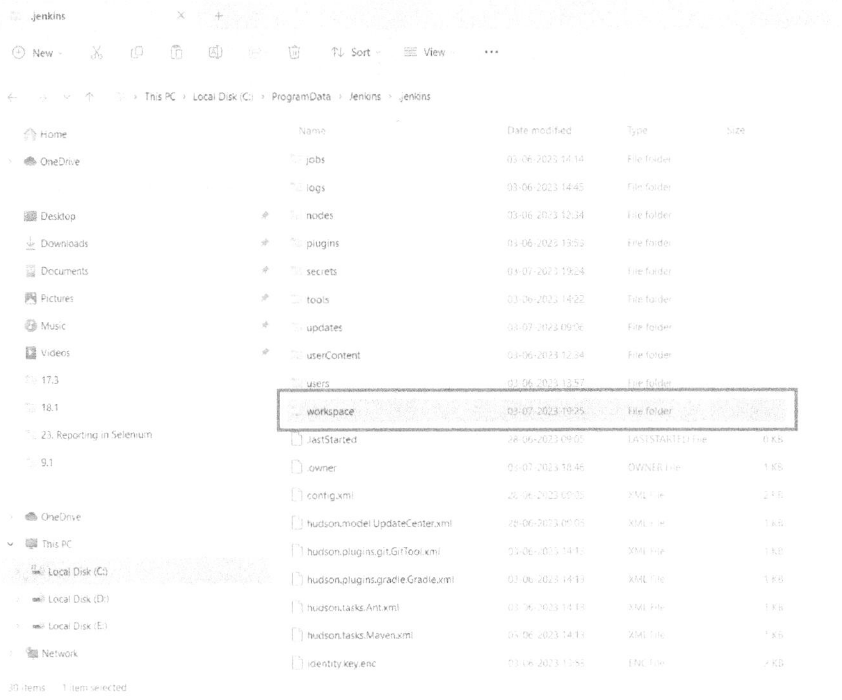

Figure 25-24- Workspace folder

14. In this workspace folder we will **copy** our project folder from location **C:\Selenium\workspace**

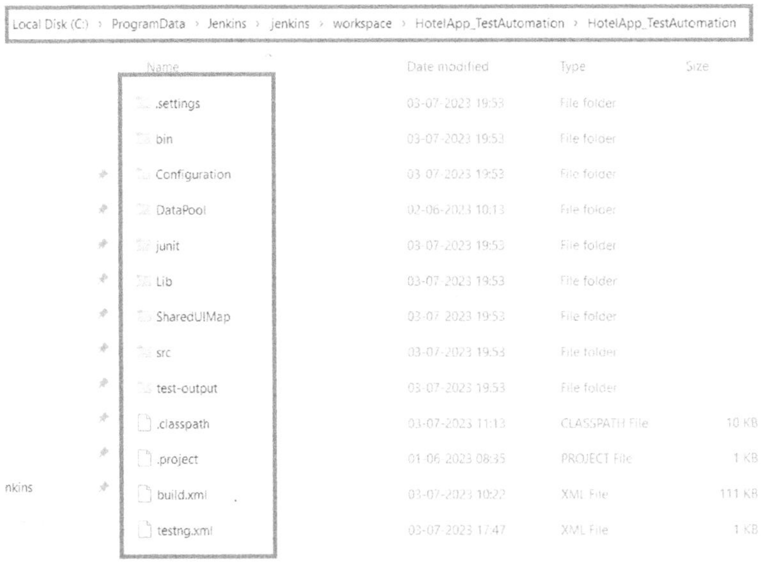

Figure 25-25- Project Folder copied in Jenkins directory

15. Verify the **build.xml** file exists in your project folder in C:\ProgramData\Jenkins\.jenkins\workspace\HotelApp_TestAutomation\HotelApp_TestAutomation

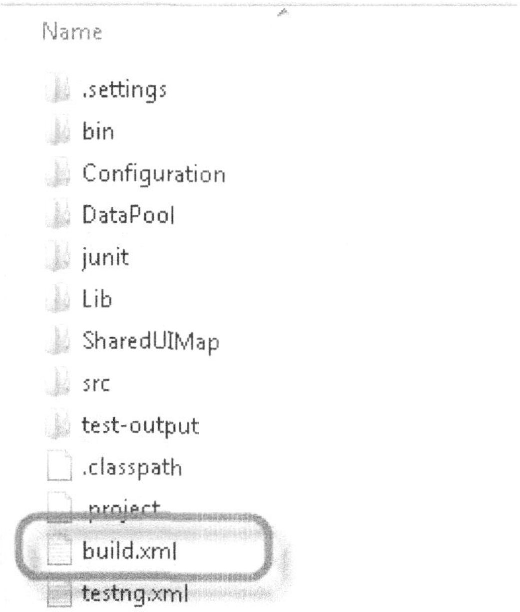

Figure 25-26- Build.xml file

Note: If you do not see the build.xml file in your project structure, then follow the instructions in the "Reporting with Selenium" chapter to get the build.xml file

16. Open the Build.xml file and select any target name you would want to run. Let us say we will run the script with the target name **VerificationPointTest** and **junitreport**

```
<target name="VerificationPointTest">
    <mkdir dir="${junit.output.dir}"/>
    <junit fork="yes" printsummary="withOutAndErr">
        <formatter type="xml"/>
        <test name="tests.VerificationPointTest" todir="${junit.output.dir}"/>
        <classpath refid="HotelApp_TestAutomation.classpath"/>
    </junit>
</target>
<target name="SharedUIMapTest">
```

Figure 25-27- Target Test

```
    <target name="junitreport">
        <junitreport todir="${junit.output.dir}">
            <fileset dir="${junit.output.dir}">
                <include name="TEST-*.xml"/>
            </fileset>
            <report format="frames" todir="${junit.output.dir}"/>
        </junitreport>
    </target>
</project>
```

Figure 25-28- Target junitreport

Note: We have this test from our previous build.xml file configured in "Reporting in Selenium" Chapter.

17. On the job page, select **HotelApp_TestAutomation → Configure link**

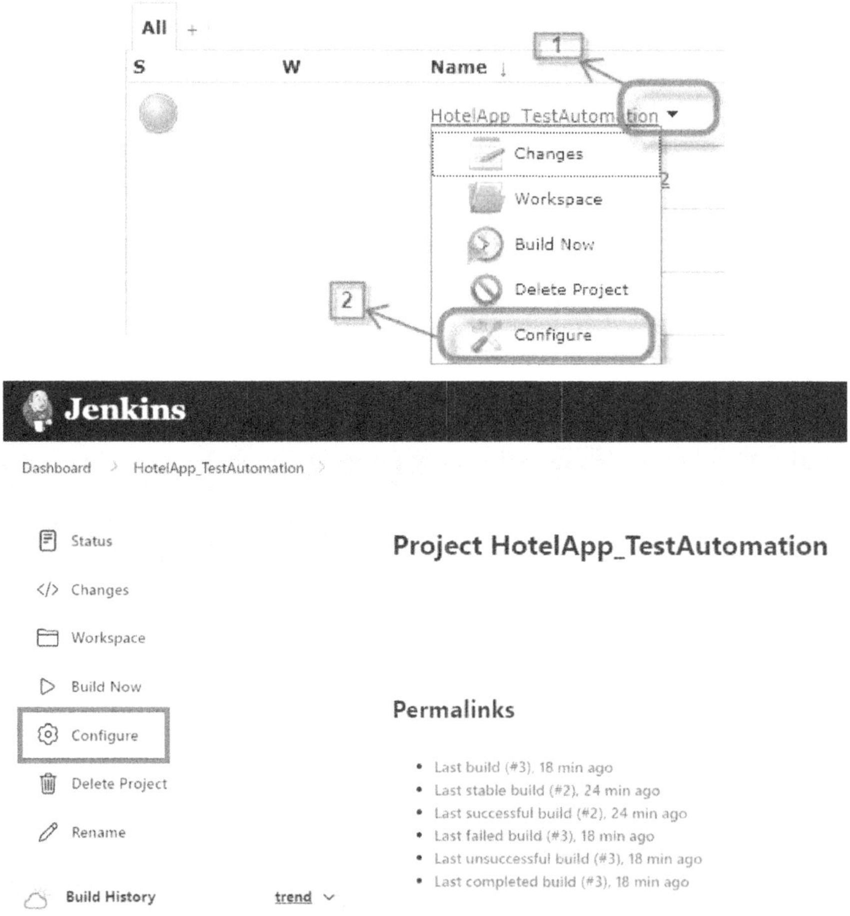

Figure 25-29- Configure Project

18. In the build section for the field Target enter the value VerificationPointTest and junitreport

Figure 25-30- Targets

19. Click on **Advanced** button in **Build section**

Figure 25-31- Advanced button

20. Enter build.xml path in **Build File** field which is the path from your Jenkins directory. In our case use C:\ProgramData\Jenkins\.jenkins\workspace\HotelApp_TestAutomation\HotelApp_TestAutomation\build.xml

Figure 25-32- Build File Path

21. In the Post Build Action section, select **Add Post-build action → Publish JUnit test result report**

Figure 25-33 – Post build action

22. In the **Test report XMLs** field enter ****\junit*.xml**

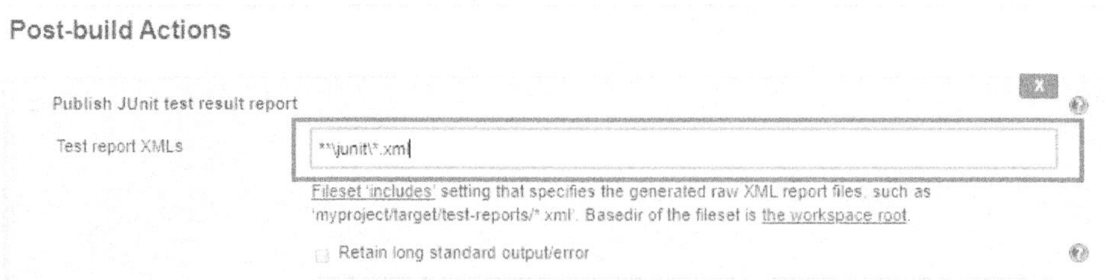

Figure 25-34 – Test report XMLs

ANT test results are stored in the junit folder under the eclipse project. Hence we are **using wildcards to find all the xml files under the junit folder.**

23. Click on the **Save** button

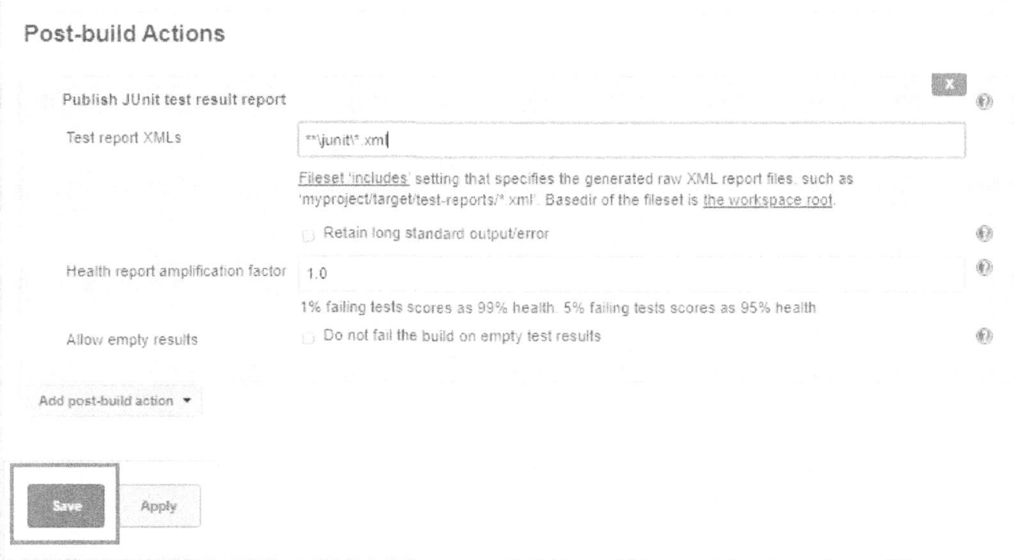

Figure 25-35 – Save Button

24. Click on **Back to Dashboard** link to see the Jobs page

Now we are ready to execute our test.

25. Click on the **Schedule run** button next to our Job **HotelApp_TestAutomation**

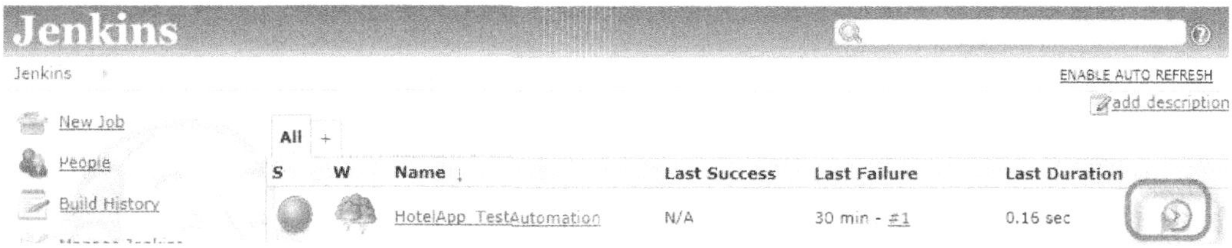

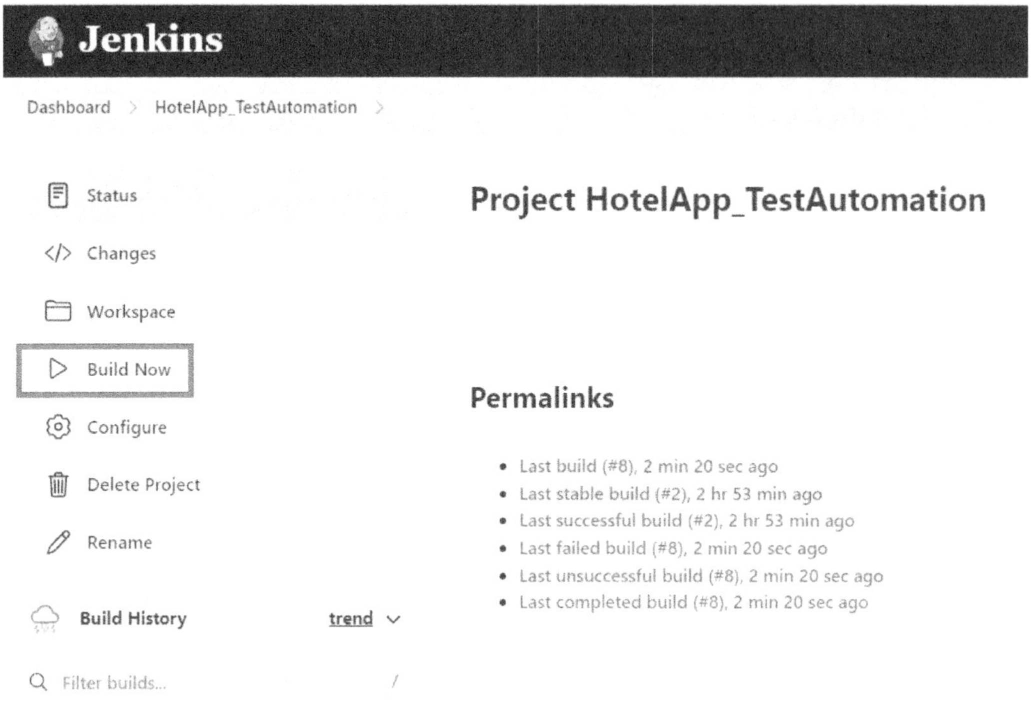

Figure 25-36 – Schedule run

26. You will notice that the Job starts execution

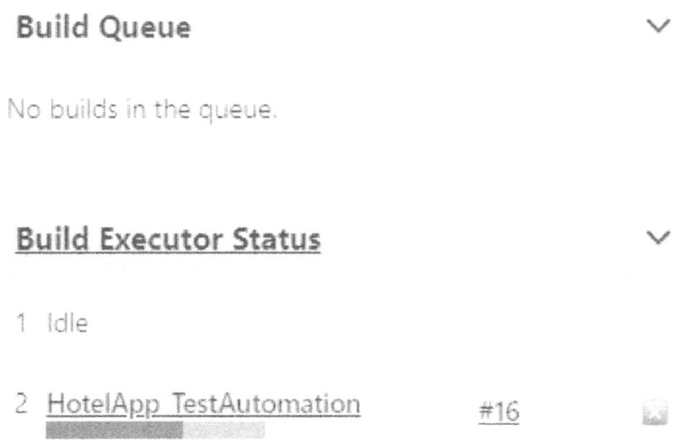

Figure 25-37 – Job Execution

Note – You will not see anything being executed on the screen. This is because execution is happening in silent mode.

27. You will notice that the Job starts executing. You can look at the console output to see the execution progress.

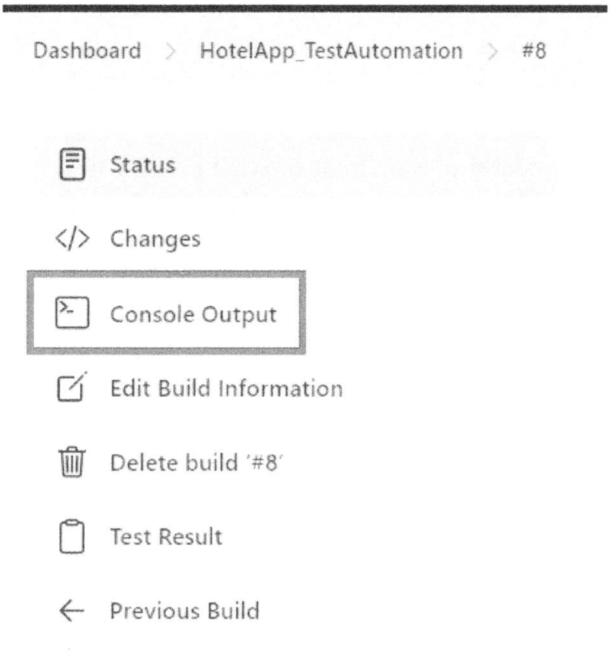

Figure 25-38 – Console Output

28. After Test Run finishes you can drill down and see the Test Results

Figure 25-39 – Test Results

Note –Make sure to add the junit.jar in ANT_HOME/Lib folder for successful execution of the item.

Note – You can run Jenkin tests from the command line too using java -jar jenkins-cli.jar -s http://localhost:8080/build HotelApp_TestAutomation

jar jenkins-cli.jar can be downloaded from http://localhost:8080/cli link.

Note - Scheduling the build for automatic execution

- Go to the Selenium Cookbook project configuration in Jenkins.
- In the Build Triggers section, select the Build periodically checkbox and enter scheduling details

Note – IF you are working within an internal proxy you can add proxy details at **Jenkins → Manage Jenkins → Plugins → Advanced Settings**

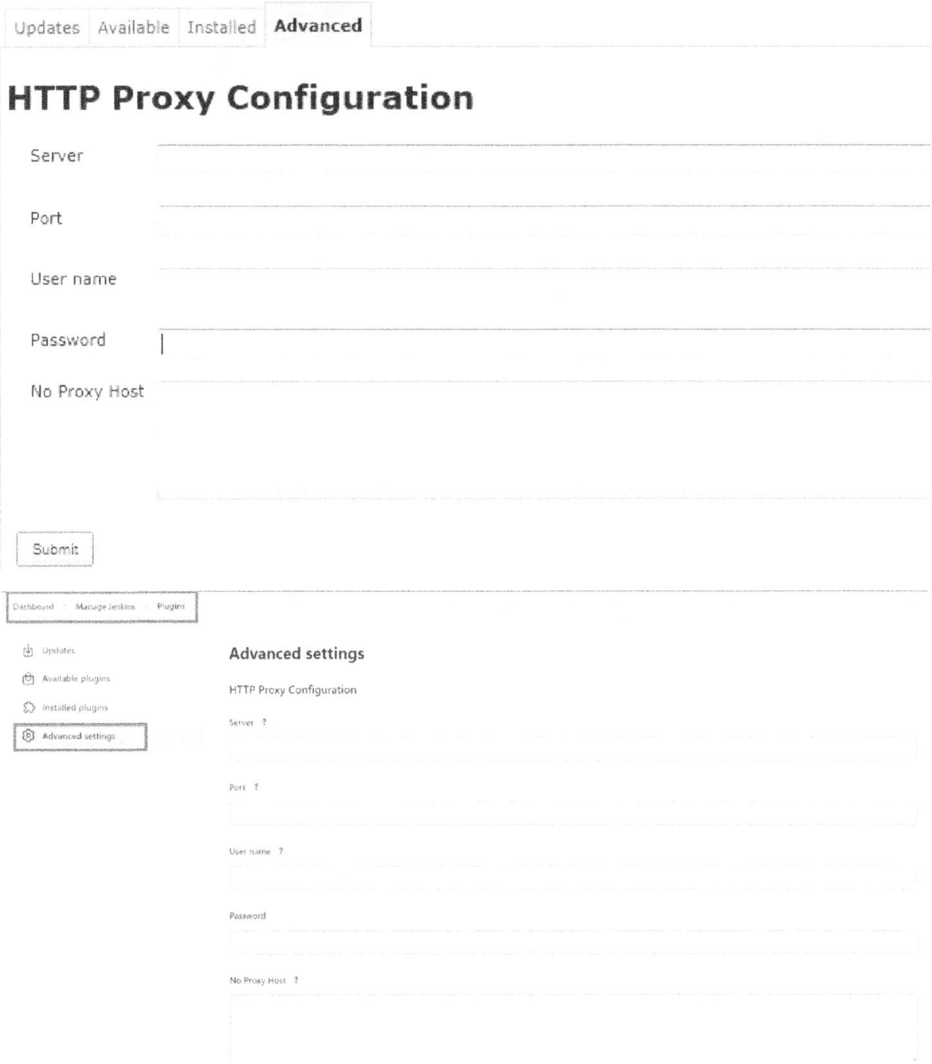

Figure 25-40 – HTTP Proxy Configuration

§

26

Automation Frameworks

Being a part of the software testing domain, we would have heard the term 'Automation Frameworks' time and again. Again, it is a very common question one encounters at interviews too. In this chapter we will try to understand the answers to three basic questions:

- Why do we need a framework? What are the advantages of frameworks?
- What exactly is an automation framework? What are the components of the framework?
- How do we implement frameworks? What are the different types of frameworks?

26.1. Why do we need Automation Frameworks?

1. Maintainability

One of the key reasons behind creating an automation framework is to reduce the cost of script maintenance. If there is any change in the functionality of the application, then we need to get our scripts fixed and working utilising the least amount of time and effort.

Ideally, there should not be too much need to update the scripts, in case the application changes. Most of the fixes should be handled at the framework level itself.

2. Productivity

If I ask how many manual test cases we can automate in a day that might be a difficult question to answer. But the important thing to ask is whether we can increase our productivity by automating more test cases per day?

Yes, we can. If we have an effective framework, we can increase the productivity manifold. In one of my previous projects, we increased the productivity from 3-4 test cases a day to 10-12 test cases a day, mainly through effective framework implementation.

3. Learning curve

If you have a new person joining your team, you would like to reduce the effort in training the person, and have him/her up and running on the framework as soon as possible.

Creating an effective framework helps reduces the learning curve.

As a best practice, I always advise my clients to keep the framework as simple as possible.

4. Make result analysis easier

Once the test cases are automated, a lot of time is spent by the testing team on analysing the results. Sometimes they are not detailed enough, which might make it hard to pinpoint the error. Most often it is not script failure but environment or data issues that turns out to be the source of problems. A better reporting format in the framework will cut down on result analysis time considerably.

26.2. What exactly is an Automation Framework?

Frameworks are a set of guidelines which define how we will structure the various components in an automation environment. These components include object repository, test data, functions, reports and batch execution script.

When the development team begins development, it creates a high level design of the application. Similarly, we, as an automation team, need to create an automation framework to define how different automation components will interact with each other. We can also call it high level design for automation components.

So what are the components of the frameworks? Let us see bel

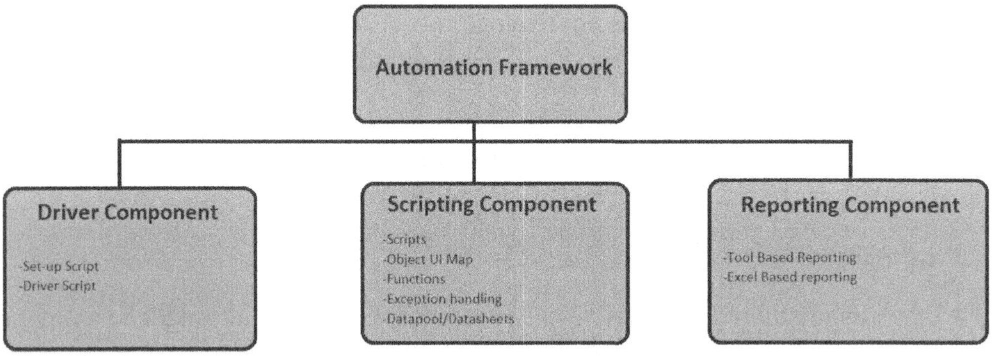

Figure 26-1 – Automation Framework Components

At a high level, Automation Framework can be divided into three components:

- **Driver Component** – How will you execute the script as a batch? What setup will you need before you start execution?
- **Scripting Component** – How will you structure all of the key components of your automation framework?
- **Reporting Component** - How will you get your results?

Together, all these components make up an **Automation Framework**.

So, let us understand what exactly is stored within these components.

- **Setup Script** – This script defines what setup you need before you can start script execution

example

In one of our projects, we had to install the new desktop thick client application, before we could kick off the automation test execution. So we created a setup script to download the latest thick client and install it on the test machines to set up the environment.

- **Driver Script** – Most of the time we need to run our scripts overnight. So we need a driver script which can run the scripts as a suite (one after another).

We discussed in previous chapters that we could use Batch Runner as the driver program or we could use Quality Center/ALM as the batch program for UFT. But if you have different automation tools, you might need to select a different batch program.

- **Scripts** – These are the actual automation programs, which execute just like manual test cases
- **Shared UI Map** – A common place where all object properties and information is stored
- **Functions** – Modular programs which can be re-used across scripts
- **Configuration File** – File in which we can set up the application URL and other variable parameters in our application
- **Exception Handling** – Exception handling scenarios in our application
- **Data Pool** – Test data that will drive our scripts
- **Tool Based Reporting** – Most of the automation tools will have their own reporting format. For instance, Selenium can run tests and create TestNG or JUnit reports. Simiilarly in OpenText UFT results can be viewed in UFT results viewer.

It is also important to think about where to store these results for future reference and the ease of drilling down to the results you are looking for.

With Selenium we can store results in database and read them using custom Web based interface. This does not come as default with Selenium but can be developed.

OpenText Quality Center/ALM as a test management tool provides a good result storage solution for OpenText UFT.

- **Excel Based Reports** – Not all tools have corresponding test management tools to store results. Quite often and as part of a framework, the automation team has to develop their own excel drivers (at times database driven) to store results in Excel Sheets, for easy viewing and analysis.

26.3. Types of Frameworks

This brings us to the next important question of how to implement frameworks and what the different types of frameworks are.

Automation developers have different reasons for following a particular framework and every framework has its own advantages and disadvantages.

We can divide frameworks into the following categories

- Level 1- Record- Replay
- Level 2- Data Driven Framework
- Level 3- Test Modularity Framework
- Level 4 - Hybrid Framework
- Level 5 - Keyword Driven Framework
- Level 6 – Page Object Model Framework

I would call Level 1 as the lowest level of framework and Level 5 as the highest.

Let's understand more about each of the different levels.

Level 1 - Record-Replay

This is not really a framework, but helps as a starting point to an introduction to frameworks.

- This framework provide the ability to capture/record the user actions and later to play them back
- Also called as Capture/Replay approach
- Enhance the recorded script with verification points, where some property or data is verified against an existing baseline
- Also note that as part of this framework, we use a shared object repository across all the scripts

Advantages

- Fastest way to generate scripts
- Automation expertise not required

Disadvantages

- Little re-use of scripts
- Test data is hard coded into the script
- New scripts always take same time to automate as previously automated scripts
- Maintenance is a nightmare

One of the key issues with this framework is that if the application workflow changes or if the test data changes, we need to go into each script and modify the script.

For instance, if you have 500 automation scripts for your application and username or password changes for your login page, you would need to go into each of the scripts and fix them, which can be a nightmare.

This leads us to our next level of framework, which resolves this issue.

Level 2 – Data Driven Framework

In this framework, while test case logic resides in test scripts, the test data is separated and kept outside the test scripts. Test data is read from the external files (Excel files, Text files, CSV files, and database) and loaded into the variables inside the test script. Variables are used both for storing input values and verification values. Test scripts themselves are prepared using the record replay approach.

Since the data is stored outside the script, if as in our previous example the username or password changes, we would need to change just one datasheet and all our 500 scripts will be fit for execution. So we avoided a huge maintenance effort using this framework.

Also we can use the same script to run multiple sets of data defined in external datasheets helping us achieve more return on investment.

Advantages

- Test data can be changed at one central place and there is no need to modify the scripts
- Changes to the Test data do not affect the Test scripts
- Test cases can be executed with multiple sets of data

Disadvantages

- If functional workflow of the application changes, it will be a maintenance nightmare
- No re-use of code

One of the key issues with the above framework is that if the workflow of the application changes, you would need to go back and fix all the scripts again.

For instance, assuming you have 500 scripts, and in each script you login to the application. Due to a new business requirement, apart from just username and password, your application now also requires your business unit name to be entered before login. This represents a change in application. Even though your data resides outside the script, you still need to go into each of the scripts and add extra lines of code to enter the business unit name. This approach is still a nightmare!

Let us look at the next level of frameworks which helps to handle this issue.

Level 3– Test Modularity Framework

As part of this framework we divide the application-under-test into libraries (Functions or Action based). These library files are then directly invoked from the test case script. This framework embodies the principle of abstraction.

In this framework, we can re-use a lot of the existing code, which helps to improve productivity.

Considering our earlier example where our login workflow has changed. We will be able to handle that issue using this framework more simplistically, as we would have created login as a separate function. This login function will be invoked from all our scripts. So we just need to add a step in the login function to enter a value in the business unit field and all our scripts should be fine.

So, as we would have understood, I would consider modular and data driven frameworks work differently; one utilizes a modular approach and the other focuses around data.

Advantages

- Higher level of code re-use is achieved in Structured Scripting compared to "Record & Playback"
- Automation scripts are less costly to develop due to higher code re-uses
- Easier Script Maintenance

Disadvantages

- Technical expertise is necessary to write Scripts using Test Library Framework
- More time is needed to plan and prepare test scripts
- Test Data is hard coded within the scripts

Level 4– Hybrid Framework

In the previous section, we saw the advantages with a data driven framework and test modularity framework. Should we not get benefits of both the data driven and modular approach?

This is exactly what we do in a hybrid framework. We keep data outside our scripts and create modular functions.

Advantages

- Higher level of code re-uses
- Test Data is at a central location and can be changed on demand
- Higher productivity and more scripts can be automated as we build the libraries
- Easier script maintenance

Disadvantages

- Technical expertise is necessary to write scripts and understanding of existing functions could take time
- More time is needed to plan and prepare test scripts
- Can be used by expert automation testers only

Note: Hybrid frameworks constitute 80-90% of the frameworks, which are highly successful.

Another common form of Hybrid Framework is Page Object Model Framework.

Page Object Model Framework

Within your web app's UI, there are areas where your tests interact with. A Page Object only models these as objects within the test code. This reduces the amount of duplicated code, which means that if the UI changes, the fix needs only to be applied in one place.

Page Object is a Design Pattern that has become popular in test automation for enhancing test maintenance and reducing code duplication. A page object is an object-oriented class that serves as an interface to a page of your AUT. The tests then use the methods of this page object class whenever they need to interact with the UI of that page. The benefit is that if the UI changes for the page, the tests themselves don't need to change, only the code within the page object needs to change. Subsequently, all changes to support that new UI are located in one place.

Advantages

- There is a clean separation between the test code and page-specific code, such as locators (or their use if you're using a UI Map) and layout.
- There is a single repository for the services or operations the page offers rather than having these services scattered throughout the tests.
- Code becomes less and optimized because of the reusable page methods in the POM classes.
- Since the test classes are separate from the classes containing the web elements and the operation on them, updating the code is very easy if any web element updates or a new one adds.

Level 5– Keyword Driven Framework

Hybrid frameworks have a lot of advantages, but the disadvantage is that they get too technical. Inherently testers are not programmers and so automation gets limited to automation testers only and cannot be done by functional testers or business analysts.

Keyword driven framework makes it easier for functional testers and business analysts to be involved in automation. Let us see how.

The keyword driven or table driven framework requires the development of data tables (usually Excel Sheets) and keywords, **independent of the test automation tool** used to execute them. Tests can be designed with or without the application.

For example, instead of recording a script to login to the application, if we had an Excel Sheet to store username and password and re-use, wouldn't that be easy enough for functional testers? See table below.

Object Object Repository	Action (KEYWORD)	TestData
WebEdit(UserName)	Set	adactin123
WebEdit(Password)	Set	Xxxxx
WebButton(Login)	Click	
Browser(Adactin.com)	Verify	Loads

But how will the script actually run?

Embedded within the back-end, there will be an intermediate component, which will translate this Excel Sheet at run time and create an automation script on the fly.

The key point to remember about this framework is that the intermediate component which will translate high level excel sheet statements written by non-programmers is the complex part and can take time. Usually we would need expert programmers to write the intermediate component.

Advantages

- Provides high code re-usability
- Test tool independent
- Independent of Application under Test (AUT), same function works for other applications (with some limitations)
- Tests can be designed with or without AUT (Application under Test)

Disadvantages

- Initial investment being pretty high, the benefits of this can only be realized if the application is considerably bigger, and the test scripts are to be maintained for a few years
- Debugging of this kind of framework can be very hard
- Test data is hard-coded within every Excel based test script, which leads to data issues
- A high level of automation expertise is required to create the keyword driven framework

Even though keyword driven framework might look like the coolest thing to work on, I have seen a lot of keyword driven frameworks fail due to their disadvantages. Most commonly, I have seen that keyword driven frameworks end up being so complicated, that it is hard for anyone to debug and isolate the problem in case the script fails.

OpenText (formerly HP) Business Process Test is one of the successful keyword driven frameworks, which comes bundled with OpenText Quality Center/ OpenText ALM (though you need to purchase a separate licence for it.).

Personally, I prefer implementing the hybrid framework since it is simple to debug and handover to functional teams. Some of the keyword driven frameworks we encountered or developed were too hard for client teams to understand and they ended up not using the framework.

§

27

Selenium Functions, Common Questions and Tips

In this chapter we will try to address a few of the important selenium functions and other common questions and tips that can be used in Selenium.

Key objectives

- How to use JavaScript?
- How to take a Screen Shot?
- How to use Keyboard or Mouse movements?
- How to read row, columns and cells data from a table?
- Working with multiple browsers
- How to maximize the Browser window
- Checking an Element's Presence
- Checking an Element's Status
- Working with drop-down lists
- Working with Radio buttons and groups
- Working with Checkboxes
- Measuring Response time for performance testing using timer
- Xpath and Properties finder in IE and Chrome browsers
- How to use WebDriver test remotely using Selenium Grid?

27.1. How to use JavaScript?

Selenium WebDriver API provides the ability to execute JavaScript code with the browser window. This is a very useful feature where tests need to interact with the page using JavaScript. Using this API, client-side JavaScript code can also be tested using Selenium

WebDriver.

For those browsers that support it, you can execute JavaScript by casting the WebDriver instance to a JavascriptExecutor .

example Below code executes javascript and returns the Web page title

```
JavascriptExecutor js = (JavascriptExecutor) driver;
// returns Web page title
String title = (String) js.executeScript("return document.title");
//returns handle to Webelement with id myid
WebElement element = (WebElement)js.executeScript("return
document.getElementById(myid)");
```

Figure 27-1 – Using JavaScript Executor

You need to return from your Javascript snippet to return a value, so:

js.executeScript("document.title");

will return null, but:

js.executeScript("return document.title");

will return the title of the document.

Note: Based on the type of return value, we need to cast the executeScript() method. For decimal values, Double can be used, for non-decimal numeric values Long can be used, and for Boolean values Boolean can be used.

example: Below code will return the count of combo boxes on the Search Hotel Page

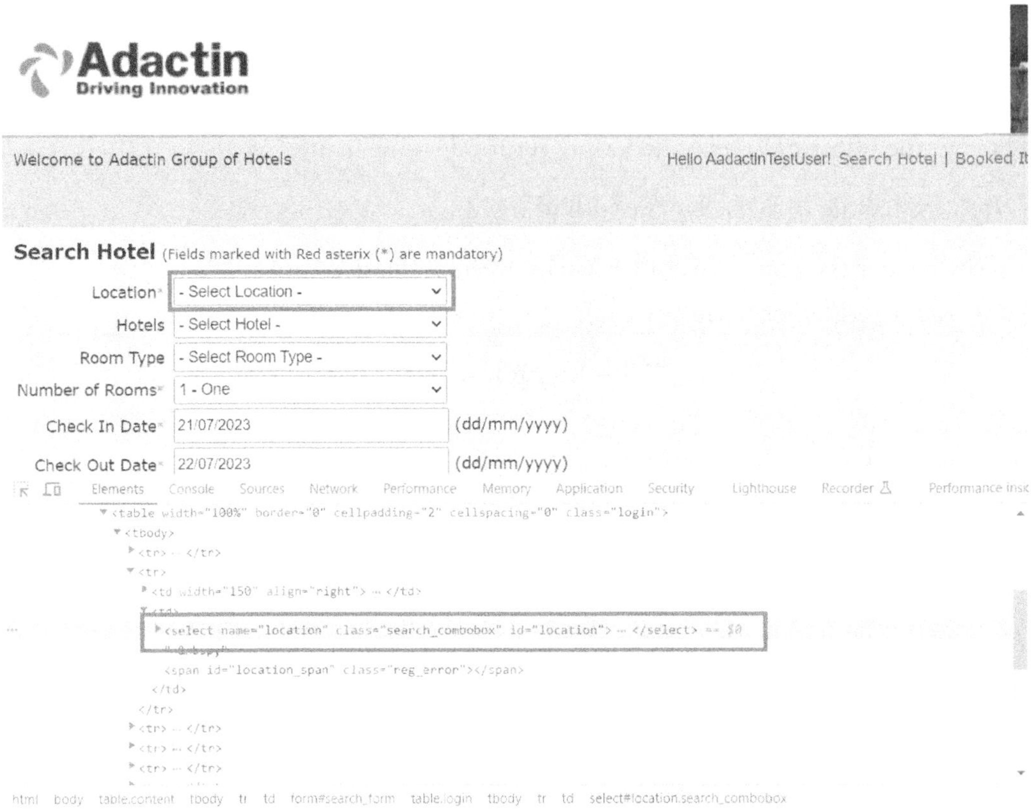

Figure 27-2 – Class of Combo boxes on search Hotel page

```
@Test
public void getComboboxcount() throws Exception
{
JavascriptExecutor js = (JavascriptExecutor) driver;

long cboxes = (Long) js.executeScript ("var cboxes =
document.getElementsByClassName ('search_combobox '); return cboxes.length");
```

Figure 27-3– Using JavaScript Executor to count comboboxes

27.2. How to take a Screenshot?

Problem – We had a client who wanted us to do take screenshots of all the critical Web pages on different browsers (Chrome, Firefox and IE) and store them at a central location which could be analysed by a business analyst. So the question was whether WebDriver can assist us in taking snapshots.

Yes, WebDriver provides a function for taking a screenshot. See below

File f = ((TakesScreenshot) driver).getScreenshotAs(OutputType.FILE);

Note: We can create it as a function which can be used as and when required to capture a screenshot.

```
//Below function takes Driver and name of screenshot file as input

public void takeScreenshot(WebDriver driver,String screenshotname) {

            String msg,path;
            String res;
            try
            {
//Take screenshot and save it in source object
          File source = ((TakesScreenshot)driver).getScreenshotAs(OutputType.FILE);

// Define path where Screenshots will be saved
            path = "./ScreenShots/" + screenshotname +".png";

//Copy the source file at the screenshot path
            FileUtils.copyFile(source, new File(path));
            msg = "Screenshot captured at"+ path;
                 res="Pass";
            }
            catch(IOException e) {
               msg = "Failed to capture screenshot: " + e.getMessage();
               res="Fail";
            }
//Reporter.log function as used in TestNG test
  Reporter.log(msg, true);
        }
```

Figure 27-4– code snippet to take screenshot

Note: For storing screenshots we can add a new folder to our framework called ScreenShots.

We can call this function as

*takeScreenshot(driver, "**TestName**")*

This will store the screenshot in our Screenshot folder (as defined in above function) with a name having our given "TestName".

27.3. How to use Keyboard or Mouse movements?

Handling special keyboard and mouse events are done using the Java's Advanced User Interactions API. It contains the Actions class that is needed when executing these events.

The following are the most commonly used keyboard and mouse events provided by the Actions class.

Method	Description
clickAndHold()	Clicks (without releasing) at the current mouse location.
contextClick()	Performs a context-click at the current mouse location
doubleClick()	Performs a double-click at the current mouse location
dragAndDrop(source, target)	Performs click-and-hold at the location of the source element, moves to the location of the target element, then releases the mouse. Parameters: source- element to emulate button down at. target- element to move to and release the mouse at.
dragAndDropBy(source, x-offset, y-offset)	Performs click-and-hold at the location of the source element, moves by a given offset, then releases the mouse. Parameters: source- element to emulate button down at. xOffset- horizontal move offset. yOffset- vertical move offset.

keyDown(modifier_key)	Performs a modifier key press. Does not release the modifier key - subsequent interactions may assume it's kept pressed. Parameters: modifier_key – any of the modifier keys (Keys.ALT, Keys.SHIFT, or Keys.CONTROL)
keyUp(modifier _key)	Performs a key release. Parameters: modifier_key – any of the modifier keys (Keys.ALT, Keys.SHIFT, or Keys.CONTROL)
moveByOffset(x-offset, y-offset)	Moves the mouse from its current position (or 0,0) by the given offset. Parameters: x-offset- horizontal offset. A negative value means moving the mouse left. y-offset- vertical offset. A negative value means moving the mouse up.
moveToElement (toElement)	Moves the mouse to the middle of the element. Parameters: toElement- element to move to.
release()	Releases the depressed left mouse button at the current mouse location
sendKeys(charsequence)	Sends a series of keystrokes onto the element. Parameters: charsequence – any string value representing the sequence of keystrokes to be sent

Figure 27-5– Keyboard and mouse events

example

Once we had an application Submit button which we were not able to identify using Selenium Webdriver. We had to click on that button to submit the form. But how could we press on that button when we were unable to recognize the control. So we used Actions class to press tab keys to bring focus on that button and then simulate Enter press. We were able to resolve our issue.

How to use Actions classes to press Tab Enter Keys

```
//Import actions class
import org.openqa.selenium.interactions.Actions;

//Instantiate a New Action Object
  Actions keyboardAction = new Actions(driver);

//Key.Enter will send Enter Key and Key.Tab will send Tab
key to the application
keyboardAction.sendKeys(Keys.Tab).perform();
keyboardAction.sendKeys(Keys.Tab).perform();
keyboardAction.sendKeys(Keys.ENTER).perform();
```

Figure 27-6– Press tab Enter keys

How to use Actions classes to drag and drop operation

Certain functionalities in our application need Drag and Drop. See below a sample code which uses the Actions class to implement Drag and Drop

```
//Import actions class
import org.openqa.selenium.interactions.Actions;

@Test
public void testDragDrop() {

      driver.get("SourceURL");
//Get Handle to Drag and Drop objects

      WebElement source = driver.findElement(By.id("TobeDragged"));
      WebElement target = driver.findElement(By.id("TobeDropped"));
//Instantiate a New Action Object

      Actions mouseAction = new Actions(driver);
//Perform Drag and Drop operation

      mouseAction.dragAndDrop(source, target).perform();
  }
```

Figure 27-7– Perform Drag and Drop

How to use Actions classes to press Double Click on element

Certain functionalities in our application need Double Click. For example some pop-up windows will open when you double click on them.

See below a sample code which uses the Actions class to implement Double click

```
//Import actions class
import org.openqa.selenium.interactions.Actions;

@Test
public void testDragDrop() {

      driver.get("SourceURL");
//Get Handle to element on which we need to perform double click
      WebElement myElement = driver.findElement(By.id("Tobedoubleclicked"));
//Instantiate a New Action Object
      Actions mouseAction = new Actions(driver);
//Perform Drag and Drop operation

      mouseAction.doubleClick (myElement).perform();
  }
```

Figure 27-8– Perform Double click

Note: Similar to the above example we can use the Action object to send mouse controls to the application too.

27.4. How to read Rows, Columns and Cell Data from Table?

A table in HTML is a collection of <tr> and <td> elements for rows and cells, respectively.

See the table in the below snapshot

Figure 27-9 – Select Hotel Table

Note: If you select every <tr> tag in chropath it will highlight the corresponding row in the browser view. It can help to find out how many rows there are in the table. Also we can figure out how many columns are available in each row.

Let's create a simple test that will print data from a table, locating its rows and columns as follows:

```
import java.util.List;

@Test
public void SelectHotelTable() {
WebElement selecthoteltable = driver.findElement(By.id("select_form"));

//Get all rows
List<WebElement> rows = selecthoteltable.findElements(By.tagName("tr"));


//Output data from each row
for (WebElement row : rows) {

// Get list of all columns within the table row and stores in cols
        List<WebElement> cols = row.findElements(By.tagName("td"));

//User Gettext method to get data from each cell in table
                for (WebElement col : cols) {
                    System.out.print(col.getText() + "\t");
                }
                 System.out.println();
        }
}
```

Figure 27-10 – Code snippet to read table cells

If you look at the logic in the table we first get the handle to the table using the statement

WebElement selecthoteltable = driver.findElement(By.id("select_form"));

Then using the handle to the table we get the list of all of the <tr> tags within the table, which gives us the count of the rows in the table

List<WebElement> rows = selecthoteltable.findElements(By.tagName("tr"));

After we get each row we can then iterate through each row and get access to data in each cell using the <td> tag

List<WebElement> cols = row.findElements(By.tagName("td"));

Once we reach each cell we use the GetText method to read data from each cell.

System.out.print(col.getText() + "\t");

27.5. Working with Multiple Browsers

So far we have used Chrome browser to run our tests. Selenium WebDriver allows users to run tests on Edge, and Firefox too..

Firefox Driver - Controls the Firefox browser using a Firefox plugin. Firefox Driver is capable of being run on Windows, Mac, and Linux. It runs in a real browser and supports JavaScript. See how we use it.

```
WebDriver driver = new FirefoxDriver();
```

Chrome Driver – WebDriver works with Chrome through the Chromedriver binary.

```
WebDriver driver = new ChromeDriver();
```

Edge Driver - This driver is controlled by a .dll and is thus only available on Windows OS.

```
WebDriver driver = new EdgeDriver ();
```

27.6. How to Maximize Browser Window?

It is always good to maximize the browser window before you start executing your test.

Use manage method to maximize the browser window

```
driver = new ChromeDriver();

driver.manage().window().maximize();
```

27.7. Checking an Element's Presence

Many a time there would be a need to check if the Web element is present on the page or not. We might even want to wait some extra time to confirm whether the Web element exists on the required page.

example - In one of the applications we tested, it had a + icon to add bookings. But the maximum number of bookings per user was 5. So, after 5 bookings the "+" icon would disappear. One of our test cases had to verify that the "+" icon does not appear after 5 bookings. How can we verify that element exists on a page or not?

See below a sample of code which helps us to check whether an element exists within the stipulated timeout

```
public boolean HotelApp_WebElementExist(WebDriver webDriver, By by, int iTimeOut)
        {
                int iSleepTime = 5000;
                for(int i = 0; i < iTimeOut; i += iSleepTime)
                {
// tries to match element based on locator
                        List<WebElement> oWebElements = WebDriver.findElements(by);
                        if(oWebElements.size()>0)
                        {
// element found is size is greater than 0 and return true
                                return true;
                        }
                        else
                        {


//element not found so wait for isleeptime and continue search again.
Thread.sleep(iSleepTime);
System.out.println(String.format("Waited for" +( i+ iSleepTime) + " milliseconds");
                        }


                                }
                // Can't find 'by' element. Therefore return false.

                return false;
        }
```

Figure 27-11 – Code snippet to check element's presence

In the above code we first try to see if we can locate the element based on locator

List<WebElement> oWebElements = WebDriver.findElements(by);

If the size of number of items is greater than 0 it means we have found the element and returned true or else we sleep for the default time and again re-verify if we can find the element.

27.8. Checking an Element's Status

Many a time there is a need to check an element's status to see if the element is enabled, selected or displayed.

The WebElement class provides the following methods to check the state of an element:

Method	Description
isEnabled()	This method checks if an element is enabled. Returns true if enabled, else false for disabled.
isSelected()	This method checks if an element is selected (radio button, checkbox, and so on). It returns true if selected, else false for deselected
isDisplayed()	This method checks if an element is displayed.

Figure 27-12 – Methods to check element status

27.9. Working with Drop-down lists

The Selenium WebDriver provides the Select class for working with Drop-down or List Controls. We can identify and locate these controls in a way that is similar to how we locate WebElements. However, we will use the Select class instead of the WebElement class.

Look back to your MyFirstWebDriverTest.java script

```
new Select(driver.findElement(By.id("location"))).selectByVisibleText("Sydney");
```

Or alternatively we could use

```
Select Location = new Select(driver.findElement(By.id("location")));

Location.selectByVisibleText("Sydney");
```

Find below a few of the Select methods we have discussed earlier in the Selenium methods chapter

Method	Purpose
selectByIndex(int index)/ deselectByIndex(int index)	Selects/deselects the option at the given index.
selectByValue(String value)/ deselectByValue(String value)	Selects/deselects the option(s) that has a value matching the argument.
selectByVisibleText(String text)/ deselectByVisibleTest(String text)	Selects/deselects the option(s) that displays text matching the argument.
deselectAll()	Deselects all options.
getAllSelectedOptions()	Returns a List<WebElement> of all selected options.
getFirstSelectedOption()	Returns a WebElement representing the first selected option.
getOptions()	Returns a List<WebElement> of all options.
isMultiple()	Returns true if this is a multi-select list; false otherwise.

Figure 27-13 – Select Methods

example See below a few alternative ways of selecting location.

```
Select Location = new Select(driver.findElement(By.id("location")));
Location. selectByIndex (2);
```

Or

```
Select Location = new Select(driver.findElement(By.id("location")));
Location. selectByValue ("Sydney");
```

example See below a sample of code to print options in the drop-down using the GetOptions method

```
Select Location = new Select(driver.findElement(By.id("location")));

for(WebElement option : Location.getOptions())
//Print options values
System.out.println(option.getText());
```

27.10. Working with Radio Buttons and Groups

Using Selenium WebDriver we can select and deselect the radio buttons using the click() method of the WebElement class and check whether a radio button is selected or deselected using the isSelected method.

Look back to your MyFirestWebDriverTest.java script. You will see this statement

```
driver.findElement(By.id("radiobutton_1")).click();
```

Alternatively we could have checked whether the Radio button was selected using the isSelected Method

```
// Select Radio button only if is not previously selected

If (!driver.findElement(By.id("radiobutton_1")).isSelected()) {

driver.findElement(By.id("radiobutton_1")).click();
}
```

Working with Radio Groups

We could also work with Radio Groups by locating all the radio buttons in the group and then select a specific Radio button based on the property match

```
// Select Radio button only if is not previously selected

List<WebElement> HotelName = driver.findElements(By.name("SelectHotel"));

for (WebElement rdbtn : HotelName)
{
//Search for Hotel Creek Radiobutton in the Radio Group and select it
if(rdbtn.getAttribute("value").equals("Hotel Creek"))
   {

        // if rdbtn is equal to Hotel Creek then select it
           rdbtn.click();
         //if found then exit from for loop
           break;
   }
}
```

Figure 27-14 – Code snippet to work with Radio Groups

27.11. Working with Checkboxes

Similar to Radio Buttons we can select or deselect a checkbox using the click() method of the WebElement class and check whether a checkbox is selected or deselected using the isSelected() method.

```
// Select Checkbox  only if is not previously selected

If (!driver.findElement(By.id("checkbox_1")).isSelected()) {

driver.findElement(By.id("checkbox _1")).click();
}
```

27.12. Measuring Response time for Performance Testing using Timer

We can measure page load or response time in the Selenium WebDriver tests. We can use timers in the test code to capture the time taken for page load, rendering of the elements and JavaScript code execution.

Problem Statement – You may have noticed that when we book our hotel on Book a Hotel page by clicking on the **Book Now** button it takes some time for the page to respond. Let us see how we can find out the time it takes to book our hotel.

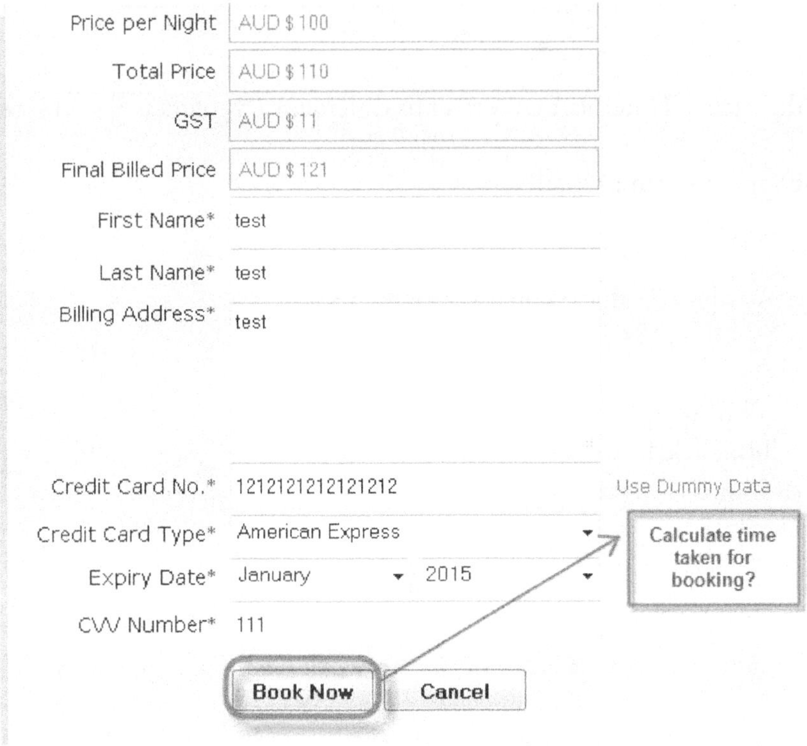

Figure 27-15 – Calculate time taken for booking

Logically there are three key steps we would need

- First, a way to capture timing before we click on "Book Now" button called **start time**
- Second a way to notify us that Booking has completed and order has been been generated
- Third to capture time again after booking called **end time**, so that we can get difference of start and end time

How to get start time

We will use the System.currentTimeMillis() method to get the start time

```
// Get the start Time
long startTime = System.currentTimeMillis();
```

How to find out if Booking has completed and order number is generated

What is the visual cue which gives us an indication that the order has been booked? We will notice that once an order has been inserted the Logout button appears on screen. We can also use other visual cues if required.

Now we can use a custom written WebElementExist method (as defined in previous sections) to return true when the Logout button appears. See below sample code

```
//calling HotelApp_WebElementExist function with timeout of 30 seconds or 30000 milliseconds

sElementExist = HotelApp_WebElementExist (driver, By.id("logout"), 30000)

public boolean HotelApp_WebElementExist(WebDriver WebDriver, By by, int iTimeOut)
        {
                //increments of 100 millseconds or .1 seconds. So out results could be out by 0.1 millseconds.
                We can reduce this further if required
                int iSleepTime = 100;
                for(int i = 0; i < iTimeOut; i += iSleepTime)
                {
// tries to match element based on locator
                        List<WebElement> oWebElements = WebDriver.findElements(by);
                        if(oWebElements.size()>0)
                        {
// element found is size is greater than 0 and return true
                                return true;
                        }
                        else
                        {


//element not found so wait for isleeptime and continue search again.
Thread.sleep(iSleepTime);
System.out.println(String.format("Waited for" +( i+ iSleepTime) + " milliseconds");
                                }


                }
                // Can't find 'by' element. Therefore return false.

                return false;
        }
```

How to get End time

We will use the System.currentTimeMillis() method to get the End time, same way we got start time

```
// Get the end Time
long endTime = System.currentTimeMillis();
```

So our final code will looks like

```
// Get the Start Time
long startTime = System.currentTimeMillis();

sElementExist =  HotelApp_WebElementExist (driver, By.id("logout"), 30000)

// Get the End Time
long endTime = System.currentTimeMillis();

long totalTime = endTime - startTime;
System.out.println("Total Booking Time: " + totalTime + " milliseconds");
```

Figure 27-16– Calculate time taken for booking

27.13. Xpath and Properties Finder in Edge and Firefox browsers

Apart from locating Xpath or DOM properties in Chrome using ChroPath plug-ins we might also need to find those properties in Edge or Firefox.

example

In one of the applications we automated we were using the getAttribute method on the property name "textcontent" to get the label value from one of the elements. When we ran the same script in Edge we found that Edge does not recognize property "textcontent". So we had to use the property "innertext" which is recognised by both Edge and Firefox DOM. So if you are running tests across multiple browsers it would be good to know how we can find Xpath and object properties in other browsers apart from Firefox.

How to see Xpath and DOM property in Firefox and Edge?

Note: There is no specific installation that is required. This feature is available by default in Firefox and Edge

1. Open the application URL (www.adactinhotelapp.com) in the Firefox/Edge Browser

2. **Right click** on the page and select **Inspect element**

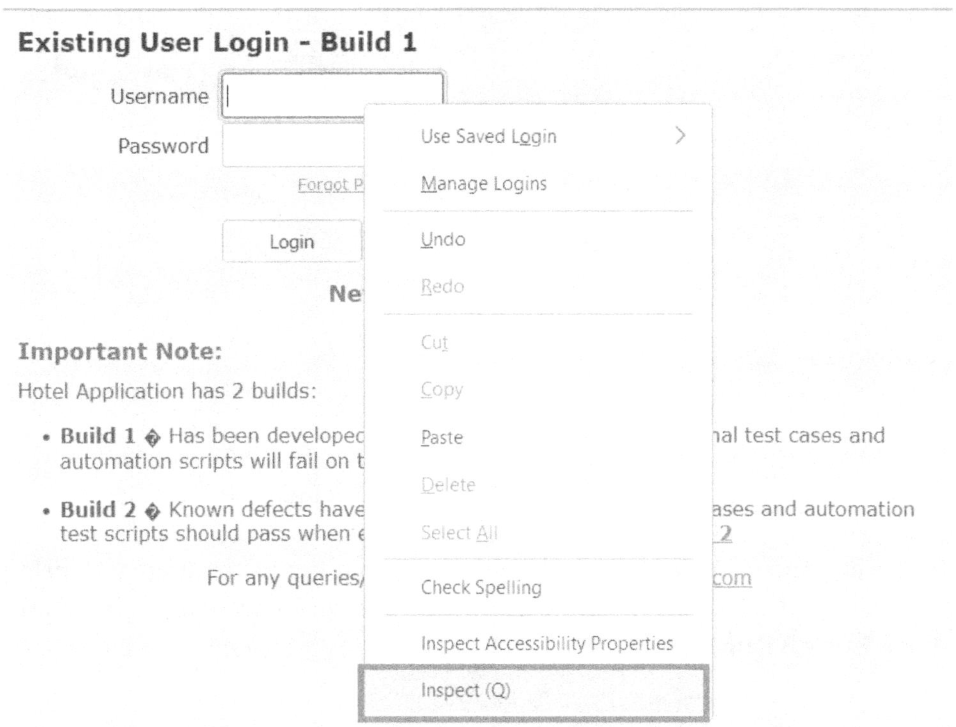

Figure 27-17 – Inspect Element

3. You will see a console open up in the bottom half of the browser

Figure 27-18 – Inspect Element Console

4. Click on **Select an inspect element** icon and click on **Username** edit field

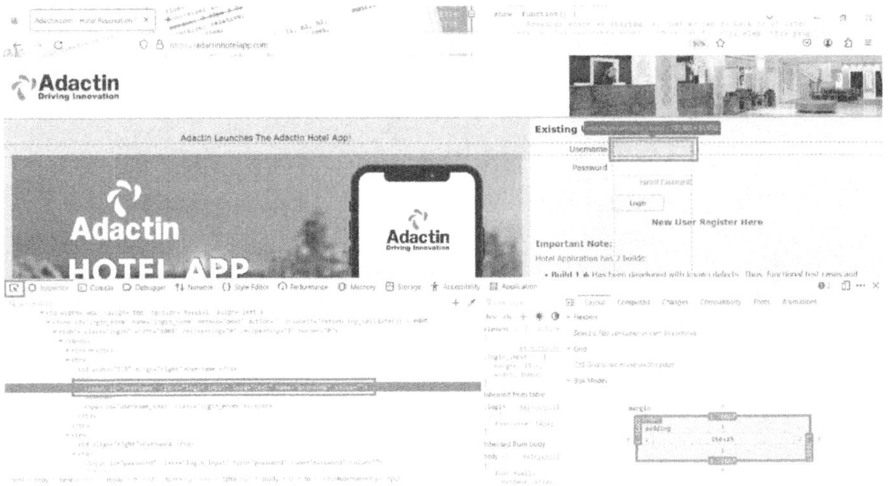

Figure 27-19 – Inspect an Element

5. You will notice that part of the DOM corresponding to the username field gets highlighted. Right click on the **highlighted text** and select **Copy → XPath (In Edge Copy XPath)**

Figure 27-20 – Copy XPath

6. **Open** a new NotePad and **paste** the XPath

Figure 27-21 – XPath Value

7. We can use this XPath value in the script if required

Note: There are quite a few extensions/plug-ins also available with Firefox to get Xpath and view Dom properties.

27.14. How to use WebDriver test remotely using Selenium Grid?

Selenium Grid is a part of the Selenium Suite that specializes in running multiple tests across different browsers, operating systems, and machines in parallel.

The latest version of Selenium Grid is Grid 4.

Selenium Grid uses a hub-node concept where you only run the test on a host machine called a **hub**, but the execution will be done by different client machines called **nodes**.

You would want to use Selenium Grid when you want to run your tests in parallel on multiple OS, multiple browsers and multiple platforms at the same time. This will save on overall execution time.

example

As part of a client project we had to run our test on these 7 OS/Browser combinations

- Windows 10 Chrome
- Windows 10 Firefox
- Windows 10 Edge
- Windows 11 Chrome
- Windows 11 Firefox

- Windows 11 Edge
- Mac Safari

Android ChromeOur test suite would take 3-4 hours of execution time per configuration. We had multiple test machines available but if we had to execute our tests sequentially it would take 3-4 working days to complete the test execution and then analyse the results.

Instead we used Selenium Grid to perform parallel execution and we could get the results in a single day.

Concept of Hub and Nodes in Selenium

A grid consists of a single hub, and one or more nodes.

The hub is a central point which initiates a test to be executed along with information on which browser and 'platform' (i.e. WINDOWS, LINUX, etc) where the test should be run. It 'knows' the configuration of each node that has been 'registered' to the hub. Using this information it selects an available node that has the requested browser-platform combination. There is only a single Hub for selenium Grid.

Once a node has been selected, Selenium commands initiated by the test are sent to the hub, which passes them to the node assigned to that test.

The node runs the browser, and executes the Selenium commands within that browser against the application under test. There can be multiple nodes for a hub.

Configure a test for Selenium Grid 4.0

In this section we will use 2 machines. One machine will act as a Hub (we will call it Machine-1) and the second machine will act as a Node (we will call it Machine-2). Let us use Machine-1 on which we have selenium project setup so far as the Hub.

Pre-requisite: Make sure you have downloaded/access to your selenium-server--4.x.x.jar file. You can download it again if required from https://www.selenium.dev/downloads/

In our case we already have the jar file at location C:\Selenium\workspace\HotelApp_TestAutomation\Lib\ selenium-server-4.9.1.jar

Key steps summary to setup Selenium Grid

1. Setup Hub – This is the central point from where tests will be initiated
2. Setup Node - This is the client machine on which tests will be executed
3. Setup Selenium Test to Run on Node

Step1 -Setup Hub

1. Go to Machine-1 and open your command prompt and use cd commands to reach location C:\Selenium\workspace\HotelApp_TestAutomation\Lib\ where selenium jar file is location.

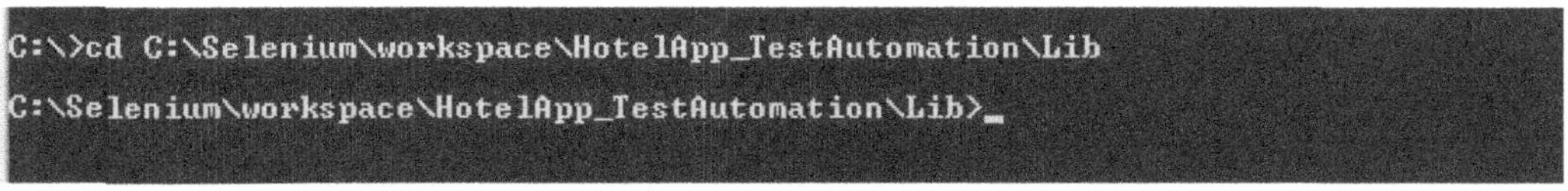

Figure 27-24– Command Prompt Setup

Note: If the jar file is not already downloaded,download the same from https://www.selenium.dev/downloads/ under Selenium Grid.

Type and Run the command java -jar selenium-server-4.9.1.jar hub The hub should successfully be launched. Your command prompt should look similar to the image below

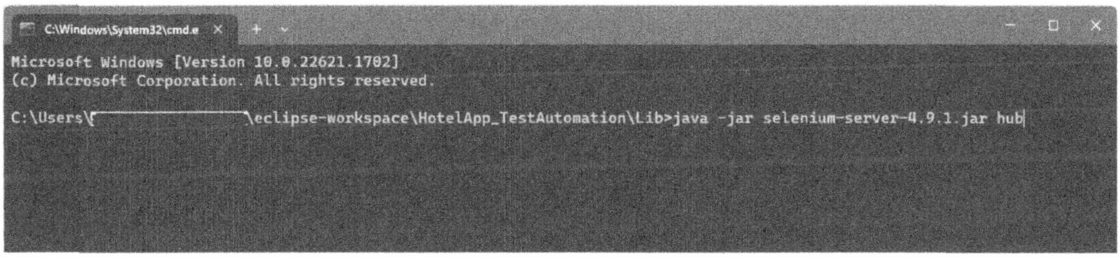

Figure 27-25– Start Hub

2. Another way to verify whether the hub is running is by using a browser. Selenium Grid, by default, uses Machine A's port 4444 for its Web interface. Simply open up a browser and go to http://localhost:4444/ui/.

Note: By default the Grid uses port 4444. It can be accessed from remote machines with URL http://Machine-1-IPAddress:4444/ui/

Figure 27-26– Grid Console

Step2- Setup Node

We are now going to setup a Node on Machine-2

Pre-requisite – Copy selenium-server-4.x.x.jar file in C:\Selenium of the Machine-2. Also get IP Address of Machine-1 so that it can be accessed from Machine-2

3. Open the **command prompt** and navigate to **C:\Selenium folder** (where selenium jar file is located) or to Users/Selenium in Mac

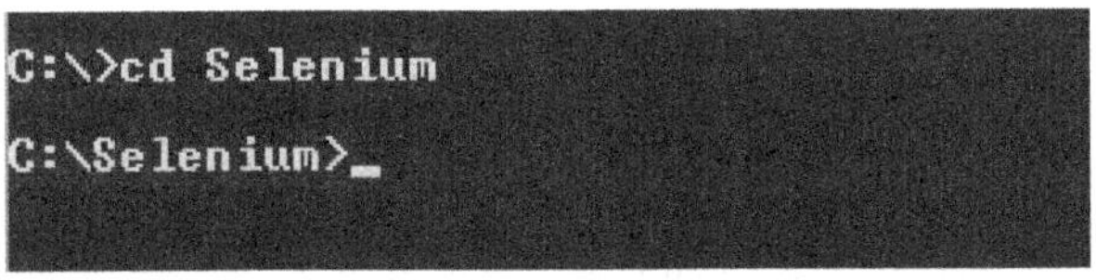

Figure 27-27– Command Prompt

4. Type command
5. java -jar selenium-server-4.9.1.jar node --selenium-manager true

Note: In the above statement 192.168.1.101 is the IP address of Machine-1. Replace it with IP address of your machine-1. Also port 6666 is a sample port. We can choose any alternate port number too but make sure you remember which port number you have used

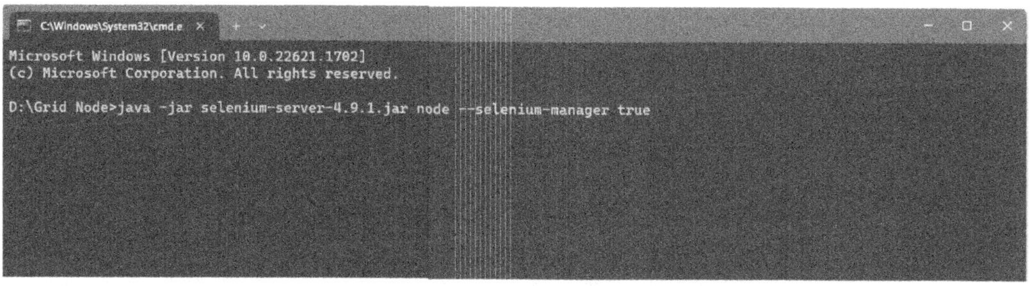

Figure 27-28– Node Command

6. You will notice that node is registered.

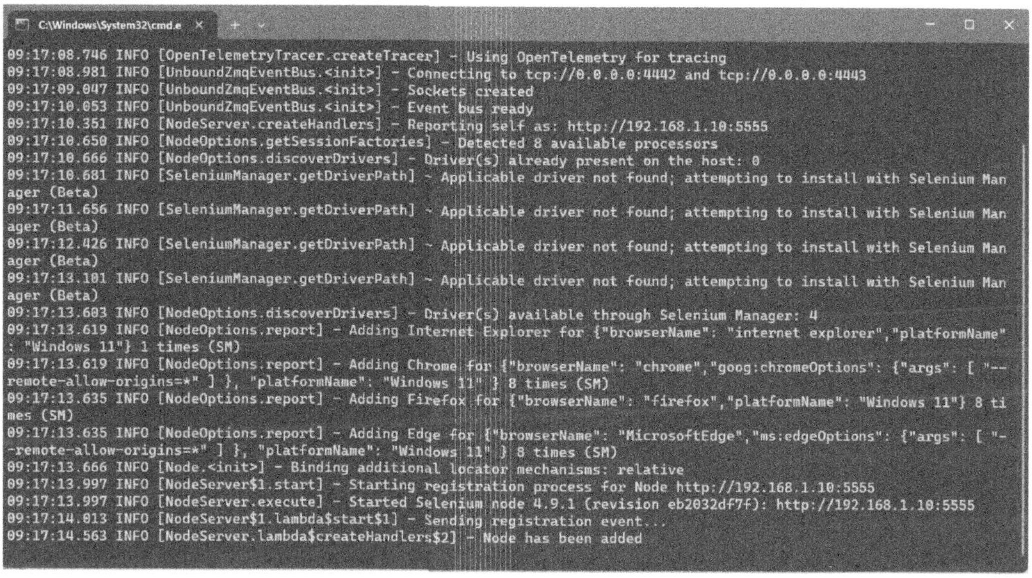

Figure 27-29– Node is registered

7. Go to Machine-1 (Hub), **Refresh** URL http://localhost:4444/grid/console.

Figure 27-30–Refreshed Grid Console

Note: Grid Console shows the IP Address and port number of the Node machines registered with Hub

Step3- Setup Selenium Test to Run on Node

1. **Right click** on your existing **MyFirstWebDriverTest.java** script, select **Copy**

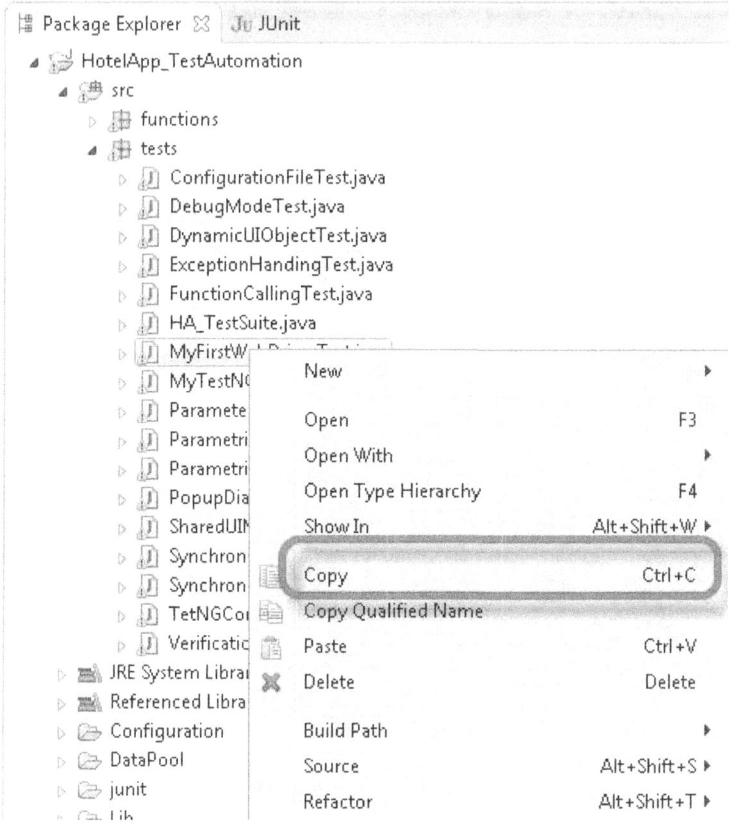

Figure 27-31 – Copy Script

2. Select the **tests** package folder, **right click** and select **Paste**
3. In the Name Conflict dialog box , enter the name of the script as **SeleniumGridTest** and click **OK**

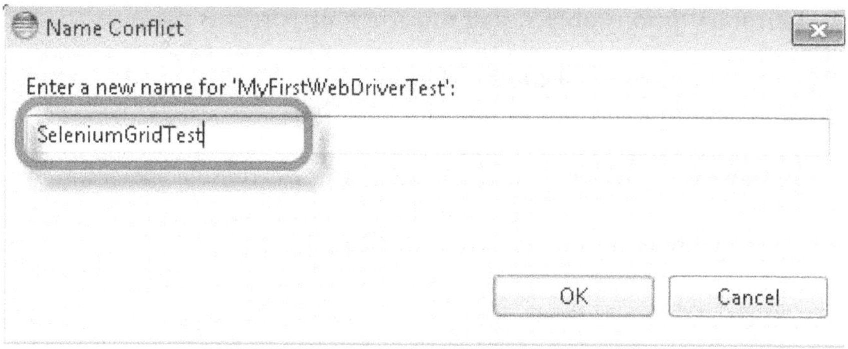

Figure 27-32 – Name Conflict dialog

4. Double click on the newly created "**SeleniumGridTest.java**" script to see the script

```
package tests;

import java.time.Duration;

public class SeleniumGridTest {

    private WebDriver driver;
    String hubURL = "http://192.168.1.10:4444";    // IP Address of hub machine
    @Before
    public void setUp() throws Exception
    {
        ChromeOptions chromeOptions = new ChromeOptions();
        chromeOptions.setCapability("platformName", "Windows");

        driver = new RemoteWebDriver(new URL(hubURL), chromeOptions);
        driver.get("https://adactinhotelapp.com/");
        driver.manage().window().maximize();
        driver.manage().timeouts().implicitlyWait(Duration.ofSeconds(20));
    }

    @Test
     public void testMyFirstWebDriver() throws Exception {

        driver.findElement(By.xpath(".//*[@id='username']")).clear();

        driver.findElement(By.xpath(".//*[@id='username']")).sendKeys("Adactin");

        driver.findElement(By.id("password")).clear();

        driver.findElement(By.id("password")).sendKeys("adactin123");

        driver.findElement(By.id("login")).click();

        new Select(driver.findElement(By.id("location"))).selectByVisibleText("Sydney");

        driver.findElement(By.id("Submit")).click();

        driver.findElement(By.id("radiobutton_1")).click();

        driver.findElement(By.id("continue")).click();

        driver.findElement(By.id("first_name")).clear();
```

Figure 27-33 – Selenium Grid Test

5. We now need to use the ChromeOptions **object** to set the type of **browser** and **OS** that we will automate.

To use the ChromeOptions **object, you must first import this package**

6. We also need to use the **RemoteWebDriver** object which is used to set which node (or client machine) that our test will run against

To use the **RemoteWebDriver** object, you must import these packages.

```
import org.openqa.selenium.remote.RemoteWebDriver;

import java.net.URL;
```

```
import java.time.Duration;
import org.junit.After;
import org.junit.Before;
import org.junit.Test;
import org.openqa.selenium.By;
import org.openqa.selenium.WebDriver;
import org.openqa.selenium.chrome.ChromeOptions;
import org.openqa.selenium.remote.RemoteWebDriver;
import org.openqa.selenium.support.ui.Select;
import java.net.URL;
```

Figure 27-34 – Imported packages

7. To use the ChromeOptions object we will need to find out the Browser and OS of node (Machine-2) which we need to set in our script. Go to Machine-1 (hub) and open the URL http://localhost:4444/ui/

Figure 27-35 – Hub Console

8. Let us assume we want to run the test on a Chrome browser. Bring your cursor on top of the Chrome icon

Figure 27-36 – Node Configuration Values

You will notice that-

Browser = chrome

OS = WIN11

Note: You will notice that maxInstances = 8 for Firefox and Chrome which means that using Selenium Grid you can run parallel tests on 8 Firefox and Chrome browsers.

9. Go back to you script and add the below code in the Setup Method and replace command driver = new ChromeDriver();

```
//setup URL to be opened on Node machine
String hubURL = "http://192.168.0.6:4444/";

//Setup Chrome options with browser and OS

ChromeOptions chromeOptions = new ChromeOptions();
chromeOptions.setCapability("platformName", "Windows");
driver = new RemoteWebDriver(new URL(nodeURL), chromeOptions);
driver.get("https://adactinhotelapp.com/");

//Using Remote WebDriver to launch the driver. Note we are not using Chrome driver as we had earlier used.
```

Figure 27-37 – code Snippet for using RemoteWebDriver

```
private WebDriver driver;
String nodeURL = "http://192.168.1.10:4444";      // IP Address of hub machine
@Before
public void setUp() throws Exception
{
    ChromeOptions chromeOptions = new ChromeOptions();
    chromeOptions.setCapability("platformName", "Windows");

    driver = new RemoteWebDriver(new URL(nodeURL), chromeOptions);
    driver.get("https://adactinhotelapp.com/");
    driver.manage().window().maximize();
    driver.manage().timeouts().implicitlyWait(Duration.ofSeconds(20));
}
```

Figure 27-38 – Using Remote Driver

In the above script we are using Remote WebDriver instead of Chrome driver (driver = new ChromeDriver();). Also we are using node IP addres in node URL for driver.

10. Ctrl + S or File → Save your script

11. Make sure that both Hub and Node servers are still running. You can verify that by looking at the command prompt where we had started them in the previous step.

12. Perform Project clean-up using **Project → Clean Up…** and clean all selected projects

13. Let us now run the script. Select the script and select **Run As → JUnit Test.** You will notice that a Chrome browser opens in a node machine and runs the test to completion.

Figure 27-39– Run Test

Note: If you are wondering, would we be required to change the WebDriver everytime to RemoteWebDriver when we need to run a test using Selenium Grid or locally? This can be a daunting task in itself.

Answer is: We can define separate functions for invoking the driver, based on whether we would want to run the test locally or use Selenium Grid. We can define how we want to run the test in a configuration file. It means that if we want to use Selenium Grid it will invoke Login function which will call RemoteWebDriver and if we would want to run the test locally it will invoke Login function which will call Chrome or corresponding driver.

§

28

Creating Project using Maven

We have made the Java project in which we have added the Jar files by downloading it and uploading in the lib folder. We can skip this using maven project. In Maven we have to add dependencies so all the Jar files gets downloaded and gets applied in Project. Let us see how we can create Maven Project.

1. Open **Eclipse** and **File>>New>>Maven Project**

Figure 28.1 Create Maven Project

2. Click on Create a **simple project** checkbox and click on **Next** button.

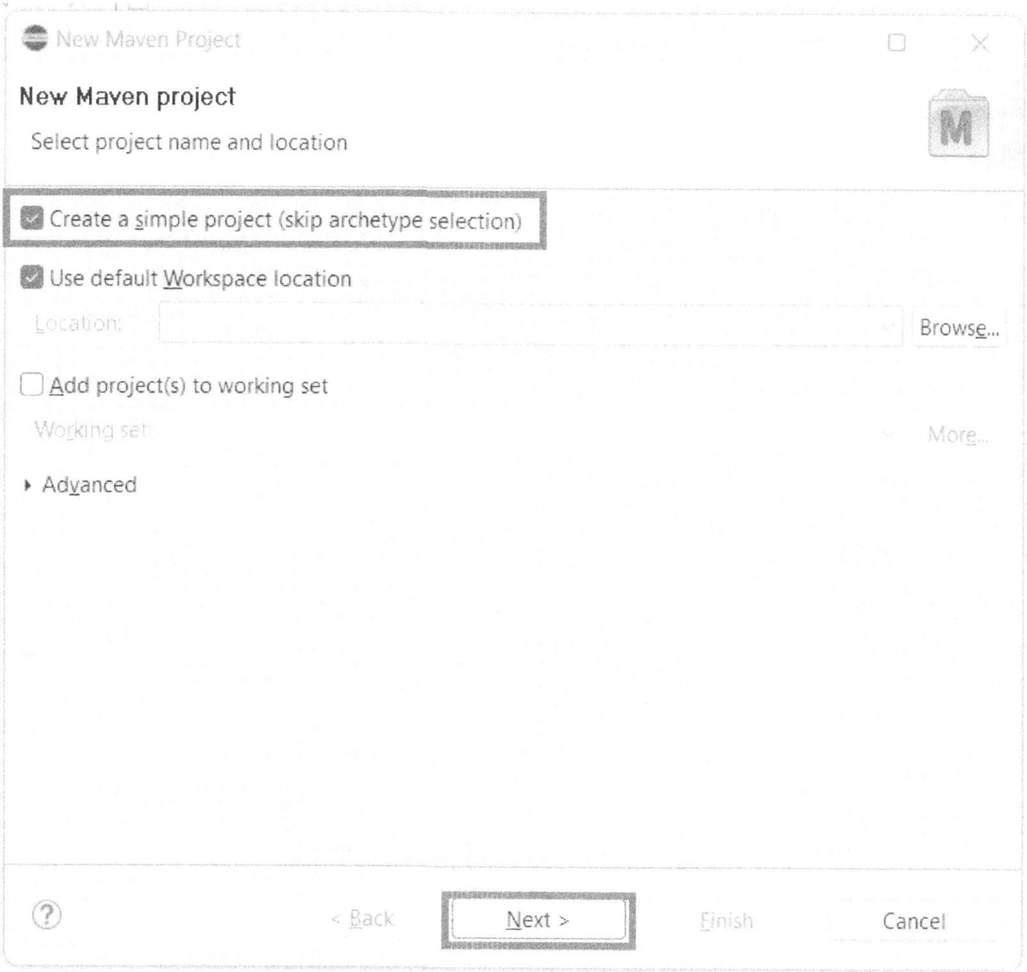

Figure 28.2 Skip archtype selection

3. Enter Group Id as '**TestProject**' and ArtifactId as '**SeleniumMavenProject**' and click on **Finish** button

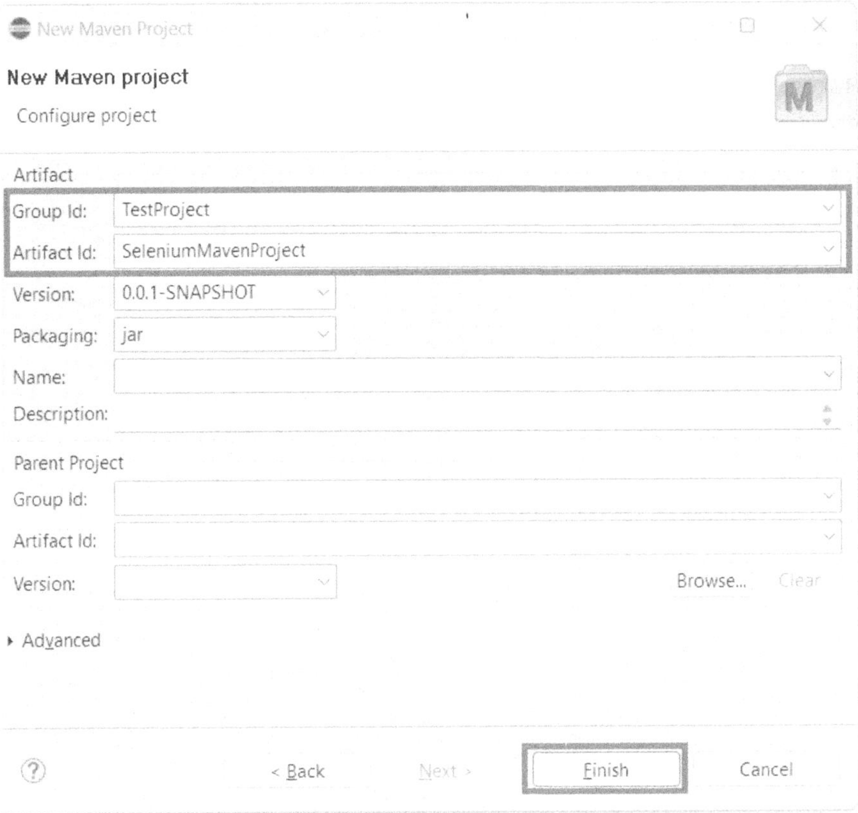

Figure 28.3 Complete Maven Project

4. New Project has created in IDE with project structure.

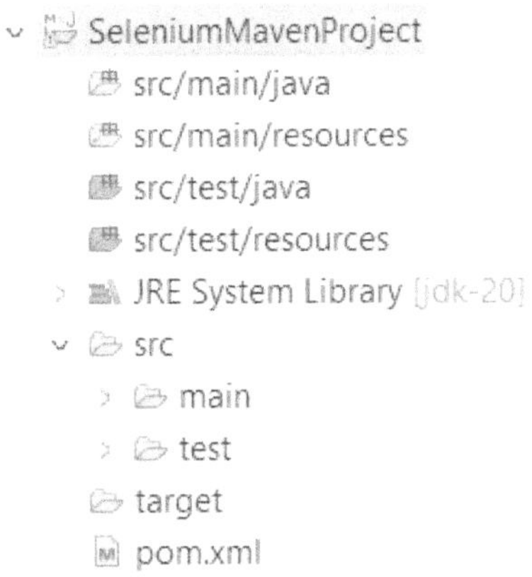

Figure 28.4 Maven Project Structure

5. Double click on **pom.xml**

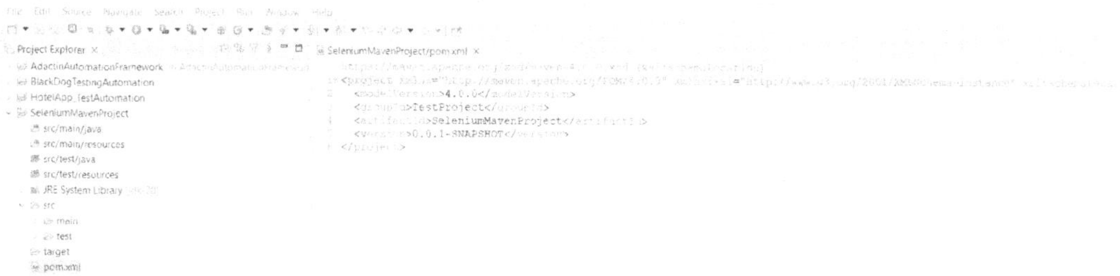

Figure 28.5 POM.xml

6. Create Dependencies Tag after version tag to add dependency for particular Jar file.

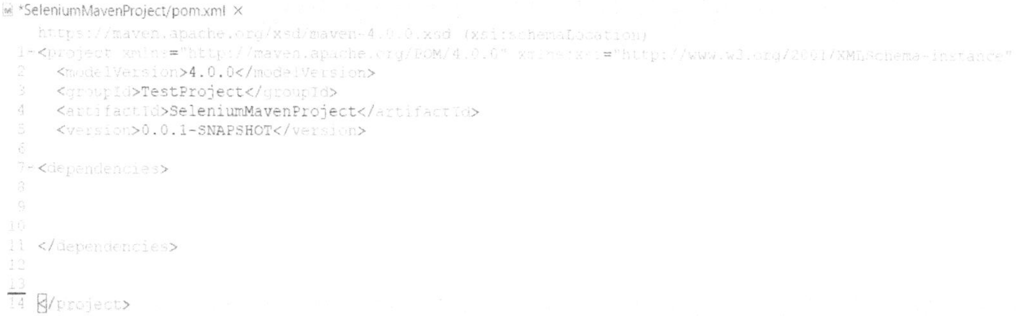

Figure 28.6 Dependencies Section

7. Let us add Jar for the Selenium. Open browser and Open URL Google and search for selenium maven dependency.

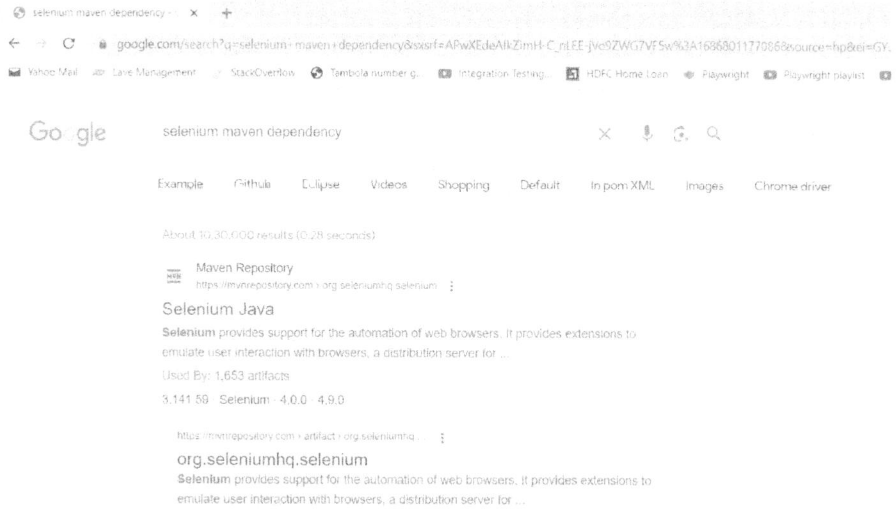

Figure 28.7 Selenium Maven Dependency Google Search

8. Click on first link of Maven Repository as Selenium Java

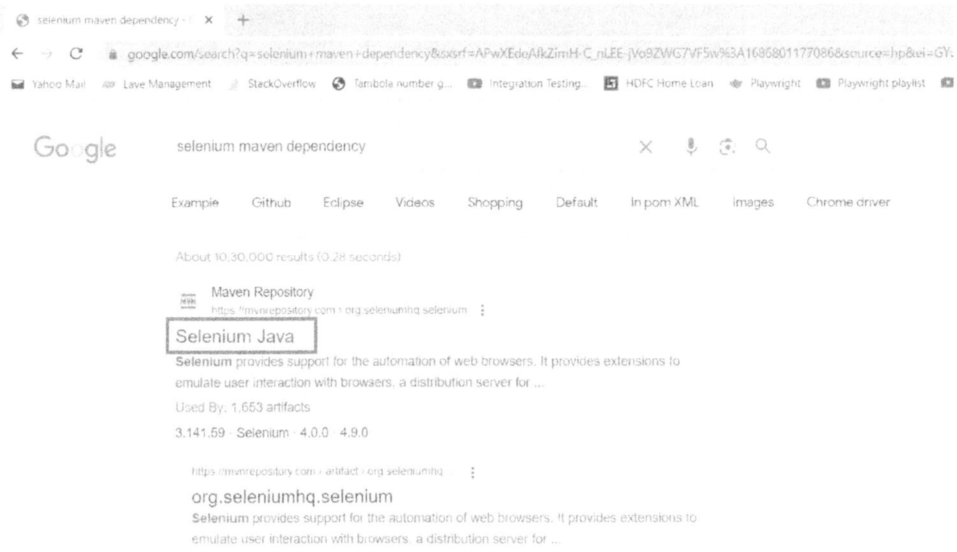

Figure 28.8 Selenium Java Link

9. Select latest version link under central tab.

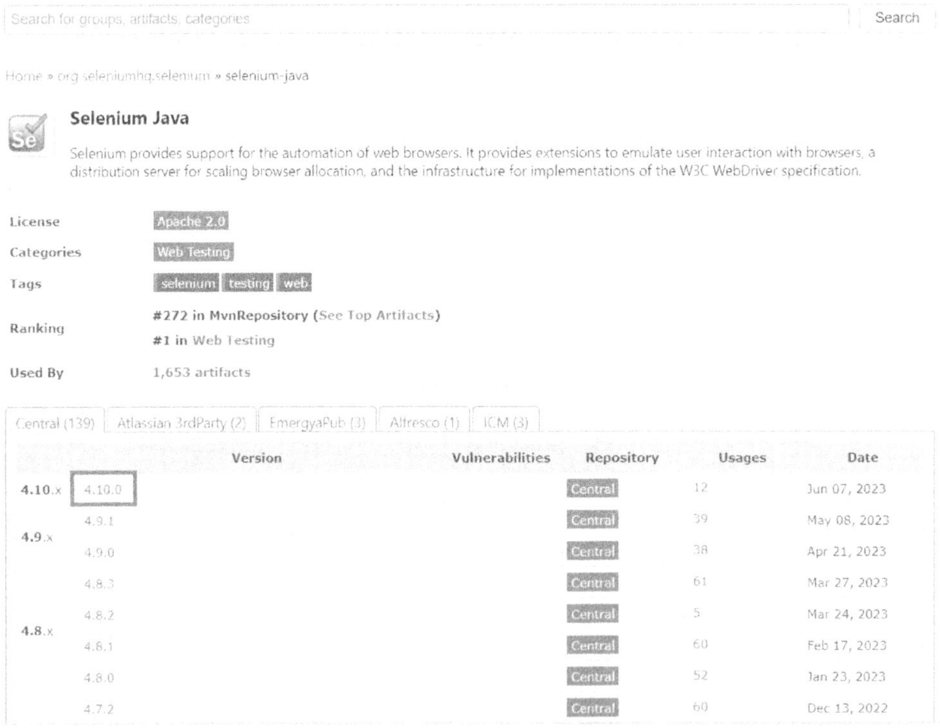

Figure 28.9 Selenium Maven Dependency version

10. Click on the Maven Section and copy all content.

Figure 28.10 Copy Selenium Maven Dependency

11. Open Eclipse and open pom.xml and paste the above data inside dependencies section and save file.

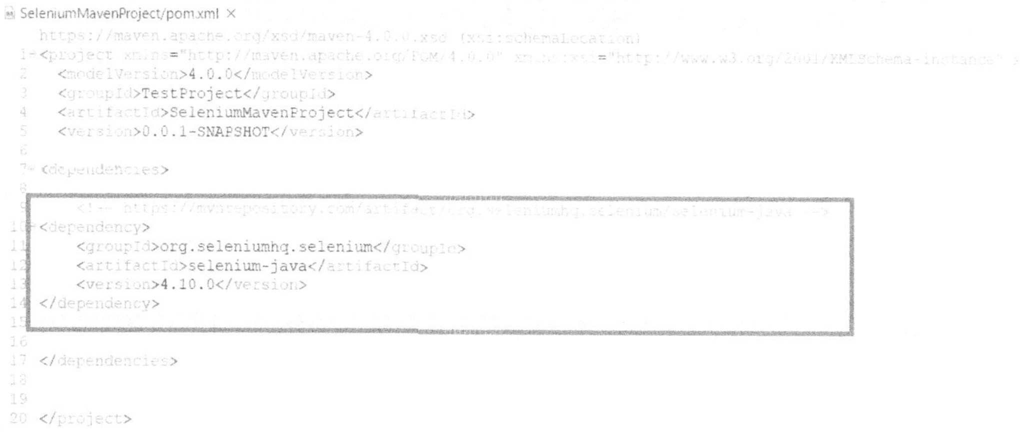

Figure 28.11 Updated POM With selenium dependency

12. Select Project and **right click** on project and **Maven → Update Project.**

Figure 28.12 Update Maven Project

13. Select **Project** and click on **Ok** Button.

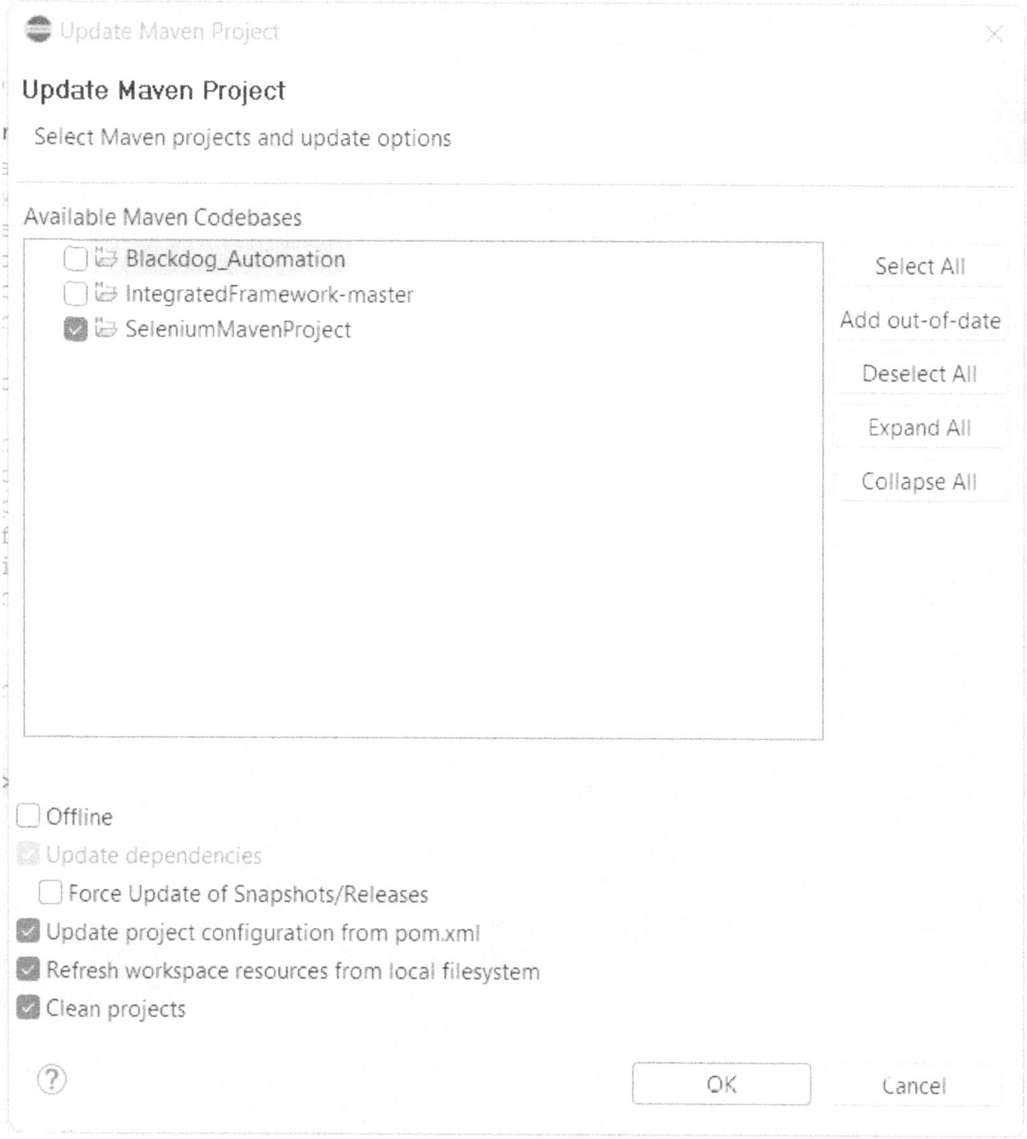

Figure 28.13 Maven Update Project window

14. Go Project section and expand the maven dependencies section here you will see all the jar files are downloaded and added in project.

Figure 28.14 Updated Maven dependencies

15. Same way we can add other Jar files by searching like JXL Maven Dependency and adding in pom.xml.

§

29

Multiple Choice Questions Set-3

1. Which of the following will pause the execution of the program at the specific step?
   A. DebugPoint
   B. Breakpoint
   C. Step Into
   D. Step Out

2. Which of the following will help us get the value of the Expression in Eclipse IDE?
   A. Variables
   B. Breakpoint
   C. Watch
   D. Commands

3. In order to handle an exception in Selenium script which of the following can be used?
   A. OnErrorGoTo method
   B. Try-Catch blocks
   C. Switch Case statement

4. Which of the following are approaches to generate reports in Selenium WebDriver?
   A. Junit HTML reports
   B. TestNG reports
   C. Excel based reports
   D. All of the above

5. What is the TestNG equivalent for Junit's @Before annotation?
   A. @BeforeClass
   B. @BeforeMethod
   C. @BeforeSuite
   D. None of the above

6. Which object and method is used to report results in TestNG based WebDriver script?
   A. Result.Reporter
   B. Log.report
   C. Reporter.log
   D. Report.ReportLog

7. Which of the below is a way to Batch execute our Selenium WebDriver scripts?
   A. JUnit with ANT
   B. TestNG Framework
   C. Master WebDriver scripting call to other scripts
   D. All of the above

8. Jenkins is an example of a
   A. Continuous Integration Tool
   B. Test Management Tool
   C. Performance Testing Tool
   D. Configuration Management Tool

9. Which Class is used to perform keyboard and mouse movements in Java?
   A. JavascriptExecutor
   B. Actions
   C. RemoteWebDriver
   D. KeyboardMouseAction

10. Which of the following components of Selenium suite can be used for running scripts on multiple browsers and OS in Parrallel?
   A. Selenium Grid
   B. Selenium RC
   C. Selenium 4.0
   D. Selenium TestNG Framework

Answers

Q1. Answer: B
Explanation – A Breakpoint will pause the execution of a program at a specific step at which it is inserted

Q2. Answer: C
Explanation – A Watch is used to return the value of an Expression

Q3. Answer: B
Explanation – Try Catch blocks are used to catch any exceptions thrown by the Selenium scripts

Q4. Answer: D
Explanation – In Selenium WebDriver we can create reports using Junit, TestNG or Excel based reports

Q5. Answer: B
Explanation – TestNG annotation @BeforeMethod is equivalent to @Before annotation in JUnit

Q6. Answer: C
Explanation – Reporter.Log (with Reporter as Object and Log() as method is used to report results in TestNG

Q7. Answer: D
Explanation – All of the methods listed can be used for Batch execution of Selenium WebDriver scripts

Q8. Answer: A
Explanation – Jenkins is an example of a continuous integration tool

Q9. Answer: B
Explanation Actions classes are used to perform keyboard and mouse operations like double click, send keys, drag and drop, etc.

Q10. Answer: A
Explanation – Selenium Grid helps in running parallel scripts on multiple OS and multiple browsers

§

30

Sample Naming and Coding Conventions

30.1. Sample Naming Conventions

Standardized naming and coding conventions ensure that automation components including names of scripts, functions, Web elements and variables are consistent throughout our framework. This reflects good coding practice and assists in code maintenance later on.

1. **Automation components naming convention** – You can follow these naming conventions for automation components

Subtype	Syntax	Example
Tests	**[Product]_ [TestCaseID]_ [TestType]_[Test Name]**	*HA_TC101_BP_FindHotel*
Functions	**[Product]_[FunctionType]_ [FunctionName]**	*HA_GBF_Hotelbooking.java*
Data Table	**[Product]_ [DataTableType]_ [Table Name]**	*HA_DE_FindHotel.xls*
Shared UI Map	**[Product]_[UI Map Name]**	*HA_SharedUIMap.properties*
Objects	**[Object Type]_[Object Description]**	*Btn_Submit*

Syntax Description

- **Product/Project [Product]**
  - **HA** - Hotel Application (Name of the project/product)
- **TestCase ID [TestCaseID]**
  - **TC101** – Represents Testcase id TC101
- **Test Types [TestType]**
  - **BP**: Business Process Script (End to end test)
  - **TC**: Test Case Script (test that maps to functional test case)

- **UT**: Utility Script (test that assists as a utility script. E.g., test data creation test)
- **UI**: User Interface Script (test that validate User Interface)

- **Test Name [TestName]**
  - **VerifyValidLogin** – Brief description of test case obejctive

- **Function Types [FunctionType]**
  - **GBF**: Global Business Function
  - **LBF**: Local Business Function
  - **VF**: Verification Function
  - **UF**: Utility Function
  - **RF**: Recovery Function

- **Function Name [Function Name]**
  - **Login** – Brief description of Function Objective

- **Data Table Types [DataTableType]**
  - **DE**: Data Entry Data table (Datasheet used with objective of reading data from a particular row of sheet. For e.g., Login with row id 3)
  - **DL**: Data Loop Data Table (Datasheet used with objective of data driving the test with multiple data values)

- **UI Map Name [UI Map Name]**
  - **SharedUIMap** – Brief description of Shared UI Map

- **Objects**
  - **Btn**: Button
  - **Ed**: Edit box
  - **Tbl**: Table
  - **Cb**: Combobox
  - **Lnk**: Link

- **Extensions:**
  - Library: **.java**
  - Datatable: **xls**
  - Shared Repository: **properties**
  - Configuration file: **properties**

2. Descriptive names

Names should use mixed case and should be as complete as necessary to describe its purpose or related objective.

30.2. Coding Conventions

1. Java Coding Conventions

Variable Naming Convention

For purposes of readability and consistency, use the prefixes listed in the following table, along with descriptive names for variables in your VBScript code.

Subtype	Prefix	Example
Boolean	bln	blnFound
Integer	int	intQuantity
Object	obj	objCurrent
String	str	strFirstName
Variant	var	varData
Array	arr	arrLocationList

Descriptive Variable and Procedure Names

The body of a variable or procedure name should use mixed case and should be as complete as necessary to describe its purpose. In addition, procedure names should begin with a verb, such as InitNameArray or CloseDialog.

For frequently used or long terms, standard abbreviations are recommended to help keep name length reasonable. In general, variable names greater than 32 characters can be difficult to read.

When using abbreviations, make sure they are consistent throughout the entire script. For example, randomly switching between Cnt and Count within a script or set of scripts may lead to confusion.

2. Code Commenting Conventions

Comments are an integral part of any programming language. They help maintenance engineers understand the objective of lines of code. As a good coding practice, key logic in the code should have comments.

Guidelines for Comments for Functions

Heading	Mandatory	Comment Contents
Function	Mandatory	Name of the function and the description.
Inputs	Mandatory	List of variables passed into the function as the parameters.
Outputs	Mandatory	List of variables as the output of the function.
Returns	Mandatory	List of variables returned by the function.
Usage	Optional	Information about specifically how the function is implemented, and how that might affect its usage in a script.
See Also	Optional	Linking to any related topic – similar / opposite functions, type definitions.

Example:

/*

****

public static String Function: String HA_GF_readXL (int introw, String strcolumnName, String strFilePath)

A sample function for demonstrating comment format.

Inputs:

introw – An integer representing the row of the Excel sheet to be read
strcolumnName– A string used to represent the name of the column to be read
strFilePath – A string path of the location of the Excel file

OutPuts:

strContent– A String value for data read from an Excel sheet and returned by the function

Returns:

A valid String for success
In case of failure will return an exception and message

History:

QA1 Create Version 1.0 06/06/2014
QA2 Update Version 1.1 07/07/2014 - Updated for error handling

Usage / Implementation Notes:

Make sure file is of extension .xls and not .xlsx
Make sure datasheet ends with extension

See Also:

<HA_GF_WriteXl>

```
****************************************************************

*/

public static String HA_GF_readXL (int row, String column, String strFilePath)
{
	// lines of code
}
```

§

31

Common Selenium Interview Questions

I have been on interview panels for quite a few IT departments of IT consulting and non-IT companies. In my personal experience, interview questions are centred on Tool technical knowledge, test automation experience, scenario and approach based techniques, which helps the interviewer judge the depth of automation experience that a candidate possesses.

This chapter discusses some of the questions I recommend that you need to be prepared for when you are shortlised for Selenium based Test Automation interviews.

31.1. Common Test Automation and Selenium Interview Questions

- Can you explain the automation framework you have developed using WebDriver in your recent project?

Answer - Refer to our automation frameworks chapters to answers this question

- What are the key challenges you think you faced doing automation?

Answer - Some of the key challenges are

  - Automation Environment unavailability
  - Application is unstable
  - Features and workflows being changed frequently leading to maintenance issues
  - Test Data is changed frequently and needs regular modification and maintenance
  - Objects and their properties are changed frequently
  - Unreal expectations from the project manager or test manager, if they expect automation to happen at the click of a button
  - Automation tool support to automate all features of an application
  - Managing test results for every build for future reference

- How comfortable are you with scripting and programming (Java or C#)?
- If you have to choose between using Shared UI Map and Local UI Map, what will you choose and why?

Answer – Shared UI Map, since if the objects/UI in the application change, there will be one central place where object properties will be modified saving huge maintenance overhead of making changes in all scripts.

Alternatively, using a local object repository would mean redundant objects across script and maintenance overhead to change in all scripts in case of object/UI changes.

- Any scenarios in your previous projects where you had worked on maintenance of scripts? What were the challenges encountered?
- How did you decide on your selection candidates (i.e., test cases to be automated) for automation in your previous projects?

Answer – Key criteria include

  - Acceptance test cases
  - High priority business requirements
  - Test cases which need to be executed multiple times with different sets of data
  - Test environment preparation scenario
  - Test cases which are complex and take a lot of time for automation
  - End-to-end business processes
  - Based on defects found earlier in the application

- So what automation process will you follow, if you would need to automate the application from the ground up?

Answer – Refer to Planning for Automation chapter to answer this question

- How many test cases can you automate in one day?

Answer – Now this really depends on how big the test case is, the type of application you are working with, and the automation framework you are using. At a very high level, if you start without any automation framework on a Web based application and a test case with 10 steps each; you can automate up to 3-4 test cases a day. But as mentioned, it can vary a lot based on different factors and your expertise on the tool.

- How do you run WebDriver scripts as a Batch?

Answer – Refer to our Batch Execution chapter to answer this

- How do you handle dynamic objects in WebDriver? Or
- How have you used Regular Expressions in WebDriver?

Answer – Refer to Working with Dynamic Objects chapter to answer this question

- How do you get properties of objects from your application, using WebDriver?

Answer – Use getAttribute method to get a property value

- How can you report a custom message from a WebDriver script to be viewed in TestNG test results after script execution?

Answer – Use Reporter.log method

- Have you used Selenium Grid? What is Selenium Grid used for?

Answer: Refer to chapter, "Selenium Functions, Common Questions and Tips" on how to implement Selenium Grid.

- Have you integrated Selenium scripts with any continuous Integration tool? How can we use Continuous integration tools

Answer – Refer to chapter, "Continuous Integration with Jenkins"

- How does Selenium WebDriver Handle Pop-up or Frames?

Answer – Refer to chapter, "Handling Pop-ups and Child Frames/Windows"

- How did you handle reporting in Selenium? What are the different ways to report results in Selenium WebDriver?

Answer – Refer to chapter, "Reporting in Selenium"

- How can you read or write data to an Excel file in Java (or Selenium WebDriver)

Answer – Refer to chapter, "Data Driven Testing and Parameterization"

- Can you use Selenium WebDriver to perform Drag and Drop operations?

Answer – Yes, using Actions object. Refer to chapter "Selenium Functions, Common Questions and Tips" on how to use the Actions object.

- Which add-ons did you use to get Xpath or DOM properties of objects while using Selenium?

Answer – Chropath add-on

- Have you executed your Selenium scripts on multiple browsers? How?

Answer – Refer to chapter "Selenium Functions, Common Questions and Tips" on how to run Selenium WebDriver scripts on multiple browsers.

- How can you check if a checkbox or Radio button in your application is currently selected?

Answer – use .IsSelected method

- How comfortable are you in debugging Selenium WebDriver scripts? What are Breakpoints, Variables, Watches used for?

Answer – Refer to chapter "Debugging Scripts"

- How do you handle exceptions in Selenium?

Answer – Refer to chapter "Exception Handling in WebDriver"

- What are the naming conventions you follow in Selenium WebDriver?

Answer – Refer to chapter "Sample Naming and Coding Conventions"

- Give a list of tools/softwares/add-ons you have used while working with Selenium WebDriver

Answer: Key tools/software include

  - Java/JDK
  - Eclipse
  - JUnit
  - Selenium WebDriver/IDE
  - ANT
  - TestNG
  - Chropath
  - Selenium Grid
  - JXL.jar
  - Jenkins

§

32

Sample Test Cases for Automation

TEST CASE ID	OBJECTIVE	STEPS	TEST DATA	EXPECTED RESULTS
TC-101	To verify valid login Details	1. Launch hotel reservation application using URL as in test data. 2. Login to the application using username and password as in test data.	URL:http://adactinhotelapp.com/index.php User:{test username} Password:{test password}	User should login to the application.
TC -102	To verify whether the check-out date field accepts a later date than check-in date.	1. Launch hotel reservation application using URL as in test data. 2. Login to the application using username and password as in test data. 3. Select location as in test data. 4. Select hotel as in test data. 5. Select room type as in test data. 6. Select no-of-rooms as in test data. 7. Enter check-in-date later than the check-out-date field as in test data. 8. Click on Search 9. Verify that system gives an error saying 'check-in-date should not be later than check-out-date'.	URL: http://adactinhotelapp.com/index.php User:{test username} Password:{test password} Location: Sydney Hotel: Hotel Creek Room type: standard No-of-rooms:1 Check-in-date: today + 7 date Checkout date:today+5 date	System should report an error message.

TC -103	To check if error is reported if check-out date field is in the past	1. Launch hotel reservation application using URL as in test data. 2. Login to the application using username and password as in test data. 3. Select location as in test data. 4. Select hotel as in test data. 5. Select room type as in test data. 6. Select no-of-rooms as in test data. 7. Enter check-out-date as in test data. 8. Click on Search 9. Verify that application throws error message	URL: http://adactinhotelapp.com/index.php User:{test username} Password:{test password} Location: Sydney Hotel: Hotel Creek Room type: standard No-of-rooms:1 Check-in-date: today's -5 date Check-out date: today's -3 date	System should report an error message 'Enter Valid dates'.
TC-104	To verify whether locations in Select Hotel page are displayed according to the location selected in Search Hotel	1. Launch hotel reservation application using URL as in test data. 2. Login to the application using username and password as in test data. 3. Select location as in test data. 4. Select hotel as in test data. 5. Select room type as in test data. 6. Select no-of-rooms as in test data. 7. Enter check-out-date as in test data. 8. Select No-of-adults as in test data. 9. Select No-of-children as in test data. 10. Click on Search button. 11. Verify that hotel displayed is the same as selected in search Hotel form.	URL: http://adactinhotelapp.com/index.php User:{test username} Password:{test password} Location: Sydney Hotel: Hotel Creek Room type: standard No-of-rooms:1 Check-in-date: today's date Checko-ut-date:today+1 date No-of-adults:1 No-of-children: 0	Location displayed in Select Hotel should be the same as location selected in search hotel form.

TC-105	To verify whether Check-in date and Check-out date are being displayed in Select Hotel page according to the dates selected in search Hotel.	1. Launch hotel reservation application using URL as in test data. 2. Login to the application using username and password as in test data. 3. Select location as in test data. 4. Select hotel as in test data. 5. Select room type as in test data. 6. Select no-of-rooms as in test data. 7. Enter check-out-date as in test data. 8. Select No-of-adults as in test data. 9. Select No-of-children as in test data. 10. Click on Search button. 11. Verify that check-in-date and check-out-dates are the same as selected in search hotel form.	URL: http://adactinhotelapp.com/index.php User:{test username} Password:{test password} Location: Sydney Hotel: Hotel Creek Room type: standard No-of-rooms:1 Check-in-date: today's date Checkoutdate:today+1 date No-of-adults:1 No-of-children:0	Check-in-date and check-out-date should be displayed according to the data entered in search hotel form.
TC-106	To verify whether no. of rooms in Select Hotel page is same as the Number of rooms selected in search hotel page	1. Launch hotel reservation application using URL as in test data. 2. Login to the application using username and password as in test data. 3. Select location as in test data. 4. Select hotel as in test data. 5. Select room type as in test data. 6. Select no-of-rooms as in test data. 7. Enter check-out-date as in test data. 8. Select No-of-adults as in test data. 9. Select No-of-children as in test data. 10. Click on Search button. 11. Verify that no-of-rooms is reflected according to the number of rooms selected in search hotel page.	URL: http://adactinhotelapp.com/index.php User:{test username} Password:{test password} Location: Sydney Hotel: Hotel Creek Room type: standard No-of-rooms:3 Check-in-date: today's date Checkoutdate:today+1 date No-of-adults:1 No-of-children: 0	No-of-rooms should be displayed and match with number of rooms in search hotel page

TC-107	To verify whether Room Type in Select Hotel page is same as Room type selected in search hotel page	1. Launch hotel reservation application using URL as in test data. 2. Login to the application using username and password as in test data. 3. Select location as in test data. 4. Select hotel as in test data. 5. Select room type as in test data. 6. Select no-of-rooms as in test data. 7. Enter check-out-date as in test data. 8. Select No-of-adults as in test data. 9. Select No-of-children as in test data. 10. Click on Search button. 11. **Verify that room type reflected** is the same as selected in search hotel page.	URL: http://adactinhotelapp.com/index.php User:{test username} Password:{test password} Location: Sydney Hotel: Hotel Creek Room type: Deluxe No-of-rooms:1 Check-in-date: today's date Checkoutdate:today+1 date No-of-adults:1 No-of-children:0	Room type displayed should be the same as selected in search hotel page
TC-108	To verify whether the total price (excl.GST) is calculated as "price per night * no. of nights* no of rooms".	1. Launch hotel reservation application using URL as in test data. 2. Login to the application using username and password as in test data. 3. Select location as in test data. 4. Select hotel as in test data. 5. Select room type as in test data. 6. Select no-of-rooms as in test data. 7. Enter check-out-date as in test data. 8. Select No-of-adults as in test data. 9. Select No-of-children as in test data. 10. Click on Search button. 11. Select the hotel and click on continue button 12. Verify that total-price(excl. GST) is being calculated as (price-per-night*no-of-nights*-no-of-rooms)	URL: http://adactinhotelapp.com/index.php User:{test username} Password:{test password} Location: Sydney Hotel: Hotel Creek Room type: standard No-of-rooms:2 Check-in-date: today's date Check-out-date:today+1 date No-of-adults:1 No-of-children: 0	Total price =125*1*2 =250$

TC-109	To verify when pressed, logout button logs out from the application.	1. Launch hotel reservation application using URL as in test data. 2. Login to the application using username and password as in test data. 3. Select location as in test data. 4. Select hotel as in test data. 5. Select room type as in test data. 6. Select no-of-rooms as in test data. 7. Enter check-out-date as in test data. 8. Select No-of-adults as in test data. 9. Select No-of-children as in test data. 10. Click on Search button. 11. Select the hotel and click on continue button. 12. Enter the details and click on book now. 13. Click on logout and verify we have been logged out of the application.	URL: http://adactinhotelapp.com/index.php User:{test username} Password:{test password} Location: Sydney Hotel: Hotel Creek Room type: standard No-of-rooms:2 Check-in-date: today's date Check-out-date:today+1 date No-of-adults:1 No-of-children:0	User should logout from the application.
TC-110	To check correct total price is being calculated as "price per night*no of days*no of rooms in Book a hotel page	1. Launch hotel reservation application using URL as in test data. 2. Login to the application using username and password as in test data. 3. Select location as in test data. 4. Select hotel as in test data. 5. Select room type as in test data. 6. Select no-of-rooms as in test data. 7. Enter check-out-date as in test data. 8. Select No-of-adults as in test data. 9. Select No-of-children as in test data. 10. Click on Search button. 11. Select the hotel and click on continue button 12. Verify that total-price is being calculated as (price-per-night*-no-of-rooms*no-of-days + 10% GST")	URL: http://adactinhotelapp.com/index.php User:{test username} Password:{test password} Location: Melbourne Hotel: Hotel Creek Room type: standard No-of-rooms:2 Check-in-date: today's date Check-out-date:today+1 date No-of-adults:1 No-of-children: 0	Total-price should be calculated as (price-per-night*no-of-rooms*no-of-days Total Price= 125*2*1 = 250$ In book a hotel page

TC-111	To check Hotel name, Location, room type, Total Day, price per night are same in Booking confirmation page as they were selected in previous screen	1. Launch hotel reservation application using URL as in test data. 2. Login to the application using username and password as in test data. 3. Select location as in test data. 4. Select Hotel as in test data. 5. Select room type as in test data. 6. Select no-of-rooms as in test data. 7. Enter check-out-date as in test data. 8. Select No-of-adults as in test data. 9. Select No-of-children as in test data. 10. Click on Search button. 11. Select the hotel and click on continue button 12. Verify Hotel name, Location, room type, Total Day, price per night are same in Booking confirmation page as they were selected in previous screen	URL: http://adactinhotelapp.com/index.php User:{test username} Password:{test password} Location: Sydney Hotel: hotel Creek Room type: standard No-of-rooms:2 Check-in-date: today's date Check-out-date:today+1 date No-of-adults:1 No-of-children: 0	Data should be same as selected in previous screen
TC-112	To check correct Final billed price is Total Price + 10% Total price in Book a Hotel page	1. Launch hotel reservation application using URL as in test data. 2. Login to the application using username and password as in test data. 3. Select location as in test data. Select Hotel as in test data. 4. Select room type as in test data. 5. Select no-of-rooms as in test data. 6. Enter check-out-date as in test data. 7. Select No-of-adults as in test data. 8. Select No-of-children as in test data. 9. Click on Search button. 10. Select the hotel and click on continue button 11. Verify that Final Billed Price is being calculated as (price-per-night*no-of-rooms*no-of-days	URL: http://adactinhotelapp.com/index.php User:{test username} Password:{test password} Location: Sydney Hotel: Hotel Creek Room type: standard No-of-rooms:2 Check-in-date: today's date Check-out-date:today+1 date No-of-adults:1 No-of-children: 0	Final billed Price =125+12.5 =137.5 in Book a Hotel page

TC-113	To verify whether the data displayed is same as the selected data in Book hotel page	1. Launch hotel reservation application using URL as in test data. 2. Login to the application using username and password as in test data. 3. Select location as in test data. 4. Select Hotel as in test data. 5. Select room type as in test data. 6. Select no-of-rooms as in test data. 7. Enter check-out-date as in test data. 8. Select No-of-adults as in test data. 9. Select No-of-children as in test data. 10. Click on Search button. 11. Select the hotel and click on continue button 12. Verify displayed data is same as the selected data in Book hotel page	URL: http://adactinhotelapp.com/index.php User:{test username} Password:{test password} Location: Sydney Hotel: Hotel Creek Room type: standard No-of-rooms:2 Check-in-date: today's date Check-out-date:today+1 date No-of-adults:1 No-of-children: 0	Hotel: hotel Creek Room type: Standard No-of-rooms:2 Check-in-date: 27/07/2012 Checkout-date: 28/07/2012 No-of-adults:1 No-of-children: 0
TC-114	Verify Order number is generated in booking **confirmation** page	1. Launch hotel reservation application using URL as in test data. 2. Login to the application using username and password as in test data. 3. Select location as in test data. 4. Select hotel as in test data. 5. Select room type as in test data. 6. Select no-of-rooms as in test data. 7. Enter check-out-date as in test data. 8. Select No-of-adults as in test data. 9. Select No-of-children as in test data. 10. Click on Search button. 11. Select the hotel and click on continue button 12. Verify Order number is generated	URL: http://adactinhotelapp.com/index.php User:{test username} Password:{test password} Location: Sydney Hotel: hotel Creek Room type: standard No-of-rooms:2 Check-in-date: today's date Check-out-date:today+1 date No-of-adults:1 No-of-children: 0	ORDER no should be generated

TC-115	To verify whether the booked itinerary details are not editable.	1. Launch hotel reservation application using URL as in test data. 2. Login to the application using username and password as in test data. 3. Select location as in test data. 4. Select Hotel as in test data. 5. Select room type as in test data. 6. Select no-of-rooms as in test data. 7. Enter check-out-date as in test data. 8. Select No-of-adults as in test data. 9. Select No-of-children as in test data. 10. Click on Search button. 11. Select the hotel and click on continue button 12. Fill the form and click on Book now button. 13. Click on My itinerary button 14. Verify that the details are not editable	URL: http://adactinhotelapp.com/index.php User:{test username} Password:{test password} Location: Adelaide Hotel: Hotel Cornice Room type: standard No-of-rooms:2 Check-in-date: today's date Check-out-date:today+1 date No-of-adults:1 No-of-children: 0	Details once accepted should not be editable
TC-116	To check whether the booked itinerary reflects the correct information in line with the booking.	1. Launch hotel reservation application using URL as in test data. 2. Login to the application using username and password as in test data. 3. Select location as in test data. 4. Select hotel as in test data. 5. Select room type as in test data. 6. Select no-of-rooms as in test data. 7. Enter check-out-date as in test data. 8. Select No-of-adults as in test data. 9. Select No-of-children as in test data. 10. Click on Search button. 11. Select the hotel and click on continue button 12. Fill the form and click on Book now button. 13. Click on My itinerary button 14. Verify that the details are reflected correctly as per the booking	http://adactinhotelapp.com/index.php User:{test username} Password:{test password} Location: Sydney Hotel: Hotel Creek Room type: standard No-of-rooms:2 Check-in-date: today's date Check-out-date:today+1 date No-of-adults:1 No-of-children: 0	Itinerary should reflect the correct information in line with the booking.

TC-117	To check whether "search order id" query is working and displaying the relevant details.	1. Launch hotel reservation application using URL as in test data. 2. Login to the application using username and password as in test data. 3. Click on booked itinerary link. 4. Enter the order id. 5. Verify that the relevant details are displayed	http://adactinhotelapp.com/index.php User:{test username} Password:{test password} Order id :pick existing order id	Search Order ID query should display the relevant details for Order ID
TC-118	Verify that all the details of newly generated order number in booked itinerary page are correct and match with data during booking.	1. Launch hotel reservation application using URL as in test data. 2. Login to the application using username and password as in test data. 3. Book an order as in previous test cases 4. Click on My itinerary button 5. Search for Order number 6. Verify all the details of order number are correct as entered during saving order	http://adactinhotelapp.com/index.php User:{test username} Password:{test password} Location: Sydney Hotel: Hotel Creek Room type: standard No-of-rooms:2 Check-in-date: today's date Check-out-date:today+1 date No-of-adults:1 No-of-children: 0	All the details in booked itinerary page should be same as those entered during booking
TC-119	To verify that the order gets cancelled after click on Cancel order number link	1. Launch hotel reservation application using URL as in test data. 2. Login to the application using username and password as in test data. 3. Book the Hotel as in previous test cases. Keep a note of order number generated 4. Click on Booked Itinerary link 5. Search for order number booked 6. Click on Cancel <Order Number> 7. Click Ok on pop-up which asks to cancel order or not 8. Verify that order number is cancelled and no longer exists in Booked Itinerary page	http://adactinhotelapp.com/index.php User:{test username} Password:{test password}	Order number should not longer be present in booked itinerary page after cancellation

TC-120	To Verify Title of every **Page reflects** what the page objective is. For example Title of Search Hotel page should have "Search Hotel"	1. Launch hotel reservation application using URL as in test data. 2. Login to the application using username and password as in test data. 3. Verify that title of each page is the same as the page objective 4. Click on Search hotel link and verify whether application directs to search hotel form 5. Click on booked itinerary link and verify that application directs to booked itinerary form	http://adactinhotelapp.com/index.php User:{test username} Password:{test password}	Title of each page should **reflect its** objective and the buttons should redirect as **specified, to** the relevant page.

Made in the USA
Las Vegas, NV
17 July 2026

51151735R00221